Pitman Research Notes in Mathematics Series

Submission of proposals for consideration

Suggestions for publication, in the form of outlines and representative samples, are invited by the Editorial Board for assessment. Intending authors should approach one of the main editors or another member of the Editorial Board, citing the relevant AMS subject classifications. Alternatively, outlines may be sent directly to the publisher's offices. Refereeing is by members of the board and other mathematical authorities in the topic concerned, throughout the world.

Preparation of accepted manuscripts

On acceptance of a proposal, the publisher will supply full instructions for the preparation of manuscripts in a form suitable for direct photo-lithographic reproduction. Specially printed grid sheets can be provided and a contribution is offered by the publisher towards the cost of typing. Word processor output, subject to the publisher's approval, is also acceptable.

Illustrations should be prepared by the authors, ready for direct reproduction without further improvement. The use of hand-drawn symbols should be avoided wherever possible, in order to maintain maximum clarity of the text.

The publisher will be pleased to give any guidance necessary during the preparation of a typescript, and will be happy to answer any queries.

Important note

In order to avoid later retyping, intending authors are strongly urged not to begin final preparation of a typescript before receiving the publisher's guidelines. In this way it is hoped to preserve the uniform appearance of the series.

Longman Scientific & Technical
Longman House
Burnt Mill
Harlow, Essex, CM20 2JE
UK
(Telephone (0279) 426721)

Titles in this series. A full list is available on request from the publisher.

201 Riemannian geometry and holonomy groups
S Salamon
202 Strong asymptotics for extremal errors and polynomials associated with Erdős type weights
D S Lubinsky
203 Optimal control of diffusion processes
V S Borkar
204 Rings, modules and radicals
B J Gardner
205 Two-parameter eigenvalue problems in ordinary differential equations
M Faierman
206 Distributions and analytic functions
R D Carmichael and D Mitrovic
207 Semicontinuity, relaxation and integral representation in the calculus of variations
G Buttazzo
208 Recent advances in nonlinear elliptic and parabolic problems
P Bénilan, M Chipot, L Evans and M Pierre
209 Model completions, ring representations and the topology of the Pierce sheaf
A Carson
210 Retarded dynamical systems
G Stepan
211 Function spaces, differential operators and nonlinear analysis
L Paivarinta
212 Analytic function theory of one complex variable
C C Yang, Y Komatu and K Niino
213 Elements of stability of visco-elastic fluids
J Dunwoody
214 Jordan decomposition of generalized vector measures
K D Schmidt
215 A mathematical analysis of bending of plates with transverse shear deformation
C Constanda
216 Ordinary and partial differential equations. Volume II
B D Sleeman and R J Jarvis
217 Hilbert modules over function algebras
R G Douglas and V I Paulsen
218 Graph colourings
R Wilson and R Nelson
219 Hardy-type inequalities
A Kufner and B Opic
220 Nonlinear partial differential equations and their applications: Collège de France Seminar. Volume X
H Brezis and J L Lions
221 Workshop on dynamical systems
E Shiels and Z Coelho
222 Geometry and analysis in nonlinear dynamics
H W Broer and F Takens
223 Fluid dynamical aspects of combustion theory
M Onofri and A Tesei
224 Approximation of Hilbert space operators. Volume I. 2nd edition
D Herrero
225 Operator theory: proceedings of the 1988 GPOTS–Wabash conference
J B Conway and B B Morrel
226 Local cohomology and localization
J L Bueso Montero, B Torrecillas Jover and A Verschoren
227 Nonlinear waves and dissipative effects
D Fusco and A Jeffrey
228 Numerical analysis 1989
D F Griffiths and G A Watson
229 Recent developments in structured continua. Volume II
D De Kee and P Kaloni
230 Boolean methods in interpolation and approximation
F J Delvos and W Schempp
231 Further advances in twistor theory. Volume I
L J Mason and L P Hughston
232 Further advances in twistor theory. Volume II
L J Mason and L P Hughston
233 Geometry in the neighborhood of invariant manifolds of maps and flows and linearization
U Kirchgraber and K Palmer
234 Quantales and their applications
K I Rosenthal
235 Integral equations and inverse problems
V Petkov and R Lazarov
236 Pseudo-differential operators
S R Simanca
237 A functional analytic approach to statistical experiments
I M Bomze
238 Quantum mechanics, algebras and distributions
D Dubin and M Hennings
239 Hamilton flows and evolution semigroups
J Gzyl
240 Topics in controlled Markov chains
V S Borkar
241 Invariant manifold theory for hydrodynamic transition
S Sritharan
242 Lectures on the spectrum of $L^2(\Gamma\backslash G)$
F L Williams
243 Progress in variational methods in Hamiltonian systems and elliptic equations
M Girardi, M Matzeu and F Pacella
244 Optimization and nonlinear analysis
A Ioffe, M Marcus and S Reich
245 Inverse problems and imaging
G F Roach
246 Semigroup theory with applications to systems and control
N U Ahmed
247 Periodic-parabolic boundary value problems and positivity
P Hess
248 Distributions and pseudo-differential operators
S Zaidman
249 Progress in partial differential equations: the Metz surveys
M Chipot and J Saint Jean Paulin
250 Differential equations and control theory
V Barbu

251 Stability of stochastic differential equations with respect to semimartingales
X Mao
252 Fixed point theory and applications
J Baillon and M Théra
253 Nonlinear hyperbolic equations and field theory
M K V Murthy and S Spagnolo
254 Ordinary and partial differential equations. Volume III
B D Sleeman and R J Jarvis
255 Harmonic maps into homogeneous spaces
M Black
256 Boundary value and initial value problems in complex analysis: studies in complex analysis and its applications to PDEs 1
R Kühnau and W Tutschke
257 Geometric function theory and applications of complex analysis in mechanics: studies in complex analysis and its applications to PDEs 2
R Kühnau and W Tutschke
258 The development of statistics: recent contributions from China
X R Chen, K T Fang and C C Yang
259 Multiplication of distributions and applications to partial differential equations
M Oberguggenberger
260 Numerical analysis 1991
D F Griffiths and G A Watson
261 Schur's algorithm and several applications
M Bakonyi and T Constantinescu
262 Partial differential equations with complex analysis
H Begehr and A Jeffrey
263 Partial differential equations with real analysis
H Begehr and A Jeffrey
264 Solvability and bifurcations of nonlinear equations
P Drábek
265 Orientational averaging in mechanics of solids
A Lagzdins, V Tamuzs, G Teters and A Kregers
266 Progress in partial differential equations: elliptic and parabolic problems
C Bandle, J Bemelmans, M Chipot, M Grüter and J Saint Jean Paulin
267 Progress in partial differential equations: calculus of variations, applications
C Bandle, J Bemelmans, M Chipot, M Grüter and J Saint Jean Paulin
268 Stochastic partial differential equations and applications
G Da Prato and L Tubaro
269 Partial differential equations and related subjects
M Miranda
270 Operator algebras and topology
W B Arveson, A S Mishchenko, M Putinar, M A Rieffel and S Stratila
271 Operator algebras and operator theory
W B Arveson, A S Mishchenko, M Putinar, M A Rieffel and S Stratila
272 Ordinary and delay differential equations
J Wiener and J K Hale
273 Partial differential equations
J Wiener and J K Hale
274 Mathematical topics in fluid mechanics
J F Rodrigues and A Sequeira
275 Green functions for second order parabolic integro-differential problems
M G Garroni and J F Menaldi
276 Riemann waves and their applications
M W Kalinowski
277 Banach C(K)-modules and operators preserving disjointness
Y A Abramovich, E L Arenson and A K Kitover
278 Limit algebras: an introduction to subalgebras of C*-algebras
S C Power
279 Abstract evolution equations, periodic problems and applications
D Daners and P Koch Medina
280 Emerging applications in free boundary problems
J M Chadam and H Rasmussen
281 Free boundary problems involving solids
J M Chadam and H Rasmussen
282 Free boundary problems in fluid flow with applications
J M Chadam and H Rasmussen
283 Asymptotic problems in probability theory: stochastic models and diffusions on fractals
K D Elworthy and N Ikeda
284 Asymptotic problems in probability theory: Wiener functionals and asymptotics
K D Elworthy and N Ikeda
285 Dynamical systems
R Bamon, R Labarca, J Lewowicz and J Palis
286 Models of hysteresis
A Visintin
287 Moments in probability and approximation theory
G Anastassiou
288 Mathematical aspects of penetrative convection
B Straughan
289 Ordinary and partial differential equations. Vol ume IV
B D Sleeman and R J Jarvis
290 *K*-theory for real *C**-algebras
H Schröder
291 Recent developments in theoretical fluid mechanics
G P Galdi and J Necas
292 Propagation of a curved shock and nonlinear ray theory
P Prasad
293 Non-classical elastic solids
M Ciarletta and D Ieșan
294 Multigrid methods
J Bramble
295 Entropy and partîal differential equations
W A Day
296 Progress in partial differential equations: the Metz surveys 2
M Chipot
297 Nonstandard methods in the calculus of variations
C Tuckey
298 Barrelledness, Baire-like- and (LF)-spaces
M Kunzinger
299 Nonlinear partial differential equations and their applications. Collège de France Seminar. Volume XI
H Brezis and J L Lions
300 Introduction to operator theory
T Yoshino

301 Generalized fractional calculus and applications
V Kiryakova
302 Nonlinear partial differential equations and their applications. Collège de France Seminar Volume XII
H Brezis and J L Lions
303 Numerical analysis 1993
D F Griffiths and G A Watson
304 Topics in abstract differential equations
S Zaidman
305 Complex analysis and its applications
C C Yang, G C Wen, K Y Li and Y M Chiang
306 Computational methods for fluid-structure interaction
J M Crolet and R Ohayon
307 Random geometrically graph directed self-similar multifractals
L Olsen
308 Progress in theoretical and computational fluid mechanics
G P Galdi, J Málek and J Necas
309 Variational methods in Lorentzian geometry
A Masiello
310 Stochastic analysis on infinite dimensional spaces
H Kunita and H-H Kuo
311 Representations of Lie groups and quantum groups
V Baldoni and M Picardello
312 Common zeros of polynomials in several variables and higher dimensional quadrature
Y Xu
313 Extending modules
N V Dung, D van Huynh, P F Smith and R Wisbauer
314 Progress in partial differential equations: the Metz surveys 3
M Chipot, J Saint Jean Paulin and I Shafrir
315 Refined large deviation limit theorems
V Vinogradov
316 Topological vector spaces, algebras and related areas
A Lau and I Tweddle
317 Integral methods in science and engineering
C Constanda
318 A method for computing unsteady flows in porous media
R Raghavan and E Ozkan
319 Asymptotic theories for plates and shells
R P Gilbert and K Hackl
320 Nonlinear variational problems and partial differential equations
A Marino and M K V Murthy
321 Topics in abstract differential equations II
S Zaidman
322 Diffraction by wedges
B Budaev
323 Free boundary problems: theory and applications
J I Diaz, M A Herrero, A Liñan and J L Vazquez
324 Recent developments in evolution equations
A C McBride and G F Roach

A C McBride and G F Roach (Editors)

University of Strathclyde

Recent developments in evolution equations

Copublished in the United States with
John Wiley & Sons, Inc., New York

Longman Scientific & Technical
Longman Group UK Limited
Longman House, Burnt Mill, Harlow
Essex CM20 2JE, England
and Associated companies throughout the world.

Copublished in the United States with
John Wiley & Sons Inc., 605 Third Avenue, New York, NY 10158

First published 1995

AMS Classifications: 20, 35, 46

ISSN 0269-3674

ISBN 0 582 24669 5

British Library Cataloguing in Publication Data

A catalogue record for this book is
available from the British Library

Library of Congress Cataloging-in-Publication Data

A catalog record for this book is available

Printed and bound in Great Britain
by Biddles Ltd, Guildford and King's Lynn

Contents

Preface

This meeting was the most recent in a series of International Seminars which have been held in the University of Strathclyde since 1975. The primary aim of the meetings has been to discuss problems which still stubbornly resisted solutions rather than problems which had been solved.

Understandably, since the first of these meetings, their scope has broadened. Certainly their topics remain very finely focussed but they have become increasingly more multidisciplinary both in emphasis and appeal. Consequently, it was decided to make the work of these meetings more readily available by publishing proceedings. This volume comprises contributions from the meeting held at the University of Strathclyde from 25-29 July, 1994. We are most grateful to Longman Scientific and Technical for agreeing to publish these Proceedings.

Meetings such as this could not take place without a considerable amount of support and assistance from a number of sources. In this connection we are most grateful to

The British Council

The City of Glasgow (Lord Provost)

The Edinburgh Mathematical Society (EMS)

The Royal Bank of Scotland

The Soros Foundation

The University of Strathclyde

for their financial support. We are particularly grateful to the EMS for endowing one of the Plenary Lectures and for allowing us to name it accordingly.

Thanks are due to our many friends who, as Advisory Editors, gave so much of their time for refereeing and selecting the contributions to this volume. Last but by no means least we record our appreciation of the work done by Drs. Bill Anderson, Wilson Lamb and Des McGhee, by the postgraduate students and by the secretarial staff of the Department of Mathematics in the University of Strathclyde, especially Mrs. Mary Sergeant whose quiet efficiency helped in so many ways to ensure the smooth running of the Conference.

A. C. McBride
G. F. Roach

University of Strathclyde
Glasgow G1 1XH
August 1994

Plenary Lectures

J A GOLDSTEIN AND G SHI

Obstacle scattering for elastic waves

§1. Introduction

We present a new approach to scattering theory for equations that are of order N in time. The main application will be to elastic waves in exterior domains, but acoustic waves, the wave equation with a potential, and other cases are covered by this approach.

Here is an outline of this paper. Section 2 gives a short description of (one and two space) scattering theory. Section 3 treats the d' Alembert formula for factored equations. Section 4 introduces the elastic wave equation. Section 5 deals with obstacle scattering. Section 6 treats the inverse problem.

§2. Scattering Theory

For $j = 0, 1$, let H_j be a selfadjoint operator on a Hilbert space $\mathcal{H}_j$. The abstract Schrödinger equation $i\frac{du}{dt} = H_j u$ is governed by the (C_0) unitary group $\left\{e^{-itH_j} : t \in \mathbb{R}\right\}$; the unique solution u of the Schrödinger equation with initial data $u(0) = f$ is given by $u(t) = e^{-itH_j} f$. (Cf. e.g. [5].)

One is to think of the subscript 1 (resp. 0) as describing "perturbed" (resp. "free") motion, and in some sense, the two groups are expected to be equivalent at $t = \pm\infty$. For definiteness, suppose $\mathcal{H}_1 = \mathcal{H}_0 = \mathcal{H}$. We suppose that the perturbed solution $e^{-itH_1} f$ looks like a free solution $e^{-itH_0} f_\pm$ as $t \to \pm\infty$ in the sense that $e^{-itH_1} f - e^{-itH_0} f_\pm \to 0$ as $t \to \pm\infty$. Then the *wave operators*

$$W_\pm g = \lim_{t\to\pm\infty} e^{itH_1} e^{-itH_0} g$$

exist (for $g = f_\pm$). Here all limits are in the norm topology of $\mathcal{H}$.

Let P_j be the orthogonal projection onto the absolutely continuous subspace $\mathcal{H}_{ac}(H_j)$ for H_j. { Thus $H_j = \int_{-\infty}^{\infty} \lambda dE_j(\lambda)$ by the spectral theorem. Say that $f \in \mathcal{H}_{ac}(H_j)$ (resp. $f \in \mathcal{H}_s(H_j)$) iff $\lambda - \|E_j(\lambda)f\|^2$ is absolutely (resp. singular) continuous on $\mathbb{R}$; here P_j is the (orthogonal) projection onto the closed subspace $\mathcal{H}_{ac}(H_j)$. }

Now suppose that the wave operators $W_\pm$ exist on all of $P_0\mathcal{H}$ $= \mathcal{H}_{ac}(H_0)$. Then their ranges satisfy Ran $(W_\pm) \subset \mathcal{H}_{ac}(H_1)$. The wave operators are called *complete* if Ran $(W_\pm) = \mathcal{H}_{ac}(H_1)$. The *scattering operator* $S = W_+^* W_- = W_+^{-1} W_-$ is then a unitary operator from $\mathcal{H}_{ac}(H_0)$ to $\mathcal{H}_{ac}(H_1,)$ sending f_- to f_+; f_- describes how the perturbed motion $e^{-itH_1}f$ looks like a free motion near $t = -\infty$. Thus f_- is the *incoming data*. Similarly, f_+ is the *outgoing data;* it describes how the perturbed motion looks free near $t = +\infty$. The scattering operator S maps f_- to f_+; this describes the result of a scattering experiment which "sees" the incoming and outgoing solutions.

When $\mathcal{H}_0$ and $\mathcal{H}_1$ are different spaces, one needs an identification operator $J : \mathcal{H}_0 \to \mathcal{H}_1$. J is to be a bounded linear operator. Then $e^{-itH_1}f_\pm$ should approximately equal $Je^{-itH_0}f$ as $t \to \pm\infty$. The *wave operators* are

$$\begin{aligned} W_\pm g &= W_\pm(H_1, H_0; J)g \\ &= \lim_{t\to\pm\infty} e^{itH_1} J e^{-itH_0} g \end{aligned}$$

(for $g = f_\pm$). Suppose $W_\pm$ exist on $\mathcal{H}_{0,ac}(H_0)$. Call $W_\pm$ *complete* if Ran$(W_\pm)$ $= \mathcal{H}_{1,ac}(H_1)$ and $W_\pm$ is injective on $\mathcal{H}_{0,ac}(H_0)$. Then, as before, the scattering operator $S = W_\pm^* W_-$ is unitary from $\mathcal{H}_{0,ac}(H_0)$ to $\mathcal{H}_{0,ac}(H_1)$, provided $W_\pm$ are isometric on $\mathcal{H}_{0,ac}(H_0)$.

As an example consider the wave equation with a potential

$$v_{tt} = \Delta v - V(x)v$$

where $x \in \mathbb{R}^n$ and $0 \le V \in L^\infty(\mathbb{R}^n)$. (The hypothesis on V can be greatly relaxed.) For the free equation, rewrite $v_{tt} = \Delta v$ as

$$u_t = \begin{pmatrix} v \\ v_t \end{pmatrix}_t = \begin{pmatrix} 0 & I \\ \Delta & 0 \end{pmatrix} \begin{pmatrix} v \\ v_t \end{pmatrix} = -iH_0 u.$$

H_0 is selfadjoint on

$$\mathcal{H}_0 = \left\{ \begin{pmatrix} f_1 \\ f_2 \end{pmatrix} : \|(-\Delta)^{\frac{1}{2}} f_1\|_2^2 + \|f_2\|_2^2 < \infty \right\},$$

which is a Hilbert space in the obvious way; here $\|\cdot\|_2$ refers to the $L^2(\mathbb{R}^n)$ norm. Similarly, $H_1 = \begin{pmatrix} 0 & I \\ \Delta - V & 0 \end{pmatrix}$ is selfadjoint on $\mathcal{H}_2$, which is normed by

$$\left\| \begin{pmatrix} f_1 \\ f_2 \end{pmatrix} \right\| = \left\{ \| (-\Delta + V)^{\frac{1}{2}} f_1 \|_2^2 + \| f_2 \|_2^2 \right\}^{\frac{1}{2}}.$$

$\mathcal{H}_0$ and $\mathcal{H}_1$ are equal as sets but they have different (but equivalent) norms. Let $J : \mathcal{H}_0 \to \mathcal{H}_1$ be the identity operator from $\mathcal{H}_0$ to $\mathcal{H}_1$. J is bounded but not unitary. This J is appropriate for scattering theory in this context.

For obstacle scattering by sound waves, let $\mathcal{O}$ be, say, a smooth star shaped bounded region in $\mathbb{R}^n$. The free group is as above. The perturbed group governs

$$v_{tt} = \Delta v \text{ for } x \in \mathbb{R}^n \backslash \mathcal{O}.$$

Here associate either Dirichlet or Neumann conditions with Δ, acting on $L^2(\mathbb{R}^n \backslash \mathcal{O})$. Then

$$\left\| \begin{pmatrix} f_1 \\ f_2 \end{pmatrix} \right\|_{\mathcal{H}_1}^2 = \| (-\Delta)^{\frac{1}{2}} f_1 \|_2^2 + \| f_2 \|_2^2$$

where $\| \cdot \|_2$ refers to the $L^2(\mathbb{R}^n \backslash \mathcal{O})$ norm. Here $J \begin{pmatrix} f_1 \\ f_2 \end{pmatrix} = \begin{pmatrix} f_1 \chi_{\mathbb{R}^n \backslash \mathcal{O}} \\ f_2 \chi_{\mathbb{R}^n \backslash \mathcal{O}} \end{pmatrix}$ has norm one. But J is not injective. Nonethless, $e^{-itH_0} f$ is transporting the wave (described by f) out to infinity since

$$\int_{|x| < \mathrm{R}} | \left(e^{itH_0} f \right) (x) |^2 dx \to 0$$

as $t \to \pm\infty$ for each $\mathrm{R} > 0$ (local energy decay.) Thus J "acts like the identity" on $e^{-itH_0} f$ for $|t|$ large. In this sense, J is "morally" injective.

Note that J can be replaced by any bounded linear operator $K : \mathcal{H}_0 \to \mathcal{H}_1$ satisfying

$$\|(J - K) e^{-itH_0} f_0 \| \to 0$$

as $t \to \pm\infty$ for any $f_0 \in \mathcal{H}_0$. In this sense, J is not uniquely determined.

For more details about scattering theory, see the books and papers listed in the Reference section. Scattering theory has a very substantial literature; we only list a few of the basic references.

§3. D' Alembert's Formula

Let $A_j = A_j^*$ be selfadjoint on $\mathcal{H}$ for $1 \le j \le N$ and suppose that $A_1, \ldots, A_N$ commute in the sense that $[e^{itA_j}, e^{isA_k}] = 0$ for all s, t, j, k. If also $A_j - A_k$ is injective for $j \ne k$, then every mild solution * of

$$\prod_{j=1}^{N} \left(\frac{d}{dt} + iA_j \right) u(t) = 0 \tag{3.1}$$

is of the form

$$u(t) = \prod_{j=1}^{N} e^{-itA_j} f_j \tag{3.2}$$

(where $f_j \in \mathcal{H}$). This is called *d'Alembert's formula.* Of course d'Alembert showed that $u_{tt} = u_{xx}$ implies $u = F(x+t) + G(x-t)$, which corresponds to $A_1 = -A_2 = i\frac{d}{dx}$ on $\mathcal{H} = L^2(\mathbb{R})$.

Note that the injectivity of $A_j - A_k$ is a necessary condition, since if $(A_j - A_k)f = 0$ (for $j \ne k$ and $f \ne 0$), then $u(t) = te^{-itA_j}f$ is a solution of (3.1) which is not fo the form (3.2).

The d'Alembert formula was proved in the context of (C_0) semigroups on a Banach space by Goldstein and Sandefur [7] under the additional assumption that Ran $(A_j - A_k)$ is sufficiently large for $j \ne k$. In our special case here involving unitary groups no additional assumption of this nature is required.

Here is the proof of d'Alembert's formula. By the spectral theorem for commuting selfadjoint operators, there is a unitary operator U from $\mathcal{H}$ to some $L^2(\Omega, \Sigma, \mu)$ and real Σ-measurable functions a_j on Ω such that UA_jU^{-1} is multiplication by a_j (with maximal domain) on $L^2(\Omega, \Sigma, \mu)$ for $j = 1, \ldots, N$. Moreover, by the injectivity hypothesis, $N_{jk} = \{\omega \in \Omega; a_j(\omega) \ne a_k(\omega)\}$ is a μ-null set whenever $j \ne k$.

---

* $e^{itA}f$ is a strong solution of $u' + iAu = 0$ if $f \in \mathfrak{D}(A)$. It is a mild solution if $f \in \overline{\mathfrak{D}(A)} = \mathcal{H}$. Similarly for higher order equations.

In the representation in $L^2(\Omega, \Sigma, \mu)$, (3.1) becomes

$$\prod_{j=1}^{N} \left(\frac{d}{dt} + ia_j(\omega) \right) v(t, \omega) = 0, \qquad \omega \in \Omega.$$

The general solution is given by

$$v(t, \omega) = \prod_{j=1}^{N} e^{-itA_j} g_j(\omega)$$

for all $t \in \mathbb{R}$ and $\omega \in \Omega \backslash N_0$, $N_0 = \bigcup_{j,k=1}^{N} N_{jk}$. Here $g_j \in L^2(\Omega, \Sigma, \mu)$.

This is because for $\omega \in \Omega \backslash N_0$, this Nth order constant coefficient ODE has distinct roots. Translating back to $u(t)$ in $\mathcal{H}$ yields (3.2) (where $f_j = U^{-1} g_j$).

§4. The Elastic Wave Equation

Let $x \in \mathbb{R}^n$; usually $n = 3$, but we allow n to be arbitrary. Let $u = (u_1, \ldots, u_n)$ represent the displacement vector for an elastic wave in $\mathbb{R}^n$; here $u = u(t, x)$ with $t \in \mathbb{R}$ and $x \in \mathbb{R}^n$. Let λ, μ be the Lamé parameters and ρ the density of the medium; these we take to be three positive constants. Then for $i = 1, \ldots, n$, u_i satisfies

$$\rho \frac{\partial^2 u_i}{\partial t^2} = \mu \Delta u_i + (\lambda + \mu) \frac{\partial}{\partial x_i} (div(u_i)). \tag{4.1}$$

This is a coupled system of second order (in t) equations. Each component $v = u_i$ satisfies

$$\prod_{j=1}^{2} \left(\frac{d^2}{dt^2} - \mu_j \Delta \right) v = 0$$

where $0 < \mu_1 = \frac{\mu}{\rho} < \mu_2 = (\lambda + 2\mu)/\rho$ (see [8]). Thus, we take as our basic equation

$$\prod_{j=1}^{4} \left(\frac{d}{dt} + ic_j A^{(k)} \right) u(t) = 0 \tag{4.2}$$

where

$$c_1 = -c_2 = \sqrt{\mu_1}, \qquad c_3 = -c_4 = \sqrt{\mu_2},$$

$$A^{(0)} = (-\Delta)^{\frac{1}{2}} \qquad \text{on } L^2(\mathbb{R}^n),$$

$$A^{(1)} = (-\Delta)^{\frac{1}{2}} \qquad \text{on } L^2(\mathbb{R}^n \backslash \mathcal{O}).$$

The superscript 0 (resp. 1) refers to the free [resp. perturbed] elastic wave equation. The obstacle $\mathcal{O}$ is assumed to be bounded and smooth (as in Section 2). Thus we view (4.1) (with superscripts $(0),(1)$) as our given pair of equations which we replace by (4.2). As in Section 2, suitable boundary conditions (e.g. Dirichlet or Neumann) must be assigned to Δ on $\mathbb{R}^n \backslash \mathcal{O}$, i.e. to $-(A^{(1)})^2$.

Lax and Phillips [10] made the first study of obstacle scattering for the (Dirichlet) wave equation (with $\mathcal{O}$ star-shaped* and n odd). They also considered general symmetric hyperbolic systems, which include the elastic wave equation. This is sketched in [10]. Wilcox [19] studied obstacle scattering for acoustic waves with the Neumann boundary condition. This work spurred much additional research, a sampling of which is covered by [12], [17].

Our approach here, based on d'Alembert's formula, is new and enables us to recover and unify old results as well as obtain new results.

§5. A Framework for Scattering

Let $A_1^{(k)}, \cdots, A_N^{(k)}$ be selfadjoint operators on $\mathcal{H}_k$ such that they all commute (i.e. $[\exp(itA_j^{(k)}), \exp(i\,sA_\ell^{(k)})] = 0$ for all t, s, j, ℓ) and $A_j^{(k)} - A_\ell^{(k)}$ is injective for $j \neq k$; here $k = 0, 1$. Of concern is

$$\prod_{j=1}^{N} \left(\frac{d}{dt} + iA_j^{(k)} \right) u(t) = 0 \tag{5.1}$$

for $k = 0, 1$. Let $J_j \in \mathcal{B}(\mathcal{H}_0, \mathcal{H}_1)$, $1 \leq j \leq N$, and let P_{kj} be the orthogonal projection onto $\mathcal{H}_{kj} := \mathcal{H}_{k,ac}(A_j^{(k)})$ for all j, k. Define

---

* Star shaped means, that for $x \in \partial\mathcal{O}$ and ν the unit outer normal to $\partial\mathcal{O}$ at x, then $\nu \cdot x \geq 0$.

$$A(t)f := \sum_{j=1}^{N} \exp(-itA_j^{(1)})f_j,$$

$$B(t)g := \sum_{j=1}^{N} J_j \exp(-itA_j^{(0)})g_j. \tag{5.2}$$

Thus

$$A(t) : \mathcal{H}_1^N \to \mathcal{H}_1,$$
$$B(t) : \mathcal{H}_0^N \to \mathcal{H}_1$$

are bounded linear operators (uniformly in $t \in \mathbb{R}$). Here we have used such obvious notations as $f = (f_1, \cdots, f_n) \in \mathcal{H}_1^N$, etc. Let

$$\mathcal{A}_\pm = \{f \in \mathcal{K}_1 := \bigoplus_{j=1}^{N} \mathcal{H}_{1j} : \lim_{t \to \pm\infty} \|A(t)f\| = 0\},$$

$$\mathcal{B}_\pm = \{g \in \mathcal{K}_0 := \bigoplus_{j=1}^{N} \mathcal{H}_{0j} : \lim_{t \to \pm\infty} \|B(t)g\| = 0\}. \tag{5.3}$$

We say that the system $\mathcal{S} = \left(\vec{A}^{(1)}, \vec{A}^{(0)}; \vec{J} \right)$

$$\left(= A_1^{(1)}, \ldots, A_N^{(1)}, A_1^{(0)}, \ldots, A_N^{(0)}; J_1, \ldots, J_N \right)$$

has the *wave operator existence property* [WOEP] iff for all $g \in \mathcal{K}_0^*$ there is an $f_\pm \in \mathcal{K}_1{}^*$ such that

$$\|A(t)f_\pm - B(t)g\| \to 0$$

as $t \to \pm\infty$. The *wave operator existence and uniqueness property* [WOEUP] for $\mathcal{S}$ means that, in addition, $f_\pm$ are unique. Similarly $\mathcal{S}$ has the *wave operator semi-completeness property* [WOSCP] means for all $f \in \mathcal{K}_1$, there exist $g_\pm \in \mathcal{K}_0$ such that $\|A(t)f - B(t)g_\pm\| \to 0$ as $t \to \pm\infty$. The *wave operator completeness property* [WOCP] means that, in addition to the above, $g_\pm$ are unique.

---

* $\mathcal{K}_0, \mathcal{K}_1$ are defined in (5.3).

It is easy to see that

(i) WOEUP <=> WOEP and $\mathcal{A}_\pm = \{0\}$;

(ii) WOCP <=> WOSCP and $\mathcal{B}_\pm = \{0\}$.

The *vector wave operators* $\vec{W}_\pm : \mathcal{K}_0 \to \mathcal{K}_1$ are defined by

$$\|A(t)\vec{W}_\pm g - B(t)g\| \to 0$$

as $t \to \pm\infty$ for $g \in \mathcal{K}_0$. Note that $\vec{W}_\pm$ exist and are unique iff $\mathcal{S}$ has the WOEP, in which case Ran $(\vec{W}_\pm) \subset \mathcal{A}_\pm^\perp$.

Prior to stating the main theorem, we define the Riemann-Lebesgue class. Let $A = A^*$ on $\mathcal{H}$. Then A is in the *Riemann-Lebesgue class* iff $e^{itA} \to 0$ as $t \to \pm\infty$ in the weak operator topology (see [6]). Writing A as $A = \int_{-\infty}^{\infty} \lambda dE(\lambda)$, then A is in the Riemann-Lebesgue class iff for all $f, g \in \mathcal{H}$, $\int_{-\infty}^{\infty} e^{it\lambda} d_\lambda \langle E(\lambda)f, g\rangle \to 0$ as $t \to \pm\infty$. By the Riemann-Legesgue lemma, this holds provided $\mathcal{H}_{ac}(A) = \mathcal{H}$. (But the converse is not true.)

Main Theorem. *For $k = 0, 1$, let $A_1^{(k)}, \cdots, A_N^{(k)}$ be commuting selfadjoint operators on $\mathcal{H}_k$ such that $A_j^{(k)} - A_\ell^{(k)}$ is injective for $j \neq \ell$ and each $A_j^{(k)}$ is in the Riemann-Lebesgue class. Let $J_j \in \mathcal{B}(\mathcal{H}_0, \mathcal{H}_1)$ and suppose the wave operators $W_{\pm,j} = W_\pm(A_j^{(1)}, A_j^{(0)}; J_j)$ exist for $j = 1, \cdots, N$. Then*

(i) $\mathcal{S} = \left(\vec{A}^{(1)}, \vec{A}^{(0)}; \vec{J}\right)$ *has the WOEP;*

(ii) $\mathcal{A}_\pm = \{0\}$; *this implies that the vector wave operators $\vec{W}_\pm$ exist and $\vec{W}_\pm g =$ Diag $(W_{\pm,j})g$ for all $g \in \mathcal{K}_0$, where Diag $(W_{\pm,j})$ is the matrix (a_{ij}) with $a_{ij} = \delta_{ij} W_{\pm,j}$;*

(iii) $\vec{W}$ *is isometric iff each $W_{\pm,j}$ is partially isometric, in which case $\mathcal{B}_\pm = \{0\}$.*

(iv) $W_{\pm,j}$ *is complete for each j implies that $\mathcal{S}$ has the WOSCP;*

(v) *Each* $W_{\pm,j}$ *is partially isometric and complete implies that* $\overrightarrow{W}_{\pm} : \mathcal{K}_0 \to \mathcal{K}_1$ *are unitary.*

The proof is too long to give here (see [15]), but we will discuss some of the key ideas. Let the hypothesis on $A_j^{(k)}$ hold. Then by unitarity and the law of cosines,

$$\begin{aligned} &\|\sum_{j=1}^{N} \exp(-itA_j^{(1)})h_j\|^2 \\ &= \sum_{j=1}^{N} \|h_j\|^2 + 2\mathrm{Re} \sum_{j<\ell} \langle \exp(-itA_j^{(1)})h_j, \exp(-itA_\ell^{(1)})h_\ell \rangle \\ &= \sum_{j=1}^{N} \|h_j\|^2 + 2\mathrm{Re} \sum_{j<\ell} \langle \exp\{-it(A_j^{(1)} - A_\ell^{(1)})\}h_j, h_\ell \rangle \to \|h\|^2 \end{aligned}$$

as $t \to \pm_\infty$ by the Riemann-Lebesgue property. This implies that $\mathcal{A}_\pm = \{0\}$. If $\overrightarrow{W}_\pm$ are partially isometric, we can easily show that $\mathcal{B}_\pm = \{0\}$.

Basically the main theorem reduces scattering theory for (5.1) to scattering theory for the pair $(A_j^{(1)}, A_j^{(0)})$ for $1 \le j \le N$. In the context of our applications, $A_j^{(k)}$ is of the form $c(-\Delta)^{\frac{1}{2}}$ for $c > 0$. The invariance principle of scattering theory says that $W_\pm(\phi(S_1), \phi(S_0); J)$ exists and is independent of ϕ provided ϕ is continuous and increasing on $\sigma_{ac}(S_0) = \sigma(S_{0,ac}) = \sigma(S_0 \mid_{\mathcal{H}_{0,ac}} (S_0))$ (assuming it exists for $\phi = id$). This result has been proved in various contexts; the original results are due to Birman and Kato. See e.g. [2], [9]. In our application, $\varphi(r) = \sqrt{r}$ on $[0, \infty)$. The idea behind the invariance principle is simple. Let $\psi : \mathbb{R} \to \mathbb{R}$ be continuous and piecewise linear with positive slope away from the corners. Then

$$\begin{aligned} &\exp(-it\psi(S_1))J\exp(it\psi(S_0)) \\ &= \exp(-i\tau S_1)J\exp(i\tau S_0) \end{aligned}$$

for a suitable $\tau = \tau(t)$, where $\tau \to \pm\infty$ as $t \to \pm\infty$. So the invariance principle holds for such ψ. The general φ can be approximated by these ψ's; thus the invariance principle is clearly plausible (even if the rigorous proof is nontrivial).

Let $\mathcal{O} \subset \mathbb{R}^n$ be smooth and bounded. The Dirichlet Laplacian Δ_D on $L^2(\mathrm{R}^n \backslash \mathcal{O})$ has domain $H^2(\Omega) \cap H_0^1(\Omega)$, $\Omega = \mathbb{R}^n \backslash \mathcal{O}$. (For Sobolev spaces, see [1], [8]). The domain of the Neumann Laplacian Δ_N is all $u \in H^1(\Omega)$ such that $\Delta u \in L^2(\Omega)$ and for all $v \in H^1(\Omega)$,

$$\int_\Omega \{(\Delta u)v + \Delta u \cdot \Delta v\} dx = 0.$$

Let $B_0 = -\Delta$ on $L^2(\mathbb{R}^n)$ and let $B_1 = -\Delta_D$ or $-\Delta_N$ on $L^2(\mathbb{R}^n \backslash \mathcal{O})$. The acoustic wave equation $u_{tt} + c^2 B_k u = 0$ takes the form (5.1) with $N = 2$, $k \in \{0, 1\}$, $A_1^{(k)} = cB_k^{\frac{1}{2}} = -A_2^{(k)}$ with $J : L^2(\mathrm{R}^n) \to L^2(\mathrm{R}^n \backslash \mathcal{O})$ defined by $Jf = f \mid_{\mathbb{R}^n \backslash \mathcal{O}} = f \chi_{\mathbb{R}^n \backslash \mathcal{O}}$. The existence and completeness of $W_{\pm,j}$ follows from Lax and Phillips [10] or Wilcox [19] according as whether $\Delta = \Delta_D$ or Δ_N (For more recent results, see e.g. [20], [17].)

§6. Inverse Problems And Final Remarks

Using the above notation, let $v(t) = \begin{pmatrix} iA_1^{(k)} u(t) \\ u'(t) \end{pmatrix}$. Then the acoustic wave equation $u_{tt} + c^2 B_k u = 0$ is equivalent to

$$\frac{d}{dt} v(t) = i \begin{pmatrix} 0 & A_1^{(k)} \\ A_1^{(k)} & 0 \end{pmatrix} v(t) \tag{6.1}$$

on

$$\mathcal{H}_k = \begin{matrix} L^2(\mathbb{R}^n)^2 & \text{if } k = 0, \\ L^2(\mathbb{R}^n \backslash \mathcal{O})^2 & \text{if } k = 1. \end{matrix}$$

Define a unitary operator $\mathcal{Q}_1$ on $\mathcal{H}_1$ by $\mathcal{Q} = 2^{\frac{-1}{2}} \begin{pmatrix} I & I \\ I & -I \end{pmatrix}$. Then

$$\mathcal{Q}_1^* \begin{pmatrix} A_1^{(k)} & 0 \\ 0 & -A_1^{(k)} \end{pmatrix} \mathcal{Q}_1 = \begin{pmatrix} 0 & A_1^{(k)} \\ A_1^{(k)} & 0 \end{pmatrix}$$

and for $Jf = f\chi_{\mathbb{R}^n \backslash \mathcal{O}}$ and $\vec{J} = \begin{pmatrix} J & 0 \\ 0 & J \end{pmatrix}$, using $\mathcal{Q}_1 \vec{J} \mathcal{Q}_1^* = \vec{J}, A_j = A_1^{(j)}$, we have

$$\begin{aligned}
&\exp\left[it\begin{pmatrix}0 & A_1\\ A_1 & 0\end{pmatrix}\right]\vec{J}\exp\left[-it\begin{pmatrix}0 & A_0\\ A_0 & 0\end{pmatrix}\right]\\
&= \mathcal{Q}_1^*\exp\left[it\begin{pmatrix}A_1 & 0\\ 0 & -A_1\end{pmatrix}\right]\mathcal{Q}_1\vec{J}\mathcal{Q}_1^*\exp\left[-it\begin{pmatrix}A_0 & 0\\ 0 & -A_0\end{pmatrix}\right]\mathcal{Q}_1\\
&= \mathcal{Q}_1^*\begin{pmatrix}e^{itA_1} & 0\\ 0 & e^{-itA_1}\end{pmatrix}\begin{pmatrix}J & 0\\ 0 & J\end{pmatrix}\begin{pmatrix}e^{-itA_0} & 0\\ 0 & e^{itA_0}\end{pmatrix}\mathcal{Q}_1\\
&= \mathcal{Q}_1^*\begin{pmatrix}e^{itA_1}Je^{-itA_0} & 0\\ 0 & e^{-itA_1}Je^{itA_0}\end{pmatrix}\mathcal{Q}_1\\
&\to \mathcal{Q}_1^*\Omega_{1\pm}\mathcal{Q}_1 \text{ strongly as } t\to\pm\infty, \text{ where}
\end{aligned}$$

$$\Omega_{1\pm} = \begin{pmatrix}W_\pm(A_1, A_0; J) & 0\\ 0 & W_\mp(A_1, A_0; J).\end{pmatrix}$$

Let $W_\pm = \mathcal{Q}_1^*\Omega_{1\pm}\mathcal{Q}_1$. Then the corresponding (acoustic) scattering operator satisfies

$$\begin{aligned}
S_{1a} = W_+^*W_- &= \mathcal{Q}_1^*\Omega_{1+}^*\mathcal{Q}_1\mathcal{Q}_1^*\Omega_{1-}\mathcal{Q}_1\\
&= \mathcal{Q}_1^*(\Omega_{1+}^*\Omega_{1-})\mathcal{Q}_1\\
&= \mathcal{Q}_1^*S_1Q_1
\end{aligned}$$

where $S_1 = \Omega_{1+}^*\Omega_{1-}$. Since $\mathcal{Q}_1$ is known explicitly, S_{1a} and S_1 determine one another; similarly for $\Omega_\pm$ and $W_\pm$. Thus with J fixed, the inverse problem is solvable for the pair A_1, A_0 iff it is solvable for the pair

$$\begin{pmatrix}0 & A_1\\ A_1 & 0\end{pmatrix}, \begin{pmatrix}0 & A_0\\ A_0 & 0\end{pmatrix}.$$

The latter pair corresponds to the equations

$$u'' + A_k^2u = \prod_{j=1}^{2}\left(\frac{d}{dt} + i(-1)^jA_k\right)u = 0,\ k = 0, 1.$$

Similarly the fourth order equation

$$\prod_{j=1}^{2}\left(\frac{d^2}{dt^2}+c_j^2B_k\right)u(t)=0 \tag{6.2}$$

is equivalent to $\frac{d}{dt}v(t)=iH_kv(t)$ where

$$H_k=\frac{1}{2}\begin{pmatrix}0 & (c_1+c_2)B_k^{\frac{1}{2}} & 0 & (c_1-c_2)B_k^{\frac{1}{2}}\\ (c_1+c_2)B_k^{\frac{1}{2}} & 0 & (c_1-c_2)B_k^{\frac{1}{2}} & 0\\ 0 & (c_1-c_2)B_k^{\frac{1}{2}} & 0 & (c_1+c_2)B_k^{\frac{1}{2}}\\ (c_1-c_2)B_k^{\frac{1}{2}} & 0 & (c_1+c_2)B_k^{\frac{1}{2}} & 0\end{pmatrix},$$

$k=0,1$. (Cf. [14].)

Let $\vec{J}=$ $\text{Diag}_{4\times 4}(J)$ and $\mathcal{Q}_2=\frac{1}{2}\begin{pmatrix}\mathcal{Q}_1 & \mathcal{Q}_1\\ \mathcal{Q}_1 & -\mathcal{Q}_1\end{pmatrix}$. Then $W_\pm$ $=W_\pm(H_1,H_0;\vec{J})$ is the limit of

$$\begin{aligned}&\exp(itH_1)\vec{J}\exp(-itH_0)\\ &=\mathcal{Q}_2^*\begin{pmatrix}W_{1\pm} & & & 0\\ & W_{1\mp} & & \\ & & W_{2\pm} & \\ 0 & & & W_{2\mp}\end{pmatrix}\mathcal{Q}_2\\ &=\begin{pmatrix}\mathcal{Q}_1^*\Omega_{1\pm}\mathcal{Q}_1 & 0\\ 0 & \mathcal{Q}_1^*\Omega_{2\pm}\mathcal{Q}_1\end{pmatrix}\end{aligned}$$

where $W_{j\pm}=W_\pm(c_kB_1^{\frac{1}{2}},c_kB_0^{\frac{1}{2}};J)$, $j=1,2$, $\Omega_{2\pm}=\begin{pmatrix}W_{2\pm} & 0\\ 0 & W_{2\mp}\end{pmatrix}$, and $\mathcal{Q}_1,\Omega_{1\pm}$ are defined as before.

Thus the elastic scattering operator S_e of the pair of the fourth order equation is given by

$$S_e=W_+^*W_-=\begin{pmatrix}\mathcal{Q}_1^*S_{1a}\mathcal{Q}_1 & 0\\ 0 & \mathcal{Q}_1^*S_{2a}\mathcal{Q}_1\end{pmatrix}.$$

Hence the scattering operator for the pair (6.2) can be expressed in terms of the scattering operator for the pair (6.1) (For details see [14], [16].)

Thus if the map $\mathcal{O} \to S_a$ from the obstacle (modulo congruence) to the scattering operator for the acoustic wave equation is injective, then so is the map $\mathcal{O} \to S_e$, where S_e is the scattering operator for the elastic wave equation (with the same boundary condition).

For more general domains the inverse problem for obstacle scattering by sound waves was solved by the efforts of many. Principal contributors were A. Majda, R. Melrose and M. Taylor.

After this paper was completed we learned from Rainer Picard and Albert Milani about the recent PhD Thesis of Rainer Picard's PhD student, Nikos Kondoyannidis [21]. Kondoyannidis treats factored problems with commuting normal operators in a general context, using multiparameter spectral theory and distribution theory as in Picard [22]. He also develops a Birman-type trace class scattering theory for these higher order equations. Thus [21] is a valuable contribution to the theory of well-posedness and asymptotics for higher ordered factored equations in Hilbert space.

References

1. R. A. Adams, *Sobolev Spaces,* Academic Press, New York, 1975.

2. W.O. Amrein, J.M. Jauch, and K.B. Sinha, *Scattering Theory in Quantum Mechanics,* Benjamin, Reading, Mass., 1977.

3. M.S. Birman, A criterion for the existence of the complete wave operators in the theory of scattering with two spaces, in *Topics in Mathematical Physics, Vol. 4,* M.S. Birman ed., Consultants Bureau, New York, 1971.

4. H.L. Cycon, R.G. Froese, W. Kirsch, B. Simon, *Schrödinger Operators,* Springer, Berlin, 1987.

5. J. A. Goldstein, *Semigroups of Linear Operators and Applications,* Oxford U. Press, New York and Oxford, 1985.

6. J. A. Goldstein, An asymptotic property of solutions of wave equations, II, *J. Math. Anal. Appl.* 32 (1970), 392-399.

7. J.A. Goldstein and J.T. Sandefur, Jr., An abstract d'Alembert formula, *SIAM J. Math. Anal.* 18 (1987), 842-856.

8. F. John, *Partial Differential Equations* (3rd. ed.) Springer, New York, 1978.

9. T. Kato, *Perturbation Theory for Linear Operators*, Berlin, Springer 1966.

10. P.D. Lax and R. Phillips, *Scattering Theory,* Academic, New York, 1967.

11. R. Leis, *Initial Boundary Value Problems in Mathematical Physics,* Wiley, New York, 1986.

12. R. Leis and G. Roach, A transmission problem for the plate equation, *Proc. Roy. Soc. Edinburgh* 99A (1985), 285-312.

13. P. Perry, *Scattering Theory by the Enss Method,* Harwood, New York, 1983.

14. D. Pickett, Scattering theory for higher order equations, *Diff. Int. Eqns.* 3 (1990), 161-173.

15. M. Reed and B. Simon, *Methods of Modern Mathematical Physics. Vol. III: Scattering Theory,* Academic Press, New York, 1979.

16. G. Shi, *Mathematical Problems in Elasticity and Quantum Theory,* Ph.D. Thesis, Louisiana State University, expected in 1995.

17. R. Weder, *Spectral and Scattering Theory for Wave Propagation in Perturbed Stratified Media,* Springer, Berlin, 1991.

18. R. Weder, Spectral and scattering therory in perturbed stratified fluids. II. Transmission problems and exterior domains, *J. Diff. Eqns.* 64 (1986), 109-131.

19. C. Wilcox, *Scattering Theory for the d'Alembert Equation on Exterior Domains,* Lecture Notes in Math. N 442. Springer, Berlin, 1975.

20. D. R. Yafaev, *Mathematical Scattering Theory,* Translations of Mathematical Monographs, Vol. 105, Amer. Math. Soc., Providence, R.I. 1992.

21. N. Kondoyannidis, *Multiparameter Spectral Theory and Factored Hyperbolic Initial Value Problems*, Ph.D.Thesis, Univ. Wisconsin-Milwaukee, 1993.

22. R. Picard, Evolution equations as space-time operator equations, *J. Math. Anal. Appl.* 173 (1993), 436-458.

Department of Mathematics
Louisiana State University
Baton Rouge, LA 70803, USA

Y SAITO

Radiation condition method in spectral and scattering theory

1. Introduction

In this article we would like to present and review a method in spectral and time-independent scattering theory which has been successful for treating some Schrödinger-type operators in the framework of spectral and scattering theory. The method may be called the radiation condition method, since its starting point is a priori estimates for the radiation term and most of the important results are obtained through these estimates.

In §2 we shall explain the radiation condition method by using a very simple operator. The 2-body Schrödinger operators with short-range or long-range potentials will be discussed in §3. We could say that the radiation condition method made some contribution for studying these operators. Next in §4 we shall review several results related to the modified radiation condition which appears when we try to study the Schrödinger-type operators which is too singular at infinity to be controlled by the ordinary Sommerfeld radiation condition. As an application, we shall present some results on the inverse scattering problem for Schrödinger operators in §5. In the following two sections we shall review two new results on Dirac operators and the reduced wave operator with two unbounded media. Concluding remarks will be given in §8.

2. An Example

In this section we are going to give a brief explanation on the radiation condition method by looking at a very simple example of an ordinary differential operator

$$H = -\frac{d^2}{dr^2} + q(r) \qquad (r \geq 0) \tag{2.1}$$

in $L_2((0,\infty))$, where q is a real-valued, measurable function on $[0,\infty)$ satisfying

$$|q(r)| \leq C(1+r)^{-1-\epsilon} \qquad (r \geq 0) \tag{2.2}$$

with positive constants C and ϵ. Define the domain $D(h)$ as the set of all u such that

$$(2.3) \qquad \begin{cases} u \in L_2((0,\infty)), \\ u \in C^1([0,\infty)), \\ u(0) = 0, \\ u \ \text{ is locally absolutely continuous on } [0,\infty), \textit{and} \\ -\dfrac{d^2}{dr^2}u + q(r)u \in L_2((0,\infty)). \end{cases}$$

Then H is a selfadjoint operator. We denote the resolvent of H by $R(z)$, i.e., $R(z) = (H - x)^{-1}$. Let $t \in \mathbf{R}$. Then define the weighted Hilbert space $L_{2,t}((0,\infty))$ by

$$(2.4) \qquad L_{2,t}((0,\infty)) = \{f : (1+r)^t f \in L_2((0,\infty))\}.$$

The norm of $L_{2,t}((0,\infty))$ will be denoted by $\| \ \|_t$, i.e.,

$$(2.5) \qquad \|v\|_t = \Big[\int_0^\infty (1+r)^{2t}|v(r)|^2\,dr\Big]^{1/2}.$$

Let δ be a constant such that

$$(2.6) \qquad \delta > 1/2.$$

Let K be a bounded set of

$$(2.7) \qquad \mathbf{C}^+ = \{k = k_1 + ik_2 : k_1 \neq 0, \ k_2 > 0\},$$

and set $u = u(\cdot, k, f) = R(k^2)f, \ k \in K$. Then u satisfies the equation

$$(2.8) \qquad -u'' + qu - k^2u = f,$$

which can be rewritten as

$$(2.9) \qquad -(u' - iku)' - ik(u' - iku) = f.$$

Let $\rho(r) = \phi(r)(1+r)^{2\delta-1}$, where $\phi(r)$ is a real-valued, continuous, piecewise smooth and bounded function. Multiply (1.8) by $\rho\overline{(u' - iku)}$, integrate over (r, R) and take the real part. Then we have

(2.10)

$$\int_r^R (b\rho + \frac{1}{2}\rho')|u' - iku|^2\,dr$$

$$= \mathrm{Re}\int_r^R \rho f(\overline{u' - iku})\,dr + \int_r^R \rho q|u' - iku|^2\,dr + \left[\frac{1}{2}\rho|u' - iku|^2\right]_r^R.$$

It follows from the above equality that we obtain the following a priori estimate on the radiation condition term $u' - iku$:

$$\|u' - iku\|_{\delta-1} \le C\{\|u\|_{-\delta} + \|f\|_\delta\} \quad (u = u(\cdot,k,f),\ k \in K,\ f \in L_{2,\delta}((0,\infty)),\ \rho \ge 0) \tag{2.11}$$

with a positive constant $C = C(K)$ depending only on K. Another important fact is the uniqueness of the equation, i.e., if u satisfies the homogeneous equation $-u'' + qu - k^2u = 0$, the boundary condition $u(0) = 0$, and the radiation condition

$$\|u' - iku\|_{\delta-1} < \infty, \tag{2.12}$$

then u is identically zero. At the same time it follows from (2.11) and the equation (2.8) itself that

(2.13)

$$\|u\|^2_{-\delta,(\rho,\infty)} \le C(1+\rho)^{-(2\delta-1)}\{\|u\|^2_{-\delta} + \|f\|^2_\delta\}$$
$$(u = u(\cdot,k,f),\ k \in K,\ f \in L_{2,\delta}((0,\infty)),\ \rho \ge 0)$$

with a positive constant $C = C(K)$ depending only on K, where

$$\|u\|^2_{-\delta,(\rho,\infty)} = \int_{r>\rho} (1+r)^{-2\delta}|u(r)|^2\,dr. \tag{2.14}$$

Then we can prove the limiting absorption principle; *there exists the limits*

$$\lim_{\epsilon\downarrow 0} u(\cdot, k \pm i\epsilon, f) = u_\pm(\cdot,k,f) \qquad (k > 0) \tag{2.15}$$

in $L_{2,-\delta}((0,\infty))$. *Furthermore* $u = u_\pm(\cdot,k,f)$ *satisfy the radiation condition*

$$\|u' \mp iku\|_{\delta-1} < \infty, \tag{2.16}$$

Also the resolvent $R(z)$, $z = \lambda + i\eta \in \mathbf{C} - \mathbf{R}$, *has the boundary value* $R^\pm(\lambda)$, *which is called the extended resolvent, i.e.,*

$$\lim_{\eta\to 0,\,\pm\eta>0} R(\lambda + i\eta) = R^\pm(\lambda). \tag{2.17}$$

The simplest topology for the convergence of the limits (2.17) is the operator norm topology in $\mathbf{B}(L_{2,\delta}((0,\infty)), L_{2,-\delta}((0,\infty))$.

The radiation condition method is also useful to discuss the asymptotic behavior of $u = u_{\pm}(\cdot, k, f)$ at ∞ (or, in our example, as $r \to \infty$). In fact, it follows from (1.13) that $u_{\mp}(r) = u_{\pm}(r, k, f)$ satiny

$$u'_{\pm} \mp iku(r) = \text{small} \qquad (r \to \infty)$$
$$\Longrightarrow \ (e^{\mp ikr} u_{\pm})' = \text{small} \qquad (r \to \infty)$$
$$\Longrightarrow \ u_{\pm} \sim c_{\pm} e^{\pm ikr} \qquad (r \to \infty) \tag{2.18}$$

with constants $c_{\pm}$, although there are several technical difficulties and the proof of (2.18) is not very straightforward. The asymptotic limit of $u_{\pm}$ is the central ingredient of the spectral and scattering theory. It is well-known that the S matrix for the Schrödinger operator is obtained using the generalized eigenfunction. Also we can construct the spectral representation for the operator H from the asymptotic limit of $u_{\pm}(\cdot, k, f)$ (see §3). Thus the radiation condition method can cover most of the issues in spectral theory and stationary scattering theory.

Eidus [4] seems to be among early work to discuss the limiting absorption principle. The idea of radiation condition method including the spectral representation theory can be found in Jäger [12]. Through the 70's it was shown that the radiation condition method is succesfully applied to two body Schrödinger operators with short-range or long-range potentials, which we shall discuss in the following section.

3. 2-body Schrödinger Operators in $\mathbf{R}^n$

Consider the Schrödinger operator

$$H = -\Delta + Q(x) \qquad (x \in \mathbf{R}^n). \tag{3.1}$$

Here a real-valued function $Q(x)$ is either short-range

$$|Q(x)| \le C(1+|x|)^{-1-\epsilon} \qquad (x \in \mathbf{R}^n), \tag{3.2}$$

or long-range

$$|D^\alpha Q(x)| \le C(1+|x|)^{-|\alpha|-\epsilon} \qquad (x \in \mathbf{R}^n, |\alpha| = 0, 1, 2, \cdots), \tag{3.3}$$

ϵ and C being positive constants, and

$$\begin{cases} D^\alpha = \partial_1^{\alpha_1} \partial_2^{\alpha_2} \cdots \partial_N^{\alpha_N}, \\ \partial_j = \partial/\partial x_j \qquad (j = 1, 2, \cdots, N), \\ |\alpha| = \alpha_1 + \alpha_2 + \cdots + \alpha_N. \end{cases} \tag{3.4}$$

The Schrödinger operator H with its domain $H^2(\mathbf{R}^n)$, the second order Sobolev space, is a selfadjoint operator in $L_2(\mathbf{R}^n)$. When the radiation condition method is applied to the Schrödinger operator H, one way is to look at H as a second order ordinary differential operator with operator-valued coefficients. Let U be a unitary operator from $L_2(\mathbf{R}^n)$ onto $L_2((0,\infty), L_2(S^{n-1}), dr)$ given by

$$U : L_2(\mathbf{R}^n) \ni f(x) \longmapsto r^{(n-1)/2} f(r\omega) \in L_2(S^{n-1}), dr), \tag{3.5}$$

where $r = |x|$ and $\omega = x/|x| \in S^{n-1}$. Then we have

$$L = UHU^{-1} = -\frac{d^2}{dr^2} + B(r) + C(r) \tag{3.6}$$

with

$$\begin{cases} B(r) = r^{-2}\{-\Lambda_n + (n-1)(n-3)/4\}, \\ C(r) = Q(r\omega)\times, \end{cases} \tag{3.7}$$

where Λ_n is the Laplace-Beltrami operator on S^{n-1}. [12] started the study of the operator L and his work was extended by [25] and [26] to be applied to the Schrödinger operator with a short-range potential. Later the theory was further extended to discuss the Schrödinger operator with a long-range potential ([29], [30], [31], [32], [33]).

Another direction to study the Schrödinger operator H is to start with the n-dimensional versions of the formulas of (2.10) and (2.11). We are going to define the radiation condition terms: Let $\in \mathbf{C} - \mathbf{R}$. We set

(1) $k = k(z) = \sqrt{z}$, where the branch is taken so that $\operatorname{Im} k \geq 0$;
(2) $b = b(z) = \operatorname{Im} k(z)$;
(3) $\mathcal{D}_j u = \partial_j u + \{(n-1)/(2r)\}\tilde{x}_j u - ik(x)\tilde{x}_j u$, where $\tilde{x}_j = x_j/r$, $r = |x|$, $j = 1, 2, \cdots, n$;
(4) $\mathcal{D}u = \nabla u + \{(n-1)/(2r)\}\tilde{x}u - ik\tilde{x}u$, where $\tilde{x} = x/r$;
(5) $\mathcal{D}_r u = \mathcal{D}u \cdot \tilde{x} = \partial u/\partial r + \{(n-1)/(2r)\}u - iku$.

Also we introduce the weighted Hilbert space $L_{2,t}(\mathbf{R}^n)$ which is given by (2.4) with $(0,\infty)$ replaced by $\mathbf{R}^n$.

Let $u = R(z)f = (H-z)^{-1}f$ with $f \in L_2(\mathbf{R}^n)$. Then the n-dimensional version of (2.10) has the form

$$-\sum_{j=1}^{n} \partial_j \mathcal{D}_j u + \{\frac{n-1}{2r} - ik\}\mathcal{D}_r u + \frac{c_n}{r^2} u = f \tag{3.8}$$

with $c_n = (n-1)(n-3)/4$. Suppose that the potential Q is a sum of a long-range potential Q_1 and a short-range potential Q_2, i.e., $Q = Q_1 + Q_2$. Then the n-dimensional version of (2.11) is written as

(3.9)

$$\int_{B_{rR}} (b\varphi + \frac{1}{2}\frac{\partial\varphi}{\partial r})|\mathcal{D}u|^2\, dx$$
$$+ \int_{B_{rR}} (\frac{\varphi}{r} - \frac{\partial\varphi}{\partial r})(|\mathcal{D}u|^2 - |\mathcal{D}_r u|^2)\, dx$$
$$+ \int_{B_{rR}} c_n \left\{ \frac{b\varphi}{r^2} - 2^{-1}\frac{\partial}{\partial r}\left(\frac{\varphi}{r^2}\right)\right\} dx$$
$$= \mathrm{Re}\left[\int_{B_{rR}} \varphi f \overline{\mathcal{D}_r u}\, dx\right] - \mathrm{Re}\left[\int_{B_{rR}} \varphi Q_2 u \overline{\mathcal{D}_r u}\, dx\right]$$
$$+ \int_{B_{rR}} \left\{\frac{1}{2}\left(\frac{\partial\varphi}{\partial r}Q_1 + \varphi\frac{\partial Q_1}{\partial r}\right) - b\varphi Q_1\right\}|u|^2\, dx$$
$$- \frac{1}{2}\left[\int_{S_R} - \int_{S_r}\right]\varphi\left\{|\mathcal{D}u|^2 - 2|\mathcal{D}_r u|^2 + \left(Q_1 + \frac{c_n}{r^2}\right)|u|^2\right\} dS$$

where $0 < r < R < \infty$,

$$\begin{cases} B_{rR} = \{x \in \mathbf{R}^n \ : \ r < |x| < R\}, \\ S_t = \{x \in \mathbf{R}^n \ : \ |x| = t\}, \end{cases} \tag{3.10}$$

and $\varphi(x) = \varphi(|x|)$ is a real-valued, C^1 function on $\mathbf{R}^n$. By starting with the identity (3.10), Ikebe-Saiō [10] proved the limiting absorption principle for the Schrödinger operator H when the potential is a sum of a long-range potential and a short-range potential. [27] and [28] discussed the case where the potential is complex-valued and showed an asymptotic estimate for the extended resolvent $R^{\pm}(\lambda)$

$$\|R^{\pm}(\lambda)\|_{(0,\delta)}^{(0,-\delta)} = O(1/\sqrt{\lambda}) \qquad (\lambda \to \infty), \tag{3.11}$$

where $\delta > 1/2$ and $\| \ \|_{(0,s)}^{(0,t)}$ means the operator norm in $\mathbf{B}(L_{2,s}(\mathbf{R}^n), L_{2,t}(\mathbf{R}^n))$. This estimate was later used when we discussed the inverse scattering problem for Schrödinger operators with short-range potential by the high-energy method (see §5).

For $\lambda > 0$ and $f \in L_{2,\delta}(\mathbf{R}^n)$ let $u_{\pm}(k,f) = u_{\pm}(\cdot,k,f)$, $k = \sqrt{\lambda}$, be the solution of the equation with radiation condition

$$\|\nabla u_{\pm} \mp ik\tilde{x}u_{\pm}\|_{\delta-1} < \infty. \tag{3.12}$$

As was mentioned in §2, the asymptotic behavior of $u_{\pm}(x,k,f)$ as $x \to \infty$ is obtained ([12], [26], [31], [33]). In the short-range case the asymptotic formula is as follows:

$$(3.13)\qquad u_\pm(r\omega,k,f)\sim c_\pm(\omega,k,f)r^{-(n-1)/2}e^{\pm irk}\qquad (r\to\infty),$$

where $\omega\in S^{n-1}$ (for the long-range case, see the following section). The coefficient $c_\pm(\omega,k,f)$ is important in spectral representation theory for the Schrödinger operator. Consider the free Schrödinger operator $H_0 = -\Delta$ in $\mathbf{R}^3$. Then it is well-known that $u_\pm(x,k,f)$ has the form

$$(3.14)\qquad u_\pm(x,k,f)=\text{const.}\int_{\mathbf{R}^3}\frac{e^{\pm ik|x-y|}}{|x-y|}f(y)\,dy,$$

and it is easy to see that we have the asymptotic formula

$$(3.15)\qquad u_\pm(r\omega,k,f)\sim \text{const.}r^{-1}\int_{\mathbf{R}^3}e^{\pm ik\omega\cdot y}f(y)\,dy$$

as $r\to\infty$, i.e., the coefficient $c_\pm(\omega,k,f)$ in this case is essentially the Fourier transform of f. This is generally true for the 2-body Schrödinger operator H with a short-range potential in $\mathbf{R}^n$, and we can develop spectral representation theory by defining the generalized Fourier transform by the asymptotic coefficients $c_\pm$ ([12], [26]). We can develop a similar theory for the long-range case. However the situation is more complicated ([32], [33]), which we shall discuss in the next section.

The 2-body Schrödinger operator has been studied by more traditional method which comes from theory of liner partial differential equations. Here we refer to the celebrated work by S. Agmon (Agmon [1]).

4. Modified Radiation Conditions

The classical form of the radiation condition (3.12) works only when the potential decays sufficiently fast and uniformly at infinity. Even when the long-range case, although the radiation condition (3.12) is sufficient to obtain the limiting absorption principle, it is not the "right" radiation condition to obtain the asymptotic formula for the solution $u_\pm(\cdot,k,f)$. In fact, we have the following ([31], [33]): *Let H be the Schrödinger operator with the potential $Q=Q_1+Q_2$, where Q_1 and Q_2 are long-range and short-range, respectively. Then there exists a real-valued function $\alpha(x,k)$, $x\in\mathbf{R}^n, k>0$, such that*

$$(4.1)\qquad u_+(x,k,f)\sim c_+(\omega,k,f)r^{-(n-1)/2}e^{i(rk-\alpha(r\omega,k))}\qquad (r\to\infty),$$

(a similar asymptotic formula holds for $u_-(\cdot,k,f)$). For each $k>0$, the (stationary) modifier $\alpha(x,k)$ is obtained as a solution or an approximate solution of the equation

$$|\nabla\alpha|^2 = 2k\frac{\partial\alpha}{\partial|x|} - Q_1(x). \tag{4.2}$$

As for the (time-dependent or stationary) modifiers for Schrödinger operators with long-range potentials, cf., e.g., [8], [9]. The starting point of the proof of the asymptotic formula (4.1) is the estimate

$$\|\nabla u_+ - i(k - \nabla\alpha)u\|_\beta \leq C\|f\|_{1+\beta}, \tag{4.3}$$

where β is a nonnegative constant and C is independent of f. Thus the modified radiation condition is naturally introduced. Proceeding as in the preceding section, we can construct the generalized Fourier transform from the asymptotic coefficients $c_\pm$ to develop spectral representation theory. As an application we can express the S-matrix of the Schrödinger operator with a long-range potential by using the eigenoperator and the potential ([34]). This is an extension of the well-known formula for the S-matrix for the Schrödinger operator with a short-range potential.

Suppose that the potential decays slower than a long-range potential or the potential does not decay at all. There a few example has been known in which we need modified radiation conditions even to prove the limiting absorption principle. Among others, one example is oscillating long-range potentials such as

$$Q(x) = \frac{\cos|x|}{|x|}. \tag{4.4}$$

As for Schrödinger operators with oscillatory long-range potentials, see Mochizuki-Uchiyama [14], [15], [16]. Also cf. [3]. Another example is potentials which satisfy (3.3) with $\epsilon = 0$, i.e.,

$$\begin{aligned} |D^\alpha Q(x)| \leq & C(1 + |x|)^{-|\alpha|} \\ & (x \in \mathbf{R}^n, |\alpha| = 0, 1, 2, \cdots), \end{aligned} \tag{4.5}$$

such as

$$Q(x) = \cos(x_1/|x|) \qquad (x = (x_1, x_2, \cdots x_n)). \tag{4.6}$$

For this type of potentials see [39], especially for the homogeneous potential see [7].

5. Inverse Scattering Problem

As an application of the radiation condition method, we studied the inverse scattering problem for Schrödinger equations and plasma wave equations with short-range potentials. Let $S(k)$ be the S-matrix for the Schrödinger operator with a short-range potential Q, and set

$$F(k) = -2\pi i k^{-(n-2)}(S(k) - I) \qquad (k > 0). \tag{5.1}$$

For each $k > 0$, $F(k)$, which is called the scattering amplitude, is a bounded operator on $L_2(S^{n-1})$. Let $\phi_{k,x}(\omega) \in L_2(S^{n-1})$ be defined by

$$\phi_{k,x}(\omega) = e^{-ikx\cdot\omega}, \tag{5.2}$$

where $k > 0$ and $x \in \mathbf{R}^n$ are parameters. Then our fundamental relationship between the S-matrix $S(k)$ and the potential Q is given by

$$\lim_{k\to\infty}(F(k)\phi_{k,x}, \phi_{k,x})_{S^{n-1}} = -2\pi \int_{\mathbf{R}^n} \frac{Q(y)}{|y-x|^{n-1}}\,dy \tag{5.3}$$

for any $k > 0$ and $x \in \mathbf{R}^n$, where $(\ )_{S^{n-1}}$ is the inner product of S^{n-1} ([35], [37]). The inequality (3.11) played an important role to obtain (5.3). Staring with the asymptotic formula (5.3), we can show the uniquness of the inverse scatterin problem. We also presented formulas to reconstruct the potential from the scattering data and discussed how the formula approximates the potential as the energy parameter k increases ([36], [38], [40]).

Recently V. Enss and R. Weder [6] has shown the uniquness and presented a reconstruction formula for the short-ramge potential of a N-body Schrödinger operator where the potential of the N-body Schrödinger opertor is a sum of a short-range potential and a long-range potential and the long-range potential is assumed to be known. Their method is the high energy method, too, and first they gave an asymptotic formula similar to (5.3). Obviously the inverse scattering problem for long-range potential seems to be an interesting problem.

6. Resolvent of Dirac Operators

In this and the following sections we shall discuss some rore recents results. First consider the Dirac operator

$$H = -i\sum_{j=1}^{3} \alpha_j \frac{\partial}{\partial x_j} + \beta + Q(x) \tag{6.1}$$

in $[L_2(\mathbf{R}^3)]^4$, where

$$\begin{cases} x = (x_1, x_2, x_3) \in \mathbf{R}^3, \\ \alpha_j \ (j = 1,2,3,4,\ \alpha_4 = \beta) : 4 \times 4 \text{ constant Hermitian symmetric matrices} \\ \qquad\qquad \text{satisfying the anticommutation relations} \\ \qquad\qquad\qquad\qquad \alpha_j \alpha_k + \alpha_k \alpha_j = 2\delta_{jk} I, \qquad\qquad (j,k = 1,2,3,4), \\ Q(x) = (q_{jk}(x)) : 4 \times 4 \text{ Hermitian matrix-valued function on } \mathbf{R}^3. \end{cases}$$

Let $R(z)$ be the resolvent of H and let $R_\pm(\lambda)$, $|\lambda| > 1$, be the extended resolvent of H (as for the spectral and scattering theory for Dirac operators see, e.g., Yamada [44], [45]). The work [19] $\sim$ [21] studied the extended resolvent $R_\pm(\lambda)$ under the assumption that the potential $Q(x)$ is short-range. we proved that, among others, the operator norm $\|R_\pm(\lambda)\|$ bounded as $|\lambda| \to \infty$, where $\|R_\pm(\lambda)\|$ is the operator norm in $\mathbf{B}(L_{2,\delta}, L_{2,-\delta})$. This is the best possible result since Yamada [46] has recently proved that the operator norm of the extended resolvent of the Dirac operator cannot decay, although (3.11) shows that the operator norm of the extended resolvent of the Schrödinger operator does decay as the energy parameter λ increases. Also we proved the formulas for the extended resolvent by the use of the extended resolvent of $-\Delta$ (cf. [2] where similar idea is developed). In the case of the extended resolvent $R_{0\pm}(\lambda)$ of the free Dirac operator H_0, for example, we have, for $\lambda > 1$,

$$R_{0\pm}(\lambda) = \Gamma_{0\pm}(\lambda^2 - 1) A_\lambda + B_\lambda, \tag{6.2}$$

where $\Gamma_{0\pm}$ is the extended resolvent of the free Schrödinger operator $-\Delta$, and the operators A_λ and B_λ are explicitly given. These formulas show the strong relationship between the Dirac operator and the Schrödinger operator, and we could prove that $R_\pm(\lambda)$ satisfies the radiation condition. We have used the theory of pseudodifferential operators. After these works, using the Mourre method (Mourre [18]), H. Ito [11] showed that the boundedness of $\|R_\pm(\lambda)\|$ hold even for the Dirac operator with a long-range potential (cf. [17]). These results may enable us to investigate the inverse scattering problem for Dirac operators by using the known methods and results for Schrödinger operators.

7. Reduced Wave Operator in Two Unbounded Media

Let

$$H = -\mu(x)^{-1}\Delta. \tag{7.1}$$

in $\mathbf{R}^n$. Here $\mu(x)$ is a positive, simple function on $\mathbf{R}^n$ given by

$$\mu(x) = \mu_j \qquad (x \in \Omega_j,\ j = 1,2), \tag{7.2}$$

where $\mu_1, \mu_2 > 0$, $\mu_1 \neq \mu_2$, and Ω_ℓ, $\ell = 1,2$, are open sets of $\mathbf{R}^n$ such that

$$
\left\{
\begin{array}{l}
\Omega_1 \cap \Omega_2 = \emptyset, \\
\overline{\Omega_1} \cup \Omega_2 = \Omega_1 \cup \overline{\Omega_2} = \mathbf{R}^n.
\end{array}
\right.
\tag{7.3}
$$

First D. Eidus [5] proved the limiting absorption principle for the case that the separating surface

$$
S = \partial\Omega_1 = \partial\Omega_2 \tag{7.4}
$$

has a cone-like shape. His results was improved by [41]. Then in the papers [22] ~ [24], G. Roach and B. Zhang, under the same assumption on the separating surface S, showed that the modified radiation condition

$$
\lim_{R\to\infty} R^{-1} \int_{|x|<R} |\nabla u - i\sqrt{\lambda\mu(x)}u|^2 \, dx = 0. \tag{7.5}
$$

does guarantee the uniqueness, and they developed spectral and scattering theory for the operator H. Recently Jäger-Saitō [13] proved the limiting absorption principle for the case that S can have a cylinder-like shape or a plane-like shape. This result has been obtained by the radiation condition method. Also we can give another proof of the results in [22] ~ [24] in quite a similar manner. Thus we are now able to treat the both cases by the radiation condition method.

8. Concluding remarks

We have seen that the radiation condition method can handle various Schrödinger-type operators in the framework of spectral and scattering theory. It seems that the "reasonable" solutions of Schrödinger-type equations satisfy a kind of radiation condition at infinity. Further it seems that, if we can find the "right" radiation condition for a given Schrödinger-type operator, we can get very useful information on the operator through the radiation condition term.

Another powerful method in spectral theory is the Mourre method (E. Mourre [18]). One of the strong points of Mourre method is that it can be applied to the many-body Schrödinger operator. So far the radiation condition method has not been successfully applied to the many-body Schrödinger operator. However, the recent work of D. Yafaev [42], [43] may give a good starting point when we try to apply the radiation condition method to the many-body Schrödinger operator.

References

[1] Spectral properties of Schrödinger operators and scattering theory, Ann. Scuola Nor. Sup. Pisa **2** (1975), 151-218.

[2] E. Baleslev and B. Helffer, *Limiting absorption principle and resonances for the Dirac operator*, Advances in Applied Mathematics **13** (1992), 186-215.

[3] A. Devinatz, R. Mocckel and P. Rejto, *A limiting absorption princple for Schrödinger operators with von Nuemann-Wigner type potentials*, Integral Equations and Operator Theory **14** (1991), 13-68.

[4] D. Eidus, *The principle of limiting absorption*, Amer. Math. Soc. Translations **47** (1965), 157-191 (Mat. Sb. (N.S.) 57 (99) (1965)).

[5] D. Eidus, *The limiting absorption and amplitude problems for the diffraction problem with two unbounded media*, Comm. Math. Phys. **107** (1986), 29-38.

[6] V. Enss and R. Weder, *Uniqueness and reconstruction formula inverse N-particle scattering*, To appear in the Proceeding of the Conference on Differential Equations and Mathematical Physics, Birmingham, AL, March 1994.

[7] I. Herbst, *Spectral and scattering theory for Schrödinger operators with potentials independent of $|x|$*, Amer. J. Math. **113** (1991), 509-565.

[8] L. Hörmander, *The existence of wave operators in scattering theory*, Math. Z. **145** (1976), 61-91.

[9] T. Ikebe and H. Isozaki, *Completeness of modified wave operators for long-range potentials*, Publ. RIMS, Kyoto Univ. **15** (1979), 679-718.

[10] T. Ikebe and Y. Saitō, *Limiting absorption method and absolute continuity for the Schrödinger operator*, J. Math. Kyoto Univ. **12** (1972), 513-612.

[11] H. Ito, *Talks at the operator theory seminar of Kyoto University, Japan, 1993.*

[12] W. Jäger, *Ein gewöhnlicher Differentialoperator zweiter Ordnung für Funktionen mit Werten in einem Hilbertraum*, Math. Z. **113** (1970), 68-98.

[13] W. Jäger and Y. Saitō, *The Limiting absorption principle for the reduced wave operator with cylindrical discontinuity. Preprint.*

[14] K. Mochizuki and J. Uchiyama, *Radiation conditions and spectral theory for 2-body Schrödinger operator with "oscillating" long-range potentials I – the principle of the limiting absorption –*, J. Math. Kyoto Univ. **18** (1978), 377-408.

[15] K. Mochizuki and J. Uchiyama, *Radiation conditions and spectral theory for 2-body Schrödinger operator with "oscillating" long-range potentials II – spectral representation –*, J. Math. Kyoto Univ. **19** (1979), 40-70.

[16] K. Mochizuki and J. Uchiyama, *Radiation conditions and spectral theory for 2-body Schrödinger operator with "oscillating" long-range potentials III*, J. Math. Kyoto Univ. **21** (1981), 605-618.

[17] A. Monvel-Berthier, D. Manda and R. Purice, *Limiting absorption principle for the Dirac operator*, Ann. Inst. Henri Poincarë **58** (1993), 413-431.

[18] E. Mourre, *Absence of singular continuous spectrum for certain self-adjoint operators*, Commun. Math. Phys. **78** (1991), 391-408.

[19] C. Pladdy, Y. Saitō and T. Umeda, *Resolvent estimate for Dirac operators*, to appear in Analysis.

[20] C. Pladdy, Y. Saitō and T. Umeda, *Radiation condition for Dirac operators*, Submitted (1993).

[21] C. Pladdy, Y. Saitō and T. Umeda, *Asymptotic behavior of the resolvent of the Dirac operator*, Operator Theory: Advances and Applications, Birkhäuser **70** (1993), 45-54.

[22] G. Roach and B. Zhang, *The limiting amplitude principle for the wave propagation problem with two unbounded media*, Math. Proc. Camb. Phil. Soc. **112** (1992), 207-223.

[23] G. Roach and B. Zhang, *The limiting amplitude principle for the wave propagation problem with two unbounded media*, Math. Proc. Camb. Phil. Soc. **112** (1992), 207-223.

[24] G. Roach and B. Zhang, *Spectral representation and scattering theory for the wave equation with two unbounded media*, Math. Proc. Camb. Phil. Soc. **113** (1993), 423-447.

[25] Y. Saitō, *The principle of limiting absorption for second-order differential equations with operator-valued coefficients*, Publ. RIMS, Kyoto Univ. **7** (1972), 518-619.

[26] Y. Saitō, *Spectral and scattering theory for second-order differential operators with operator-valued coefficients*, Osaka J. Math. **9** (1972), 463-498.

[27] Y. Saitō, *The principle of limiting absorption for the non-selfadjoint Schrödinger operator in R^N $(N \neq 2)$*, Publ. RIMS, Kyoto Univ. **9** (1974), 397-428.

[28] Y. Saitō, *The principle of limiting absorption for the non-selfadjoint Schrödinger operator in R^2*, Osaka J. Math. **11** (1974), 295-306.

[29] Y. Saitō, *Spectral theory for second-order differential operators with long-range operator-valued coefficients I*, Japan J. Math., New Ser. **1** (1975), 311-349.

[30] Y. Saitō, *Spectral theory for second-order differential operators with long-range operator-valued coefficients II*, Japan J. Math., New Ser. **1** (1975), 351-385.

[31] Y. Saitō, *On the asymptotic behavior of the solution of the Schrödinger equation* $(-\Delta + Q(y) - k^2)V = F$, Osaka J. Math. **14** (1977), 11-35.

[32] Y. Saitō, *Eigenfunction expansions for the Schrödinger operators with long-range potentials* $Q(y) = O(|y|^{-\epsilon})$, $\epsilon > 0$, Osaka J. Math. **14** (1977), 37-52.

[33] Y. Saitō, *Spectral Representations for Schrödinger Operators with Long-Range Potentials*, Lecture Notes in Mathematics 727, Springer, 1979.

[34] Y. Saitō, *On the S-matrix for Schrödinger operators with long-range potentials*, J. Reine Angew. Math. **324** (1980), 99-116.

[35] Y. Saitō, *Some properties of the scattering amplitude and the inverse scattering problem*, Osaka J. Math. **19** (1982), 527-547.

[36] Y. Saitō, *An inverse problem in potential theory and the inverse scattering problem, J. Math. Kyoto Univ.*.

[37] Y. Saitō, *An asymptotic behavior of S-matrix and the inverse scattering problem*, J. Math. Phys. **25** (1984), 3105-3111.

[38] Y. Saitō, *An approximation formula in the inverse scattering problem*, J. Math. Phys. **27** (1986), 1145-1153.

[39] Y. Saitō, *Schrödinger operators with a nonspherical radiation condition*, Pacific J. Math. **126** (1987), 331-359.

[40] Y. Saitō, *Inverse scattering for the plasma wave equation starting with large-t data*, J. Phys. A: Math. Gen. **21** (1988), 1623-1631.

[41] Y. Saitō, *A remark on the limiting absorption principle for the reduced wave equation with two unbounded media*, Pacific J. Math. **136** (1989), 183-208.

[42] D. Yafaev, *Radiation condition and scattering theory for three-particle Hamiltonians*, Preprint 91-01. Nantes University (1992).

[43] D. Yafaev, *Radiation condition and scattering theory for N-particle Hamiltonians*, Commun. Math. Phys **154** (1993), 523-554.

[44] O. Yamada, *On the principle of limiting absorption for the Dirac operator*, Publ. RIMS Kyoto Univ. **8** (1973), 557-577.

[45] O. Yamada, *Eigenfunction expansions and scattering theory for Dirac operators*, Publ. RIMS Kyoto Univ. **11** (1976), 651-689.

[46] O. Yamada, *A remark on the limiting absorption method for Dirac operators*, Proc. Japan Acad. **69, Ser A** (1993), 243-246.

N SAUER

Implicit evolution equations and empathy theory

1. Introduction

Evolution equations of the form

$$\begin{aligned} \frac{d}{dt}[Bu(t)] &= Au(t) \\ \lim_{t \to 0+} Bu(t) &= y \in Y \end{aligned} \tag{1}$$

with $A, B : \mathfrak{D} \subset X \to Y$ unbounded linear operators defined on a domain $\mathfrak{D}$ in a Banach space X with values in a Banach space Y often occur in applications such as non-Newtonian fluid mechanics (with nonlinear terms omitted) and dynamical boundary conditions. It frequently turns out that the operator B is not closeable so that it becomes impossible to transform the equations in (1) in such a way that $u(t)$ appears explicitly. For that reason it is virtuous to study (1) as it is [Showalter (1988)]. Study of abstract evolution equations in the form (1) have also been studied by [Sauer (1983), Sauer & Singleton (1987)] with the aid of the concept of *B-evolution.* That concept was applied to problems of 'parabolic' type by [Sauer (1983), Grobbelaar & Sauer (1989, 1993), Van der Merwe (1988)], but applications to wave-like phenomena seem to be extremely tedious. The notion of *empathy* was introduced by [Sauer & Singleton (1989)] as a generalization of B-evolution and studied by [Conradie & Sauer (1994)] in a restricted setting. In this paper a much more general approach to the notion of empathy will be put forth. The results obtained make applications to wave problems quite straightforward.

2. Heuristics

Since the state $u(t)$ is in a different space (X) than the data ($y \in Y$), we must imagine a system in which *cause* (data) and *effect* (solution) 'live in different worlds'. In order to obtain a progression in time, it becomes necessary to think of a curve $v : t \to v(t); t \geq 0$ in the space Y, emanating from the initial state y which is such that every point $v(t)$ on this curve becomes a 'cause' for subsequent effects $u(t+s) \in X$. This framework is best explained graphically as in Figure 1. Let us represent the points $u(t)$ and $v(t)$ on the two curves by means of two families of linear operators $S(t) : Y \to X$ and $E(t) : Y \to Y$ which relate the initial state y to the points on the curves according to $u(t) = S(t)y$ and $v(t) = E(t)y$. Suppose that the family $\{S(t) : t > 0\}$ has the *evolution property* $u(t+s) = S(s)v(t) = S(s)E(t)y$. Then the *empathy relation* $S(t+s) = S(s)E(t) = S(t)E(s)$ becomes a natural property.

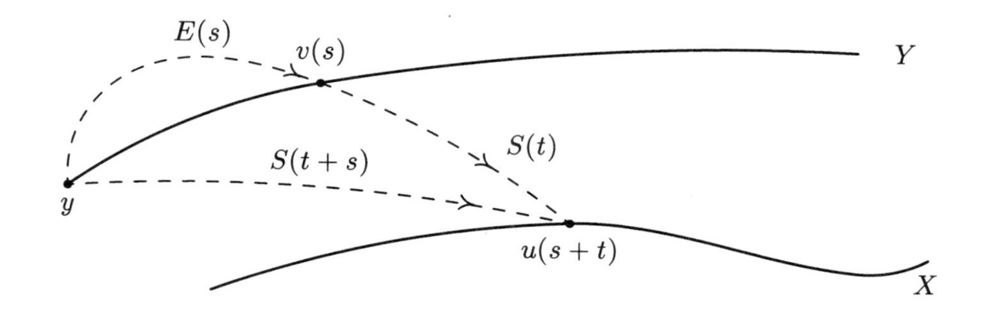

Figure 1. Heuristic setting

If we suppose, in addition, that $u(t)$ and $v(t)$ depend continuously on the initial state y, we must assume that the operators $S(t)$ and $E(t)$ are bounded. Additionally, we are interested in obtaining the operators A and B in (1) as a 'generator' of the 'double family' $\langle S(t), E(t)\rangle$, the possibility of defining an infinitesimal generator in accordance with (1) appears to be somewhat artificial. Instead, we shall use a Laplace transform approach, which gives with (1) in mind, the Laplace transform of $S(t)$ as the inverse of the operator $\lambda B - A$.

3. Empathy theory

Let X and Y be Banach spaces and $\mathcal{E} = \{E(t) : t > 0\}$ and $\mathcal{S} = \{S(t) : t > 0\}$ be two families of bounded linear operators such that $E(t) : Y \to Y$ and $S(t) : Y \to X$. We shall assume that for every $y \in Y$ and $\lambda > 0$ the Laplace transforms

$$R(\lambda)y = \int_0^\infty e^{-\lambda t} E(t)y\, dt; \quad P(\lambda)y = \int_0^\infty e^{-\lambda t} S(t)y\, dt$$

exist as Lebesgue integrals. $\langle \mathcal{E}, \mathcal{S}\rangle$ is called *an empathy* if the following conditions hold:

$$S(t+s) = S(t)E(s) \tag{2}$$

The operators $R(\lambda)$ *and* $P(\lambda)$ *are bounded* (3)

For some $\xi > 0$, $P(\xi)$ *is invertible.* (4)

Note that there are no requirements on the behaviour of $E(t)y$ or $S(t)y$ as $t \to 0+$.

PROPOSITION 3.1. *The family* $\mathcal{E}$ *is a semigroup.*

Proof. From (2) we see that for every $y \in Y$, $S(t)[E(r+s)y - E(r)E(s)y] = 0$. According to (4), $E(r+s)y = E(r)E(s)y$ for positive r and s. □

PROPOSITION 3.2. *The norms $\|E(t)\|$ and $\|S(t)\|$ are uniformly bounded on compact intervals $I \subset (0, \infty)$ and $\mathcal{E}$ and $\mathcal{S}$ are strongly continuous on $(0, \infty)$.*

Proof. Since $E(t)$ is a semigroup and $t \to E(t)y$ is strongly measurable, the statement regarding $E(t)$ is well-known [Miyadera (1951), Hille-Phillips (1957)]. Strong continuity of $S(t)$ now follows from the empathy relation (2) and the boundedness of $S(t)$. Having proved strong continuity, the local boundedness of $\|S(t)\|$ follows from the uniform boundedness principle. □

The following result is proved by taking Laplace transforms in the empathy relation and the semigroup relation:

PROPOSITION 3.3. (The pseudo-resolvent equations). *For all positive λ, μ*

$$R(\lambda) - R(\mu) = (\mu - \lambda)R(\lambda)R(\mu) \tag{5}$$

$$P(\lambda) - P(\mu) = (\mu - \lambda)P(\lambda)R(\mu). \tag{6}$$

From (5) and (6) it is a simple matter to derive the following:

PROPOSITION 3.4. *For any positive λ, μ it is true that $R(\lambda)[Y] = R(\mu)[Y] = \mathfrak{D}_E$ and $P(\lambda)[Y] = P(\mu)[Y] = \mathfrak{D}$. Moreover, $E(t)[\mathfrak{D}_E] \subset \mathfrak{D}_E$ and $S(t)[\mathfrak{D}_E] \subset \mathfrak{D}$.*

The proof of the inclusions is based on the expressions

$$R(\lambda)E(t) = E(t)R(\lambda) \tag{7}$$

$$P(\lambda)E(t) = S(t)R(\lambda) \tag{8}$$

which are derived from the identities $E(t)E(s) = E(s)E(t)$ and $S(t)E(s) = S(s)E(t)$.

PROPOSITION 3.5. *For any $\lambda > 0$ the operator $P(\lambda)$ is invertible.*

Proof. Since $R(\lambda)$ is a pseudo-resolvent, it is well-known that $N_E = \text{Ker } P(\lambda)$ is independent of λ. It is easily proved, making use of (6), that $N_E \cap \text{Ker } P(\lambda) = N_E \cap \text{Ker } P(\mu)$ for any positive λ, μ. Suppose that $y \in \text{Ker } P(\lambda)$. Then, from (6), $P(\xi)y = (\lambda - \xi)P(\xi)R(\lambda)y$. Since $P(\xi)$ is invertible, it follows that $y = (\lambda - \xi)R(\lambda)y$. Therefore, $R(\xi)y = (\lambda - \xi)R(\xi)R(\lambda)y$. Comparison with (5) gives $R(\lambda)y = 0$ which implies that $y \in N_E \cap \text{Ker } P(\lambda) = N_E \cap \text{Ker } P(\xi)y = \{0\}$. Thus $y = 0$. □

We now proceed to define the concept of generator of an empathy by introducing the operators $A_\lambda = (\lambda R(\lambda) - I)P^{-1}(\lambda) : \mathfrak{D} \to Y$ and $B_\lambda = R(\lambda)P^{-1}(\lambda) : \mathfrak{D} \to \mathfrak{D}_E$.

PROPOSITION 3.6. *The operators A_λ and B_λ do not depend on λ.*

Proof. This follows from the inversion

$$P^{-1}(\mu) - P^{-1}(\lambda) = (\mu - \lambda)R(\mu)P^{-1}(\mu) = (\mu - \lambda)R(\lambda)P^{-1}(\lambda) \tag{9}$$

proving the statement first for B_λ. □

In accordance with this result we set $A = A_\lambda$ and $B = B_\lambda$. The operator pair $\langle A, B \rangle$ will be called the *generator* of the empathy $\langle \mathcal{E}, \mathcal{S} \rangle$. We collect a number of important properties in the following result.

PROPOSITION 3.7. *Let $\langle A, B\rangle$ be the generator of an empathy $\langle \mathcal{E}, \mathcal{S}\rangle$. Then $P(\lambda) = (\lambda B - A)^{-1}$, $B[\mathfrak{D}] = \mathfrak{D}_E$, $E(t)y = BS(t)y$ for every $y \in \mathfrak{D}_E$ and $S(t+s)y = S(t)BS(s)y$ for $y \in \mathfrak{D}_E$*

The proof follows directly from the definitions of A and B and equations (7) and (8).

Next we obtain some representations with important consequences. If $y \in \mathfrak{D}_E$, we may write $y = R(\lambda)y_\lambda$ for some $y_\lambda \in Y$.

LEMMA 3.8. *If $y \in \mathfrak{D}_E$ then*

$$E(t)y = e^{\lambda t}[y - \int_0^t e^{-\lambda s} E(s)y_\lambda \, ds] \tag{10}$$

$$S(t)y = e^{\lambda t}[P(\lambda)y_\lambda - \int_0^t e^{-\lambda s} S(s)y_\lambda \, ds]. \tag{11}$$

For any $y \in Y$, $\int_0^t e^{-\lambda s} S(s)y \, ds \in \mathfrak{D}$ and

$$E(t)y = e^{\lambda t}[y - P^{-1}(\lambda) \int_0^t e^{-\lambda s} S(s)y \, ds]. \tag{12}$$

(10) and (11) are obtained by direct calculation. (12) may obtained in a similar way, making use of (8).

COROLLARY 3.9. *For $y \in \mathfrak{D}_E$, $\lim_{t\to 0+} E(t)y = y$. There exists a linear operator $C : \mathfrak{D}_E \to \mathfrak{D}$ such that*

$$Cy = \lim_{t \to 0+} S(t)y = P(\lambda)y_\lambda. \tag{13}$$

We are now set to obtain the relationship between the generator of an empathy and the Cauchy-problem (1):

THEOREM 3.10. *Let $\langle A, B\rangle$ be the generator of an empathy $\langle \mathcal{E}, \mathcal{S}\rangle$. For $y \in \mathfrak{D}_E$, $u(t) = S(t)y$ solves the Cauchy-problem (1).*

Proof. According to Proposition 3.4, $u(t) \in \mathfrak{D}$. Let $v(t) = Bu(t) = BS(t)y = E(t)y$ (Proposition 3.7). By (10) and Proposition 3.2, $v(t)$ is differentiable. Indeed, differentiation of (10) gives

$$\begin{aligned}
\frac{d}{dt}[Bu(t)] = \frac{d}{dt}E(t)y &= \lambda v(t) - E(t)y_\lambda \\
&= (\lambda E(t)R(\lambda)y_\lambda - E(t)y_\lambda \\
&= (\lambda R(\lambda) - I)P^{-1}(\lambda)P(\lambda)E(t)y_\lambda \\
&= AP(\lambda)E(t)y_\lambda = AS(t)R(\lambda)y_\lambda = AS(t)y = Au(t).
\end{aligned}$$

That the initial condition is satisfied, follows from Corollary 3.9. □

REMARK 3.11. It is seen from Lemma 3.8 that $u(t)$ itself is differentiable, but differentiation of it, does not lead to any particularly interesting equation.

Finally, we investigate the relationship between the operators B and C.

THEOREM 3.12. *$C[\mathfrak{D}_E] = \mathfrak{D}$. C is invertible with inverse B. The operators $R(\lambda)$ are resolvents of $\mathcal{A} = AC : \mathfrak{D}_E \to Y$.*

Proof. If $x = P(\lambda)z_\lambda \in \mathfrak{D}$, then for $y = R(\lambda)z_\lambda$, $Cy = P(\lambda)z_\lambda = x$. Next, if $y = R(\lambda)y_\lambda$, then $BCy = R(\lambda)P^{-1}(\lambda)P(\lambda)y_\lambda = R(\lambda)y_\lambda = y$. Thus C is invertible with inverse B. Once this is known, it follows from the expression $R(\lambda) = BP(\lambda)$ that the $R(\lambda)$ are invertible, and are therefore resolvent operators. The same expression implies that $R(\lambda) = B(\lambda B - A)^{-1} = (\lambda - AC)^{-1}$. □

4. Uniformly bounded empathies.

If for all $t > 0$, there exist $M > 0$ and $N > 0$ such that $\|E(t)\| \leq M$ and $\|S(t)\| \leq N$ the empathy $\langle \mathcal{E}, \mathcal{S} \rangle$ is said to be *uniformly bounded.* For such empathies it is not necessary to require that the resolvents $R(\lambda)$ and $P(\lambda)$ be bounded. Indeed, we have

THEOREM 4.1. *If $\langle \mathcal{E}, \mathcal{S} \rangle$ is a bounded empathy, then*

$$\begin{aligned} \|R^k(\lambda)\| &\leq \frac{M}{\lambda^k}; \quad k > 0 \\ \|P(\lambda)R^k(\lambda)\| &\leq \frac{N}{\lambda^{k+1}}; \quad k \geq 0. \end{aligned} \tag{14}$$

Moreover, the operator C is bounded.

Proof. These inequalities follow from the identities

$$R^{k+1}(\lambda)y = \frac{1}{k!}\int_0^\infty e^{-\lambda t}t^k E(t)y\,dt; \quad P(\lambda)R^k(\lambda)y = \frac{1}{k!}\int_0^\infty e^{-\lambda t}t^k S(t)y\,dt$$

which are derived by making use of the convolution theorem for the Laplace transform. The boundedness of C follows from Corollary 3.9 and the Banach-Steinhaus theorem. □

Next we consider the *Widder operators* associated with $R(\lambda)$ and $P(\lambda)$. First we define the *symbols* of these operators as

$$\mathcal{L}_k R(\lambda) = \frac{(-1)^k}{k!}\lambda^{k+1}R^{(k)}(\lambda); \quad \mathcal{L}_k P(\lambda) = \frac{(-1)^k}{k!}\lambda^{k+1}P^{(k)}(\lambda)$$

for $k = 0, 1, 2 \ldots$. The Widder operators L_k are then defined as

$$L_k R(t)y = \mathcal{L}_k R(k/t)y; \quad L_k P(t)y = \mathcal{L}_k P(k/t)y$$

for $t > 0$. Differentiation of the resolvents are carried out with the resolvent equations (5) and (6) in mind. We obtain $R^{(k)}(\lambda) = (-1)^k k! R^{k+1}(\lambda)$ and $P^{(k)}(\lambda) =$

$(-1)^k k! P(\lambda) R^k(\lambda)$, with the derivatives converging in the strong operator topology. The symbols may now be expressed differently:

$$\mathcal{L}_k R(\lambda) = \lambda^{k+1} R^{k+1}(\lambda); \quad \mathcal{L}_k P(\lambda) = \lambda P(\lambda) \lambda^k R^k(\lambda) \tag{15}$$

Theorem 4.1 may be interpreted in terms of the Widder operators:

$$\|L_k R(t) y\| \leq M \|y\|$$
$$\|L_k P(t) y\| \leq N \|y\|$$

for every $y \in Y$. The Post-Widder inversion theorem [Widder (1946), Theorem 6a., p.288] which is true in Banach spaces [Miyadera (1956)], claims that $\|L_k R(t) y - E(t) y\| \to 0$ as $k \to \infty$, and the same holds for $L_k P(t) y$ which converges to $S(t) y$ on the Lebesgue sets of $E(t) y$ and $S(t) y$ respectively.

We now turn to the problem of constructing an empathy with a given operator pair as generator. Let A and B be given linear operators defined on a common domain $\mathfrak{D} \subset X$ with ranges in Y. Suppose that for all positive λ, $\lambda B - A$ has a bounded inverse $P(\lambda) = (\lambda B - A)^{-1} : Y \to \mathfrak{D}$. Set $R(\lambda) = BP(\lambda)$ and $\mathfrak{D}_E = B[\mathfrak{D}]$. It follows that $R(\lambda)[Y] = \mathfrak{D}_E$ for every positive λ. It is a simple matter to prove that the pseudo-resolvent equations (5) and (6) hold. The following result gives a condition under which the operator B, and consequently the $R(\lambda)$, will be invertible.

LEMMA 4.2. *Suppose that for every $y \in Y$, $P(\lambda) y \to 0$ as $\lambda \to \infty$. Then B is invertible. If $\|\lambda P(\lambda)\| \leq N$ as $\lambda \to \infty$, then the inverse of B is bounded.*

Proof. Let $y \in \mathfrak{D}_E$ be represented in the form $y = R(\mu) y_\mu$ with $y_\mu \in Y$. Then, by (6),

$$\lambda P(\lambda) y = P(\mu) y_\mu + P(\lambda)[\mu y - y_\mu].$$

From the hypothesis it is clear that the limit

$$Cy = \lim_{\lambda \to \infty} \lambda P(\lambda) y = P(\mu) y_\mu. \tag{16}$$

exists. We show that the range of the linear operator C defined by (16) is $\mathfrak{D}_E$. Indeed, if $x = P(\mu) z_\mu \in \mathfrak{D}$ for some $z_\mu \in Y$, and $y = R(\mu) z_\mu \in \mathfrak{D}_E$, then $Cy = P(\mu) z_\mu = x$. Now, if $y = R(\mu) y_\mu \in \mathfrak{D}_E$, then $BCy = BP(\mu) y_\mu = R(\mu) y_\mu = y$. Hence C is the inverse of B. The statement on the boundedness of C follows from the Banach-Steinhaus theorem applied to the closure Y_E of $\mathfrak{D}_E$ in Y. □

THEOREM 4.3. *Suppose that the spaces X and Y have the Radon-Nikodym property. If there exist positive constants M and N such that*

$$\begin{aligned} \|\lambda P(\lambda)\| &\leq N \\ \|[\lambda R(\lambda)]^k\| &\leq M \end{aligned} \tag{17}$$

for $\lambda > 0$ *and* $k = 1, 2 \ldots$, *then there exist for every* $t > 0$ *bounded linear operators* $E(t) : Y \to Y$ *and* $S(t) : Y \to X$ *such that* $R(\lambda)y$ *and* $P(\lambda)y$ *are the Laplace transforms of* $E(t)y$ *and* $S(t)y$ *respectively. It is also true that*

$$\|E(t)\| \le M \text{ and } \|S(t)\| \le MN.$$

For given $y \in Y$ *there exists a set* N_y *of measure zero such that for* $s, t \notin N_y$, $S(t+s)y = S(t)E(s)y$.

Proof. From (17) it follows that Widder's theorem [Widder (1946), Theorem 16a, p.315] holds and the first statement is proved [Arendt (1987)]. The second statement follows from (5) and (6) by applying the uniqeness theorem for Laplace transforms [Arendt (1987)]. □

If it could be proved that the empathy relation holds for all s and t, the characterization of the generator of a uniformly bounded empathy would be over. It turns out that we have to settle for somewhat less as the discussion in the rest of the paper will indicate.

PROPOSITION 4.4. *If* t *is in the Lebesgue set of* $E(t)y$, *then then* $E(t)y \in Y_E$.

Proof. It is a straightforward (but not simple) matter to show that the Post-Widder inversion theorem [Widder (1946), Theorem 6a, p.288] holds in Banach spaces [Miyadera (1956)]. Thereby, $E(t)y$ is the limit as $k \to \infty$ of $L_k R(t)y$ and the result follows. □

It turns out that the space $Y_E = \mathrm{Cl}\,\mathfrak{D}_E$ is central in the analysis of the situation. The following result is used to show that strong continuity of $E(t)$ and $S(t)$ holds on the space Y_E.

LEMMA 4.5. *For* $y \in \mathfrak{D}_E$ *and any* $\lambda > 0$

$$\int_t^\infty e^{-\lambda s} S(s)y\,ds \in \mathfrak{D}; \quad \int_t^\infty e^{-\lambda s} E(s)y\,ds \in \mathfrak{D}_E$$

and

$$E(t)y = e^{\lambda t}[y - P^{-1}(\lambda)\int_0^t e^{-\lambda s} S(s)y\,ds] = e^{\lambda t}\int_t^\infty e^{-\lambda s} E(s)R^{-1}(\lambda)y\,ds. \quad (18)$$

The proof which makes essential use of the Post-Widder inversion theorem, the Widder representation theorem [Widder (1946), Theorem 11a, p.303] and the identities $P^{-1}(\lambda)R(\mu) = [I - (\mu - \lambda)R(\mu)] = R^{-1}(\lambda)R(\mu)$, is too long and detailed to be presented here and will be given in a forthcoming publication.

From this result, by making use of the identity $S(t)R(\lambda)y = P(\lambda)E(t)y$ which holds almost everywhere, we can prove

THEOREM 4.6. *For $y \in Y_E$ the mappings $t \to E(t)y$ and $t \to S(t)y$; $t > 0$ are continuous and consequently the empathy relation holds.*

It is now a simple matter to show that the Cauchy problem (1) can be solved for $y \in \mathfrak{D}_E^1 = R(\lambda)[Y_E]$.

REFERENCES

ARENDT, W. (1987), Vector-valued Laplace transforms and Cauchy problems, *Israel J. of Math.*, **59**, 327–352.

CONRADIE, W.L. & SAUER, N. (1994), Empathy, C-semigroups and Integrated semigroups, In: *Evolution Equations*, pp. 123–132. Marcel Dekker. Edited by Ferreyra, Goldstein & Neubrander.

GROBBELAAR-VAN DALSEN, M. & SAUER, N. (1989), Dynamic boundary conditions for the Navier-Stokes equations, *Proc. Royal Soc. Edinburgh*, **113A**, 1–11.

GROBBELAAR-VAN DALSEN, M. & SAUER, N. (1993), The solutions in Lebesgue spaces of the Navier-Stokes equations with dynamic boundary conditions, *Proc. Royal Soc. Edinburgh*, **123**, 745–761.

HILLE, E. & PHILLIPS, R.S. (1957), *Functional analysis and semi-groups*, American Math. Soc. Colloquium Publications, Vol. 31. Providence, Rhode Island.

MIYADERA, I. (1951), On one-parameter semi-groups of operators, *J. Math. Tokyo*, **1**, 23–26.

MIYADERA, I. (1956), On the representation theorem by the Laplace transformation of vector-valued functions, *Tohuku Math. J.*, **8**, 170–180.

SAUER, N. (1982), Linear evolution equations in two Banach spaces, *Proc. Royal Soc. Edinburgh*, **91A**, 287–303.

SAUER, N. & SINGLETON, J.E. (1987), Evolution operators related to semigroups of class(A), *Semigroup Forum*, **35**, 317–335.

SINGLETON, J.E. & SAUER, N. (1989), Evolution operators in empathy with a semigroup, *Semigroup Forum*, **39**, 85–94.

SHOWALTER, R.E. (1988), Implicit evolution equations, In: *Proceedings of the International Conference on Theory and Applications of Differential Equations*, pp. 404–411. Ohio University.

VAN DER MERWE, A.J. (1988), B-evolutions and Sobolev equations, *Applicable Analysis*, **29**, 91–105.

WIDDER, D.V. (1946), *The Laplace transform*, Princeton University Press, Princeton.

Faculty of Science
University of Pretoria
Pretoria 0002, South Africa
EMail: NSauer@Scinet.UP.AC.ZA

G F WEBB

Periodic and chaotic behavior in structured models of cell population dynamics

1. Introduction

In this paper we demonstrate how periodic and chaotic behavior arise in maturity structured models of cell population growth. The models we consider describe the changing structure of a proliferating cell population in which individual cells are distinguished by a maturity structure variable. The equations for these models are linear first order partial differential equations of transport form. The solutions of these equations yield semigroups of linear operators in function spaces of the maturity variable. We will show that the behavior of the solutions is connected to the spectrum of the infinitesimal generator of the semigroup of operators.

2. Exponentially Periodic Behavior

The maturity structured model of cell population dynamics was proposed by Rubinow in 1973 [15]. In the Rubinow model the maturity x of a cell corresponds to various physical properties associated with progress through the cell cycle. The values of maturity satisfy $0 < x_0 \leq x \leq 1$. Cell mitosis occurs at 1, whereupon a mother cell divides to produce two new daughter cells which enter the population with maturity x_0. All cells transport through the cell cycle with the same velocity $v(x)$, where $\int_{x_1}^{x_2} 1/v(x)dx$ is the time required to mature from x_1 to x_2. The equations of the model are

$$w_t(x,t) + (v(x)w(x,t))_x = 0 \tag{2.1}$$

$$v(x_0)w(x_0,t) = 2v(1)w(1,t), t \geq 0 \tag{2.2}$$

$$w(x,0) = \psi(x), x_0 \leq x \leq 1 \tag{2.3}$$

The balance law (2.1) accounts for transport of maturity. For simplicity we assume no mortality of cells (it could easily be incorporated into the analysis). The boundary condition (2.2) accounts for the mitotic process and the initial condition (2.3) accounts for the initial maturity distribution of cells. We specialize the problem to the case that $v(x) = x$. We simplify the analysis by transforming the independent variable $x \to -\ell n x$. The resulting equations are

$$u_t(x,t) = u_x(x,t) - u(x,t) \tag{2.4}$$

$$u(b,t) = cu(0,t), t \geq 0 \tag{2.5}$$

$$u(x,0) = \phi(x), 0 \leq x \leq b \tag{2.6}$$

where $c = 2/x_0$, $b = -\ell n x_0$, $\phi(x) = \psi(e^{-x}), w(x,t) = u(-\ell n x, t)$.

The problem (2.4)-(2.6) is easily solved by the method of characteristics. The solutions may be represented as $u(x,t) = B(x+t)e^{-t}$ for an appropriately prescribed scalar function B. Define the Banach space $X = \{\phi \in C[0,b] : \phi(b) = c\phi(0)\}$. Define $B(t) = \phi(t), 0 \leq t < b$, $B(t) = c\phi(t-b), b \leq t < 2b$, $B(t) = c^2\phi(t-2b)$, $2b \leq t < 3b, \cdots$. It is easily seen that $u(x,t) = B(x+t)e^{-t}$ satisfies (2.4)-(2.6) for ϕ continuously differentiable. For $\phi \in X$ define $(T(t)\phi)(x) = c^k\phi(x+t-kb)e^{-t}, 0 \leq x \leq b$, $kb \leq t < (k+1)b, k = 0,1,\cdots$.

Proposition 2.1. $T(t), t \geq 0$ is a strongly continuous semigroup in X. The infinitesimal generator is $A\phi = \phi' - \phi, D(A) = \{\phi \in X : \phi' \in X\}$ and $\sigma(A) = P\sigma(A) = \{\lambda \in \mathbb{C} : \lambda = (\ell nc)/b - 1 \pm 2n\pi i/b, n = 0,1,\cdots\}$. $T(t), t \geq 0$ is exponentially periodic in the sense that $e^{t(1-(\ell nc/b)}T(t)$ is periodic in t with period b.

Proof. The first claims of the proposition are proved in [17]. To prove the exponential periodicity it suffices to show $e^{b-\ell nc}T(b) = I$. For $\phi \in X, x \in [0,b]$, $(e^b/c)(T(b)\phi)(x) = (e^b/c)[c\phi(x+b-b)]e^{-b} = \phi(x)$. □

The solution to (2.1)-(2.3) is $w(x,t) = (T(t)\phi)(-\ell n x)$, $x_0 \leq x \leq 1$, where $\phi(x) = \psi(e^{-x})$, $0 \leq x \leq -\ell n x_0, \phi \in D(A)$. Since $(\ell nc)/b - 1 = -\ell n 2/\ell n x_0 > 0$, the solution grows exponentially in time. The exponential periodic behavior of the model (2.1)-(2.3) is accurate only for a few generations. It is commonly observed that proliferating cell populations grow asynchronously in that their initially synchronized structure is dispersed as the total population grows exponentially in time. The model (2.1)-(2.3) allows no mechanism for structure dispersion and the initial structure is effectively doubled again and again through the overlapping generations. More realistic maturity structured models which do exhibit asynchronous exponential growth have been treated in [1], [5], [7], and [17].

3. Extinction in Finite Time

Another version of a maturity structured model of cell population dynamics allows cells to enter the population with maturity at any value x, $0 < x_0 \leq x \leq 1$. The maturity variable x distinguishes primitive cell types and mature cell types. The division process is not modeled directly, but new cells enter the population at a rate proportional to the structure density of existing cells. The equations of the model are

$$w_t(x,t) + (v(x)w(x,t))_x = \beta w(x,t) \tag{3.1}$$

$$w_x(x_0,t) = 0,\ t \geq 0 \tag{3.2}$$

$$w(x,0) = \psi(x),\ x_0 \leq x \leq 1 \tag{3.3}$$

where $v(x)$ is the maturation velocity and $\beta > 1$. We again take $v(x) = x$ and $x \to -\ell n x$. The new equations are

$$u_t(x,t) = u_x(x,t) + \alpha u(x,t) \tag{3.4}$$

$$u_x(b,t) = 0,\ t \geq 0 \tag{3.5}$$

$$u(x,0) = \phi(x),\ 0 \leq x \leq b \tag{3.6}$$

where $b = -\ell n x_0$, $\alpha = \beta - 1 > 0$, $\phi(x) = \psi(e^{-x})$, and $w(x,t) = u(-\ell n x, t)$.

Let $X = C[0,b]$. For $\phi \in X$, $0 \leq x \leq b$, and $t \geq 0$ define

$$(T(t)\phi)(x) = \begin{cases} e^{\alpha t}\phi(x+t), & x+t \leq b \\ e^{\alpha t}\phi(b), & x+t > b \end{cases}$$

The following proposition is easily proved:

Proposition 3.1. $T(t), t \geq 0$ is a strongly continuous semigroup in X. The infinitesimal generator is $A\phi = \phi' + \alpha\phi$, $D(A) = \{\phi \in X : \phi'^{-}(b) = 0\}$ and $\sigma(A) = P\sigma(A) = \{\alpha\}$. $T(b)\phi = 0$ for $\phi \in X$ such that $\phi(b) = 0$.

The solution of (3.1)-(3.3) in the case $v(x) = x$ is

$$w(x,t) = \begin{cases} e^{(\beta-1)t}\psi(xe^{-t}), & x_0 \leq xe^{-t} \\ e^{(\beta-1)t}\psi(x_0), & xe^{-t} < x_0 \end{cases}$$

for $\psi' \in C[x_0,1]$. The ultimate behavior of the cell population depends on the initial maturity distribution of cells. If cells of the most primitive type x_0 are present initially, then the population grows exponentially in time. If no cells of the most primitive type x_0 are present initially, then the population extinguishes by time $t = -\ell n x_0$. A sufficient supply of primitive cells must be present initially to sustain population growth. An imbalance in the initial maturity distribution results in collapse of the population.

4. Chaotic Behavior

A maturity structured model of the blood cell production system has been studied by Lasota [9], Brunovsky [3], Brunovsky and Komornik [4], Lasota and Mackey [10], [11], Rudnicki [16], and Rey and Mackey [13]. In this model the maturity variable distinguishes primitive and mature cell types. The maturity values x satisfy $0 \leq x \leq 1$ and the maturation velocity is $v(x) = x$. The division process is not modeled directly, but there is a proportional production of new cells of all maturity values. The equations are

$$w_t(x,t) + (xw(x,t))_x = \beta w(x,t) \tag{4.1}$$

$$w(x,0) = \psi(x), 0 \leq x \leq 1 \tag{4.2}$$

where $\beta > 1$.

The behavior of the solutions has been analyzed in [3], [4], [9], [10], [11], [13], [16] (even in some more general nonlinear cases). If $\psi(0) > 0$, then $\lim_{t\to\infty} e^{(1-\beta)t}$ $w(x,t) = \psi(0) > 0$. If $\psi(0) = 0$, then the solution exhibits chaotic behavior. We will show that this chaotic behavior is connected to the spectrum of the infinitesimal generator of the associated semigroup. We again make the substitution $x \to -\ell n x$ to obtain

$$u_t(x,t) = u_x(x,t) + \alpha u(x,t) \tag{4.3}$$

$$u(x,0) = \phi(x), \ 0 \leq x < \infty \tag{4.4}$$

where $\alpha = \beta - 1 > 0$, $\phi(x) = \psi(e^{-x})$, and $w(x,t) = u(-\ell n x, t)$.

Let $X = \{\phi \in C[0,\infty) : \lim_{x\to\infty} \phi(x) \text{ exists}\}$ with supremum norm and let $X_0 = \{\phi \in X : \lim_{x\to\infty} \phi(x) = 0\}$. For $\alpha > 0$, $\phi \in X$ define

$$(T(t)\phi)(x) = e^{\alpha t}\phi(x+t), \ x \geq 0, \ t \geq 0 \tag{4.5}$$

The following proposition is easily proved:

Proposition 4.1. $T(t), t \geq 0$ is a strongly continuous semigroup in X and in X_0. The infinitesimal generator in X is $A\phi = \phi' + \alpha\phi, D(A) = \{\phi \in X : \phi' \in X\}$ with $\sigma(A) = P\sigma(A) = \{\lambda \in \mathbb{C} : \text{Re}\lambda \leq \alpha\}$. For $\lambda \in \mathbb{C}$ with $\text{Re}\lambda < \alpha$, $A\phi_\lambda = \lambda\phi_\lambda$, where $\phi_\lambda(x) = \exp[(\text{Re}\lambda - \alpha)x + i(\text{Im}\lambda)x]$, $x \geq 0$. The infinitesimal generator in X_0 is the same except that $\{\lambda \in \mathbb{C} : \text{Re}\lambda = \alpha\}$ lies in its continuous spectrum.

The solution of (4.1), (4.2) is $w(x,t) = e^{(\beta-1)t}\psi(xe^{-t})$ for $x\psi'(x) \in C[0,1]$. If $\psi(0) > 0$, then $\lim_{t\to\infty} e^{(1-\beta)t}w(x,t) = \psi(0) > 0$ uniformly in x, which corresponds to a normal blood production system. If $\psi(0) = 0$, then the solutions of (4.1), (4.2) are chaotic in the space $C_0[0,1] = \{\psi \in C[0,1] : \psi(0) = 0\}$, which corresponds to an aplastic anemia (see [9], [10], [11], [12], [13]). The following definition of linear chaos is based on the definition of nonlinear chaos due to Auslander and Yorke [2] and Knudsen [8]:

Definition 4.1. The strongly continuous semigroup of bounded linear operators $T(t), t \geq 0$ in the Banach space X is *chaotic* provided there exists $\phi \in X$ such that $\{T(t)\phi : t \geq 0\}$ is dense in X.

It is possible to show that the semigroup $T(t), t \geq 0$ of Proposition 4.1 is chaotic in X_0 by direct construction of a dense orbit (see [3], [4], [9], [10], [11], [16]). We will show this chaotic behavior by using a general result in [18], which provides sufficient conditions for chaos in terms of $\sigma(A)$.

Let $T(t), t \geq 0$ be a strongly continuous semigroup of bounded linear operators in the Banach space with infinitesimal generator A. We require the following hypotheses:

(H.1) There exists an increasing sequence of positive numbers $\{\gamma_m\}_{m=1}^{\infty}$ such that $\gamma_m + i\mu \in P_\sigma(A)$ for $\mu \in \mathbb{R}$.

For $\gamma_m + i\mu \in P\sigma(A)$ let $A\phi_{\gamma_m+i\mu} = (\gamma_m + i\mu)\phi_{\gamma_m+i\mu}$, where $\|\phi_{\gamma_m+i\mu}\| = 1$. For positive integers m and k let $X_{m,k} = \text{span}\{\phi_{\gamma_m+n\pi i/k} : n = 0, \pm1, \pm2, \cdots\}$ (where span means all finite linear combinations). For $\phi \in X_{m,k}$, $\exp[-\gamma_m t]T(t)\phi$ is periodic in t with period $2k$, since $T(t)\phi_{\gamma_m+n\pi i/k} = \exp[\gamma_m t]\exp[n\pi ti/k]\phi_{\gamma_m+n\pi i/k}$.

(H.2) Let $\psi \in X$, let m be a positive integer, and let $\in > 0$. There exists a positive integer k' such that if $k \geq k'$, then there exists $\phi \in X_{m,k}$ such that $\|\phi - \psi\| < \in$.

(H.3) Let m_1, m_2, and k_1, be positive integers with $m_1 < m_2$, let $\phi_1 \in X_{m_1,k_1}$, and let $\in > 0$. There exists a positive integer k such that if $k_1 \geq k$, then there exists

$\phi_2 \in X_{m_2,k_1 k_2}$ such that $\|\phi_1 - \phi_2\| <\in \exp[-\gamma_{m_1 k_1 k_2}]$.

Hypothesis (H.1) requires that $P\sigma(A)$ contains an infinite number of vertical lines in the right half complex plane. Hypothesis (H.2) requires that for any one of these vertical lines $\{\gamma_m + i\mu : \mu \in \mathbb{R}\}$ the subspaces $X_{m,k}$ fill out X as k increases. Hypothesis (H.3) requires that two such sets of subspaces X_{m_1,k_1} and X_{m_2,kk_1} get sufficiently close as k increases. (Note that $X_{m_1,k_1} \cap X_{m_2,kk_2} = \{0\}$ for $m_1 < m_2$, since the eigenvectors are linearly independent). The following proposition is proved in [18]:

Proposition 4.2. Let $T(t), t \geq 0$ be a strongly continuous semigroup in the separable Banach space X and let (H.1), (H.2), (H.3) hold. Then $T(t), t \geq 0$ is chaotic in X.

Proposition 4.3. The strongly continuous semigroup $T(t), t \geq 0$ defined by (4.5) is chaotic in X_0.

Proof. Let $\{\gamma_m\}_{m=1}^{\infty}$ be an increasing sequence of positive numbers in $(0, \alpha/2)$. By virtue of Proposition 4.1 this sequence satisfies (H.1), where $\phi_{\gamma_m+i\mu}(x) = \exp[(\gamma_m - \alpha)x + i\mu x]$. To prove (H.2) let $\psi \in X_0$ such that ψ has compact support, let m be a positive integer, and let $\in > 0$. Choose k'' such that $\text{supp}\psi \subset [0, k'']$. Choose $k' > k''$ such that for $k \geq k'$.

$$\exp[(\gamma_m - \alpha)(k - k'')]\|\psi\| <\in /2 \tag{4.6}$$

Let $k \geq k'$. Let $\hat{\psi}(x)$ be the restriction of $\exp[(\alpha - \gamma_m)x]\psi(x)$ to $[0, k]$ and let $\tilde{\psi}(x)$ be the even $2k-$ periodic continuous extension of $\hat{\psi}(x)$ to $\mathbb{R}$. (Note that $\tilde{\psi}$ is real-valued if ψ is real-valued). Observe that

$$\sup_{x \in \mathbb{R}} |\tilde{\psi}(x)| = \sup_{0 \leq x \leq k} |\hat{\psi}(x)| \leq \exp[(\alpha - \gamma_m)k'']\|\psi\| \tag{4.7}$$

(since $\text{supp}\psi \subset [0, k'']$). Choose a positive integer N and Fourier coefficients $c_n, -N \leq n \leq N$, such that

$$\left| \sum_{n=-N}^{N} c_n \exp[n\pi i x/k] - \tilde{\psi}(x) \right| <\in /2 \text{ for } x \geq 0 \tag{4.8}$$

Define $\phi = \sum_{n=-N}^{N} c_n \phi_{\gamma + n\pi i/k_m} \in X_{m,k}$. (Note that ϕ is real-valued if ψ is real-valued). For $0 \leq x \leq k$, (4.8) implies

$$|\phi(x) - \psi(x)| = \exp[(\gamma_m - \alpha)x] \left| \sum_{n=-N}^{N} c_n \exp[n\pi i x/k] - \tilde{\psi}(x) \right| <\in \tag{4.9}$$

For $x > k$, (4.6), (4.7), and (4.8) imply

$$\begin{aligned} &|\phi(x) - \psi(x)| \qquad (4.10) \\ &= \exp[(\gamma_m - \alpha)x]|\sum_{n=-N}^{N} c_n \exp[n\pi i x/k]| \\ &\leq \exp[(\gamma_m - \alpha)k](|\tilde{\psi}(x)| + \in /2) \\ &\leq \exp[(\gamma_m - \alpha)k](\exp(\alpha - \gamma_m)k'']\|\psi\| + \in /2) \\ &< \in \end{aligned}$$

Then, (H.2) follows from (4.9) and (4.10).

To prove (H.3) let m_1, m_2, and k_1 be positive integers with $m_1 < m_2$, let $\phi_1 \epsilon X_{m_1,k_1}$, and let $0 < \in < 1$. Since $\phi_1 \in X_{m_1,k_1}$, there exist coefficients c_{1n}, $-N_1 \leq n \leq N_1$, such that for $x \geq 0$

$$\phi_1(x) = \sum_{n=-N_1}^{N_1} c_{1n} \exp[(\gamma_{m_1} - \alpha)x] \exp[n\pi i/k_1]$$

Observe that $\exp[(\alpha - \gamma_{m_1})x]\phi_1(x)$, $x \geq 0$, is continuous and periodic with period $2k_1$. Thus, there exists a constant M such that

$$|\phi_1(x)| \leq \exp[(\gamma_{m_1} - \alpha)x]M, x \geq 0 \qquad (4.11)$$

Let k be sufficiently large such that if $k_2 \geq k$, then

$$\exp[(\gamma_{m_1} + \gamma_{m_2} - \alpha)k_1k_2] < \epsilon/(M+1) \qquad (4.12)$$

(recall $\gamma_{m_1} < \gamma_{m_2} < \alpha/2$). Let $\hat{\phi}_1(x)$ be the restriction of $\exp[(\alpha - \gamma_{m_2})x]\phi_1(x)$ to $[0, k_1k_2]$ and let $\tilde{\phi}_1(x)$ be the even $2k_1k_2$-periodic continuous extension of $\hat{\phi}_1(x)$ to $\mathbb{R}$. (Note that $\tilde{\phi}_1$ is real-valued if ϕ_1 real-valued). For $0 \leq x \leq k_1k_2$, (4.11) implies

$$\begin{aligned} |\tilde{\phi}_1(x)| &= \exp[(\alpha - \gamma_{m_2})x]|\phi_1(x)| \qquad (4.13) \\ &\leq \exp[(\alpha - \gamma_{m_2})x] \exp[(\gamma_{m_1} - \alpha)x]M \\ &< M \end{aligned}$$

For $k_1k_2 < x \leq 2k_1k_2$, (4.11) implies

$$\begin{aligned} &|\tilde{\phi}_1(x)| = |\hat{\phi}_1(2k_1k_2 - x)| \qquad (4.14) \\ &= \exp[(\alpha - \gamma_{m_2})(2k_1k_2 - x)]|\phi_1(2k_1k_2 - x)| \\ &\leq \exp[(\alpha - \gamma_{m_2})(2k_1k_2 - x)] \exp[(\gamma_{m_1} - \alpha)(2k_1k_2 - x)]M \\ &< M \end{aligned}$$

Choose a positive integer N_2 and Fourier coefficients $c_{2n}, -N_2 \leq n \leq N_2$, such that for $x \geq 0$

$$| \sum_{n=-N_2}^{N_2} c_{2n} \exp[n\pi ix/k_1k_2] - \tilde{\phi}_1(x)| < \frac{\epsilon}{2} \exp[-\gamma_{m_1} k_1 k_2] \tag{4.15}$$

Define $\phi_2(x) = \sum_{n=-N_2}^{N_2} c_{2n}\phi_{\gamma_{m_2}} + n\pi i/k_1k_2 \in X_{m_2,k_1k_2}$. Let $\tilde{\phi}_2(x) = \sum_{n=-N_2}^{N_2} c_{2n}$ $\exp[n\pi ix/k_1k_2] = \exp[(\alpha - \gamma_{m_2})x]\phi_2(x), x \geq 0$. (Note that ϕ_2 is real-valued if ϕ_1 is real-valued). Let $M_2 = \sup_{x\geq 0} |\tilde{\phi}_2(x)| = \sup_{0\leq x\leq 2k_1k_2} |\tilde{\phi}_2(x)|$. For $0 \leq x \leq 2k_1k_2$, (4.13), (4.14), and (4.15) imply $|\tilde{\phi}_2(x)| \leq |\tilde{\phi}_1(x)| + \frac{\epsilon}{2} \exp[-\gamma_{m_1} k_1k_2] < M+1$. Thus, $M_2 \leq M + 1$ and

$$|\phi_2(x)| \leq \exp[(\gamma_{m_2} - \alpha)x](M + 1), x \geq 0 \tag{4.16}$$

For $0 \leq x \leq k_1k_2$, (4.15) implies

$$\begin{aligned} |\phi_1(x) - \phi_2(x)| &= \exp[(\gamma_{m_2} - \alpha)x]|\tilde{\phi}_1(x) - \tilde{\phi}_2(x)| \\ &< \frac{\epsilon}{2} \exp[-\gamma_{m_1} k_1k_2] \end{aligned} \tag{4.17}$$

For $x > k_1k_2$, (4.11) and (4.12) imply

$$\begin{aligned} |\phi_1(x)| &\leq \exp[(\gamma_{m_1} - \alpha)k_1k_2] \\ &= \exp[(2\gamma_{m_1} - \alpha)k_1k_2] \exp[-\gamma_{m_1} k_1k_2]M \\ &<\in \exp[-\gamma_{m_1} k_1k_2] \end{aligned} \tag{4.18}$$

For $x > k_1k_2$, (4.12) and (4.16) imply

$$\begin{aligned} |\phi_2(x)| &\leq \exp[(\gamma_{m_2} - \alpha)k_1k_2](M + 1) \\ &= \exp[(\gamma_{m_1} + \gamma_{m_2} - \alpha)k_1k_2] \exp[-\gamma_{m_1} k_1k_2](M + 1) \\ &<\in \exp[-\gamma_{m_1} k_1k_2] \end{aligned} \tag{4.19}$$

Then, (H.3) follows from (4.17), (4.18), and (4.19). □

The chaotic behavior demonstrated in Proposition 4.3 means that the solution $u(x,t) = e^{\alpha t}\phi(x + t)$ of (4.3), (4.4) depends sensitively on the initial condition $\phi(x)$, when $\lim_{x\to\infty} \phi(x) = 0$. For the model (4.1), (4.2) the solution

$w(x,t) = e^{(\beta-1)t}\psi(xe^{-t}) = u(-\ell n x, t), \psi(x) = \phi(-\ell n x)$, exhibits chaotic behavior in the space $\{\phi \in C[0,1] : \psi(0) = 0\}$. The sensitivity of the solutions to the initial values when $\psi(0) = 0$ corresponds to aplastic anemia. If $\psi(0) = 0$, then there is an insufficient supply of the most primitive cell types. If $\psi(0) > 0$, then $\lim_{t\to\infty} e^{(1-\beta)t}w(x,t) = \psi(0) > 0$ uniformly for x in $[0,1]$, which corresponds to a sufficient supply of primitive cell types and to a normal blood production system. In [6] a nonlinear maturity structured model of the blood cell production system due to Rey and Mackey [13] is analyzed and abnormal behavior is associated with instability of initial values.

References

1. O. Arino and M. Kimmel, Asymptotic analysis of a cell-cycle model based on unequal division, *SIAM J. Appl. Math.* **47**(1987), 128-145.

2. J. Auslander and J. Yorke, Intervals maps, factor of maps and chaos, *Tôhoku Math. J.* **32**(1980), 177-188.

3. P. Brunovsky, Notes on chaos in the cell population partial differential equation, *Nonlinear Anal.* **7**(1983), 167-176.

4. P. Brunovsky and J. Komornik, Ergodicity and exactness of the shift on $C[0,\infty)$ and the semiflow of a first-order partial differential equation, *J. Math. Anal. Appl.* **104**(1984), 235-245.

5. O. Diekmann, H. Heijmans, and H. Thieme, On the stability of the cell size distribution, *J. Math. Biol.* (1984), 227-248.

6. J. Dyson, R. Villella-Bressan, and G. F. Webb, A nonlinear transport equation model of aplastic anemia, to appear.

7. G. Greiner and R. Nagel, Growth of cell populations via one-parameter semigroups of positive operators, *Mathematics Applied to Science*, Academic Press, San Diego, 1988, 79-105.

8. C. Knudsen, Chaos without nonperiodicity, *Amer. Math. Mon.* **101**, No. 6 (1994), 563-564.

9. A. Lasota, Stable and chaotic solutions of a first order partial differential equation, *Nonlinear Anal.* **5**(1981), 1181-1193.

10. A. Lasota and M. C. Mackey, *Probabilistic Properties of Deterministic Systems*, Cambridge University Press, Cambridge, 1985.

11. A. Lasota and M. C. Mackey, *Chaos, Fractals, and Noise Stochastic Aspects of Dynamics*, Applied Mathematical Sciences Series Vol. 97, Springer-Verlag, New York, 1994.

12. M. C. Mackey, Unified hypothesis for the origin of aplastic anemia and periodic hematopoiesis, *Blood* **51**(1978), 941-956.

13. A. O. Rey and M. C. Mackey, Multistability and boundary layer development in a transport equation with delayed arguments, *Canad. Appl. Math. Quar.* Vol 1, No. 1 (1993), 61-81.

14. M. Rotenberg, Transport theory for growing cell populations, *J. theor. Biol.* **103** (1983), 181-199.

15. S. I. Rubinow, Mathematical problems in the biological sciences, *CBMS Regional Conf. Ser. Appl. Math.*, No. 10, *SIAM*, Philadelphia (1973), 53-73.

16. R. Rudnicki, Strong ergodic properties of a first-order partial differential equation, *J. Math. Anal. Appl.* **133**(1988), 14-26.

17. G. F. Webb, An opertor-theoretic formulation of asynchronous exponential growth, *Trans. Amer. Math. Soc.* Vol. 303, No. 2 (1987), 751-763.

18. G. F. Webb, Chaos in abstract linear differential equations, to appear.

Department of Mathematics
Vanderbilt University
Nashville, TN 37240
U.S.A.

Lectures

J BANASIAK

Singular perturbations of resonance type with applications to the kinetic theory

1 Introduction

The main topic of this paper is an asymptotic analysis of singularly perturbed equations of resonance type, especially of the type appearing in the transport theory. In particular, we are interested in procedures which, roughly speaking, show that the solution to such an evolution equation can be approximated by the solution to a suitable equation of the diffusion type.

A model for our considerations is offered by equations of the kinetic theory which can be written in an abstract form as

$$\begin{aligned} \partial_t u_\epsilon &= \epsilon^{-1}\mathcal{C}u_\epsilon + \mathcal{S}u_\epsilon, \\ u(0) &= \overset{\circ}{u}, \end{aligned} \tag{1.1}$$

where $\mathcal{C}$ and $\mathcal{S}$ denote the collision and streaming operators, respectively, and ϵ^{-1} is, roughly speaking, proportional to the mean free path of particles. In this paper we assume that the operator C is linear and has $\lambda = 0$ as a semi-simple eigenvalue. This excludes both the full nonlinear Boltzmann equation and the linearized Boltzmann equation [5] but a large class of linear Boltzmann equations and linear equations of the Fokker-Plank type with the collision operator given by second order differential operators are covered by our theory.

From the physics of the problem it follows that if ϵ is close to zero, then the so-called hydrodynamic part of the solution u should be close to a solution of some diffusion equation.

The approximation to the solution of (1.1) is usually sought in the form of a truncated power series in ϵ. The series is then inserted into (1.1) and by equating coefficients at the same powers of ϵ to zero we obtain a hierarchy of equations for the coefficients of the expansion. Unfortunately, such a straightforward method which can be dated back to Hilbert, does not provide a diffusion equation but instead it gives an open hierarchy of non-homogeneous equations [5].

There were various attempts to overcome this difficulty [7, 4, 9, 12, 13]. We present here the so-called *compressed* asymptotic procedure which is a modification of the old Chapman-Enskog method [5] and which was proposed in [8]. This method gives, in a systematic way, the diffusion equation on the first level of asymptotic expansion for a large class of singularly perturbed equations of the form (1.1). Moreover, at the same time it yields an initial value corrector and initial layer terms which improve the accuracy of the approximation. It follows also that under suitable assumptions the error of this approximation is of order of ϵ^2.

In this paper we show that all assumptions of the developed theory are satisfied when (1.1) is the linear Boltzmann equation. In [1] a similar analysis is performed for kinetic equations of the Fokker-Planck type.

2 Basic notions

In this section we introduce basic notation and assumptions which will make our asymptotic analysis possible.

To avoid additional difficulties related to a possible occurrence of a boundary layer we confine ourselves to two particular types of the problem. We assume that, with respect to x, we are either dealing with the free space or with the so-called periodic boundary conditions. The periodic boundary conditions mean that we require that the boundary values of the solution (and its derivatives, if necessary) are equal on the opposite faces of the unit cube of $\mathbb{R}^n$. In other words we have

$$u(x_1, \ldots, x_k, \ldots, x_n) = u(x_1, \ldots, x_k + 1, \ldots, x_n)$$

for $k = 1, 2, \ldots, n$. In that case the domain of x is commonly identified with n-dimensional torus.

In what follows $\mathcal{L}(X, Y)$ will denote the space of linear, bounded operators between X, Y. By $Lin\{e_1, \ldots, e_k\}$ we will denote the linear envelope of elements $e_1, \ldots, e_k$. With some abuse of notation, the function $(t, x, \xi) \to u(t, x, \xi)$ will be treated as a function $(t, x) \to u(t, x)$ with values in some Hilbert space H (a space of functions of velocity variable) and for every t the value $u(t)$ itself will be an element of $\mathcal{H} = L_2(\Omega) \otimes H = L_2(\Omega, H)$, where $\otimes$ denotes the tensor product of Hilbert spaces [14]. The set $\Omega = \mathbb{R}^n$ in a free space case and $\Omega = [0, 1]^n$ in the periodic case. We assume that $\mathcal{C}$ is independent of x, and replacing $\mathcal{S}$ with a generalized streaming operator, we can write (1.1) as

$$\partial_t u = (C \otimes I)u - (\sum_{k=1}^{n} S_{(k)} \otimes \partial_{x_k})u, \tag{2.1}$$

where $x = (x_1, \ldots, x_n) \in \Omega$. The operators C and $S_{(k)}$ are now acting in the space H.

Let us introduce the space $\mathcal{H} = L_2(\mathsf{P}^n, H)$, where $\mathsf{P} = \mathbb{R}$ in the free space case and $\mathsf{P} = \mathbb{Z}$ in the periodic case. Precisely, in the latter case the space L_2 reduces itself to the space l_2 of square-summable, H-valued, multi-indexed sequences. As this will not lead to any misunderstanding, we shall use the same labels, P and L_2, for both sets of parameters and both spaces, respectively. Also, in what follows, the phrase "for every $p \in \mathsf{P}^n$", when referred to the first case, is to be understood as "for almost every $p \in \mathbb{R}^n$".

Applying the Fourier transformation with respect to x, $u \to \hat{u}$, we obtain the unitarily equivalent problem

$$\partial_t \hat{u} = (C \otimes I)\hat{u} + (\sum_{k=1}^{n} i p_k S_{(k)})\hat{u} \tag{2.2}$$

in $L_2(\mathsf{P}^n, H)$. It follows (Theorem 2.1) that under certain assumptions the equation (2.2) can be analyzed with $p = (p_1, \dots, p_n)$ treated as a parameter. This will allow to discard restrictions caused by an unboundedness of the differential operators ∂_{x_k}.

Since from now on we shall work only with the transformed equation (2.2), in what follows we shall drop *"hat"* in the notation of the solution to (2.2).

Now we introduce basic assumptions which will be used throughout this paper. Let $(A, D(A))$ denote an unbounded operator in H with domain $D(A)$ and let P^n equal either $\mathbb{R}^n$ or $\mathbb{Z}^n$. We consider an operator $(C, D(C))$ and a family of operators $\{(S_p, D(S_p)\}_{p \in \mathsf{P}^n}$ and we assume that for $p \in \mathsf{P}^n$, $p \neq 0$ we have $D(S_p) = D_S$, D_S independent of p. Bearing in mind applications we assume that $p \to S_p$ is a linear mapping of P^n into the set of closed operators in H, so that

$$S_p = \sum_{k=1}^{n} p_k S_{(k)}, \tag{2.3}$$

where all $S_{(k)}$ are closed linear operator on D_S. Let these operators have the following properties.

P1. C is a self-adjoint operator, generating a semigroup of contractions, T_C, in H. Zero is a semisimple isolated eigenvalue of C with the eigenfunction $\mathbf{m}$ and $\sup Re\{\sigma(C) \setminus \{0\}\} = -\gamma < 0$.

P2. For every $p \in \mathsf{P}^n$, S_p generates a semigroup of isometries T_{S_p}.

P3. $D(C) \cap D_S$ is dense in H and

$$\mathbf{m} \in D_S \cap D(C). \tag{2.4}$$

The operator $(C + S_p, D(C) \cap D_S)$ is dissipative and, having dense domain, it is closable [11]. We define

$$(K_p, D(K_p)) := \overline{(C + S_p, D(C) \cap D(S_p))}. \tag{2.5}$$

We postulate that

P.4 K_p generates a strongly continuous semigroup in H, denoted by T_{K_p}.

The Trotter product formula [11] implies that T_{K_p} is a contraction semigroup.

We also require that

P.5 for $p \in \mathsf{P}^n$, $p \neq 0$

$$D_K := D(K_p) = D(C) \cap D_S. \tag{2.6}$$

We equip D_S with the norm of graph of arbitrary operator $S_{(k)}$ which appears in (2.3) and in the same way we introduce a norm $\| \cdot \|_{D_K}$ in D_K.

In what follows we quote several theorems which are similar to theorems proved in [1, 2]. Let us recall that $\mathcal{H} = L_2(\mathsf{P}^n, H)$. We have $\mathcal{C} = C \otimes I$ with the domain

$$D(\mathcal{C}) = D(C \otimes I) = D(C) \otimes L_2(\mathsf{P}^n)$$

and $(\mathcal{S}u)(p) = S_p u(p)$, treated as an operator in $\mathcal{H}$, with the domain

$$D(\mathcal{S}) = \{u \in \mathcal{H};\ u(p) \in D(S_p) \text{ for } p \in \mathsf{P}^n,\ \mathcal{S}u \in \mathcal{H}\} \tag{2.7}$$

By P.1 $(C, D(C))$ generates in H a semigroup of contractions and, since $\mathcal{C} = C \otimes I$, we have by [14] that $\mathcal{C}$ also generates a semigroup of contractions. It follows [1] that the operator $(\mathcal{S}, D(\mathcal{S}))$ generates a semigroup of isometries in $\mathcal{H}$, say $T_{\mathcal{S}}$, such that for every $p \in \mathsf{P}^n$ and $u \in \mathcal{H}$ we have

$$(T_{\mathcal{S}} u)(p) = T_{S_p} u(p). \tag{2.8}$$

Let us define the operator $(\mathcal{K}, D(\mathcal{K})$ by the formula $(\mathcal{K}u)(p) = K_p u(p)$ with

$$D(\mathcal{K}) = \{u \in \mathcal{H}; u(p) \in D(K_p) \text{ for } p \in \mathsf{P}^n,\ \mathcal{K}u \in \mathcal{H}\}.$$

We have the following theorem:

Theorem 2.1 *Operator $(\mathcal{K}, D(\mathcal{K}))$ generates a semigroup of contractions in $\mathcal{H}$, say $T_{\mathcal{K}}$, given by the Trotter formula*

$$T_{\mathcal{K}}(t)u = \lim_{n\to\infty} \left(T_{\mathcal{C}}\left(\frac{t}{n}\right) T_{\mathcal{S}}\left(\frac{t}{n}\right)\right)^n u, \quad u \in \mathcal{H}. \tag{2.9}$$

This semigroup satisfies

$$(T_{\mathcal{K}} u)(p) = T_{K_p} u(p). \tag{2.10}$$

for every $u \in \mathcal{H}$.

With that result we can carry out the asymptotic analysis in the space H, ensuring that the estimates are in suitable sense uniform in p. The main rôle in the modified Chapman-Enskog (compressed) asymptotic procedure [8, 9] is played by a decomposition of H into two subspaces and splitting (1.1) accordingly. One of these spaces is the eigenspace of C corresponding to the zero eigenvalue and the other is its orthogonal complement.

Let us consider the following family of evolution equations in H

$$\partial_t u = Cu + S_p u. \tag{2.11}$$

We define P to be the operator of the orthogonal projection onto the kernel of C, $PH = N(C) = V$, and $Q = I - P$, $QH = V^{\perp} = W$. We denote

$$v = Pu, \qquad w = Qu. \tag{2.12}$$

It is easy to see that the projection operators commute with ∂_t, so applying P and Q to both sides of (2.11) and, taking into account that $PC \equiv 0$ and $CP \equiv 0$, we obtain formally the following system of equations:

$$\begin{aligned} \partial_t v &= PS_pPv + PS_pQw \\ \partial_t w &= (QCQ + QS_pQ)w + QS_pPv, \end{aligned} \tag{2.13}$$

where we have left spurious notation for the projection operators for the sake of symmetry. Thanks to the assumptions P.1 - P.3 and P.5 we can prove [2]

Proposition 2.1 *A pair $(v, w) \in V \times (D_K \cap W)$ solves (2.13) if and only if $u = v + w \in D_K$ solves (2.11).*

To perform the asymptotic analysis we will have to solve certain equations in the subspace W. To cater for that we have the following general theorem [1].

Theorem 2.2 *If A is an m-dissipative operator in a Hilbert space H, H_1 is a subspace of H of finite codimension and Q is the orthogonal projection onto H_1, then the operator QAQ generates a semigroup of contractions in H_1.*

Let us return to the particular situation which is considered in this paper. Since all operators K_p, C, S_p are m-dissipative and W has codimension one, we can apply this theorem to each of them, obtaining the following corollary.

Corollary 2.1 *Let $K_{\epsilon,p} = \epsilon^{-1}C + S_p$. The operator $QK_{\epsilon,p}Q$ generates a semigroup of contractions in W, say $G_{\epsilon,p}(t)$, which satisfies for $t \geq 0$*

$$\|G_{\epsilon,p}(t)\|_{H_1} \leq e^{-\gamma t/\epsilon}, \tag{2.14}$$

where γ is defined in P.1.

3 Formal perturbation procedure for the operator $K_{\epsilon,p}$

The full formal compressed (or modified Chapman-Enskog) perturbation procedure for the equation of the form (1.1) can be found in [8, 2]. Here we shall quote, for further reference, equations satisfied by the terms of expansion. Let us consider the singularly perturbed system (2.13) with initial data $\mathring{v} = P\mathring{u}$ and $\mathring{w} = Q\mathring{u}$, where $\mathring{u} \in \mathcal{H}$.

$$\begin{aligned} \partial_t v &= PS_pPv + PS_pQw \\ \epsilon\partial_t w &= QCQw + \epsilon QS_pQw + \epsilon QS_pPv \\ v(0) &= \mathring{v} \\ w(0) &= \mathring{w}. \end{aligned} \tag{3.1}$$

The solution of (3.1) is looked for in the form

$$\begin{aligned} v(t) &= \bar{v}(t) + \tilde{v}(\tau) \\ w(t) &= \bar{w}(t) + \tilde{w}(\tau), \end{aligned} \tag{3.2}$$

where $\tau = t/\epsilon$. Terms $\bar{v}$, $\bar{w}$ are referred to as the bulk part of the solution and should give a good approximation to v, w for large time. Terms $\tilde{v}$, $\tilde{w}$ are called the initial layer part and are expected to improve accuracy of approximation for small t. The bulk part and the initial layer part are sought independently.

The formal compressed asymptotic expansion shows that the approximation of the solution $u = v + w$ to (3.1) should take the form

$$\mathfrak{u}(t) = \bar{\mathfrak{v}}(t) + \epsilon\tilde{v}_1(t/\epsilon) + \tilde{w}_0(t/\epsilon) + \epsilon(\bar{w}_1(t) + \tilde{w}_1(t/\epsilon)) \tag{3.3}$$

where the particular terms of the expansion satisfy the following equations:

The bulk part.

$$\begin{aligned} \partial_t\bar{\mathfrak{v}} &= PS_pP\bar{\mathfrak{v}} - \epsilon PS_pQ(QCQ)^{-1}QS_pP\bar{\mathfrak{v}} \\ \bar{\mathfrak{v}}(0) &= \overset{\circ}{v} - \epsilon PS_pQ(QCQ)^{-1}\overset{\circ}{w} \end{aligned} \tag{3.4}$$

and

$$\bar{w}_0 \equiv 0 \tag{3.5}$$

$$\bar{w}_1 = -(QCQ)^{-1}QS_pP\mathfrak{v}. \tag{3.6}$$

The initial layer part

$$\tilde{v}_0(\tau) \equiv 0 \tag{3.7}$$

$$\tilde{w}_0(\tau) = T_{QCQ}(\tau)\overset{\circ}{w} \tag{3.8}$$

$$\tilde{v}_1(\tau) = PS_pQ(QCQ)^{-1}\tilde{w}_1(\tau) \tag{3.9}$$

and

$$\begin{aligned} \partial_\tau\tilde{w}_1 &= QCQ\tilde{w}_1 + QS_pQ\tilde{w}_0 \\ \tilde{w}_1(0) &= (QCQ)^{-1}QS_pP\overset{\circ}{v} \end{aligned} \tag{3.10}$$

In the next section we shall see that the terms of the asymptotic expansion are well defined by (3.4) - (3.10) and that $\mathfrak{u}$, defined by (3.3), approximates u with an error of order of ϵ^2. It follows that the error of the approximation given by

$$y(t) = v(t) - [\bar{\mathfrak{v}}(t) + \epsilon\tilde{v}_1(t/\epsilon)] \tag{3.11}$$

$$z(t) = w(t) - [\tilde{w}_0(t/\epsilon) + \epsilon(\bar{w}_1(t) + \tilde{w}_1(t/\epsilon))] \tag{3.12}$$

satisfies (formally) the system

$$\begin{aligned}
\partial_t y &= PS_pPy + PS_pQz + \epsilon PS_pP\tilde{v}_1 + PS_pQ\tilde{w}_1,\\
\partial_t z &= QS_pPy + QS_pQz + \frac{1}{\epsilon}QCQz + \epsilon QS_pQ\tilde{w}_1 + \epsilon QS_pP\tilde{v}_1 + \epsilon QS_pQ\bar{w}_1 - \epsilon\partial_t\bar{w}_1,\\
y(0) &= 0,\\
z(0) &= \epsilon^2(QCQ)^{-1}QS_pPPS_pQ(QCQ)^{-1}\mathring{w}.
\end{aligned} \tag{3.13}$$

4 Properties of the terms of the expansion

Let us recall that $PH = V = Lin\{\mathbf{m}\}$ and $Q = I - P$. The operator S_p is in general unbounded but it follows that PSP and QSP, being defined on a one-dimensional space, are continuous. It can be checked that PS_pQ is closable and therefore bounded [6]. Moreover, a simple calculation shows that it can be extended by continuity onto H. We shall use the same symbol for the continuous extension of PSQ. As in [1] we can prove that (3.4) is indeed a diffusion equation. Precisely, we have

Proposition 4.1 *There exist functions $p \to \omega(p)$, satisfying*

$$Re\ \omega = 0, \tag{4.1}$$

and $p \to \Omega(p)$ with

$$Re\ \Omega(p) \geq \omega_1 \|QS_pP\|_H^2 \tag{4.2}$$

and

$$|\Omega(p)| \leq \|(QCQ)^{-1}\|_H \|QS_pP\|_H \|PS_pQ\|_H \leq \omega_2 |p|^2 \tag{4.3}$$

for some positive constants ω_1, ω_2, independent of p, such that the solution $\bar{\mathfrak{v}}(t)$ to (3.4) is given by the formula

$$\bar{\mathfrak{v}}(t) = e^{(\omega - \epsilon\Omega)t}\bar{\mathfrak{v}}(0). \tag{4.4}$$

The formal analysis of Section 3 shows that we must ensure that the initial layer terms decay exponentially. The exponential decay of T_{QCQ} in H is ensured by assumption P.1 of Section 2. However, equations (3.9)-(3.10) which define the initial layer terms $\tilde{v}_1, \tilde{w}_0$ and $\tilde{w}_1$ involve superpositions of (usually) unbounded operators and we shall need stronger estimates so that we shall require additional assumptions. Let us denote

$$D_{S^2} := \{u \in D_S;\ S_{p_0}u \in D_S \text{ for some } 0 \neq p_0 \in \mathsf{P}^n\} \tag{4.5}$$

Using the same notation for the restriction of T_{QCQ} to both D_S and D_{S^2}, we assume that

P.6 T_{QCQ} is a semigroup of negative type in both D_S and D_{S^2}.

In other words we have

$$\|QS_pQT_{QCQ}(t)\mathring{w}\|_H \leq q_1 e^{-\gamma_0 t}\||p|\mathring{w}\|_{D_S} \tag{4.6}$$

and

$$\|QS_pQT_{QCQ}(t)\mathring{w}\|_{D_S} \leq q_2 e^{-\gamma_0 t}\||p|\mathring{w}\|_{D_{S^2}}, \tag{4.7}$$

where q_1 and q_2 are constants and where for the further applications we put $\gamma \geq \gamma_0 > 0$ (see assumption P.1 of Section 2).

We assume also that

$$\mathbf{m} \in D_{S^2}. \tag{4.8}$$

Now we can turn to the question of existence and regularity of the solution $\tilde{w}_1$ of (3.10). To this end we have the following result.

Proposition 4.2 *If the assumption P.6 is satisfied and $QCQ\mathring{w} \in D_S$, then (3.10) has a unique classical solution satisfying*

$$\|\tilde{w}_1(\tau)\|_H \leq K e^{-\gamma_0 \tau}\|p\mathring{u}\|_{D_S} \tag{4.9}$$

for some constant K.

5 Estimates of the error of asymptotic expansion

This Section is concerned with the estimate of the error of the asymptotic expansion (3.3). The following main theorem is proved in [2].

Theorem 5.1 *If assumptions P.1 - P.5, (4.6)-(4.7) and (4.8) are satisfied, then for any T, $0 < T < \infty$, there is a constant M independent of p such that the error of the asymptotic expansion given by (3.11)-(3.12) satisfies*

$$\|y(t) + z(t)\|_H \leq \epsilon^2 \mathrm{M} \max_{0 \leq k \leq 3} \left\{\|p^k \mathring{u}\|_{D_{S^2}}\right\}. \tag{5.1}$$

uniformly for $0 \leq t \leq T$.

As an immediate consequence we obtain the estimate for the error of the approximation in $\mathcal{H}$ (and, equivalently, in $L_2(\Omega, H)$).

Theorem 5.2 *If assumptions P.1 - P.5, (4.6)-(4.7) and (4.8) are satisfied, $\mathring{u} \in D(\mathcal{C})$ and*

$$\max_{0 \leq k \leq 3} \left\{\int_{\mathsf{P}_n} \|p^k \mathring{u}\|^2_{D_{S^2}} dp\right\} < +\infty \tag{5.2}$$

(that is $\mathring{u}$ is a Fourier transform of a function from the Sobolev space $W_2^3(\Omega, D_{S^2})$), then for any T, $0 < T < \infty$ there is a constant C such that

$$\|y(t) + z(t)\|_{\mathcal{H}} \leq \mathrm{C}\epsilon^2. \tag{5.3}$$

uniformly for $0 \leq t \leq T$.

The assumption that C is a self-adjoint operator, appearing in P.1, often turns out to be too restrictive. A closer look at the proofs of the discussed results [1, 2] shows that the self-adjointness of C was used first, in Corollary 2.1 to prove the existence and estimates of the semigroup $G_{\epsilon,p}$ which helped to obtain estimates for the bulk part of the error in Theorem 5.1 and second, in Proposition 4.1 to prove the positivity of the diffusion coefficent Ω.

In many cases it is possible to establish the positivity of the diffusion coefficent by other means (see for instance Proposition 6.1). In general, it is impossible, however, to obtain the estimate (2.14) for non self-adjoint C and we need some additional assumptions to obtain a counterpart of Theorem 5.1.

If we replace the assumption P.1 by

P.1' C is the generator of a semigroup of contractions, T_C, in H. Zero is a semisimple isolated eigenvalue of C with the eigenfunction $\mathbf{m}$ and $\sup Re\{\sigma(C) \setminus \{0\}\} = -\gamma < 0$. Moreover, the spectral projections P and Q, corresponding to the eigenvalue $\lambda = 0$ are orthogonal and the semigroup T_{QCQ} is of negative type in H.

and additionally assume that

$$QS_p\mathbf{m} \in D_{S^2}, \tag{5.4}$$

then the following theorem is true [2].

Theorem 5.3 *If assumptions P.1', P.2 - P.5, (4.6)-(4.7), (5.4) are satisfied, and if the diffusion coefficent Ω satisfies (4.2), then for any T, $0 < T < \infty$ there is a constant* M *independent of p such that*

$$\|y_p(t) + z_p(t)\|_H \leq \epsilon^2 \mathrm{M} \max_{0 \leq k \leq 3} \left\{ \|p^k \mathring{u}\|_{D_{S^2}} \right\}. \tag{5.5}$$

uniformly for $0 \leq t \leq T$.

Theorem 5.2 has to be modified in obvious way.

6 Diffusion approximation of the linear Boltzmann equation

In this Section we shall apply the theory which was developed in the previous sections to a class of kinetic equations with bounded collision operators. To this class belong linear Boltzmann equations, arising in low density approximations to the full Boltzmann equation. We shall consider only operators with unbounded velocities, since in case of bounded velocities most assumptions of the theory are trivially satisfied. In Propositions 6.1 - 6.3 we collected in a unified mathematical setting the required properties of the collision and transport operators. Some of them, in one form or another, belong to the folklore of the transport theory and can be found in e.g. [5, 12, 15]. The full proofs of these propositions can be found in [2].

The results of the Theorem 6.1 are new. We note only that the corrector of the initial value for the diffusion equation appeared in [15] where the diffusion approximation of the transport equation with bounded velocity range was analysed.

Let us consider the following initial value poblem for a transport equation:

$$\begin{aligned}
\partial_t u(t,x,\xi) &= -\xi\partial_x u(t,x,\xi) - \nu(\xi)u(t,x,\xi) + \int\limits_{\mathbb{R}^3} k(\xi,\xi')u(t,x,\xi)d\xi' \\
u(0,x,\xi) &= \mathring{u}(x,\xi).
\end{aligned} \tag{6.1}$$

We assume that we deal either with the free space case and then $x \in \Omega = \mathbb{R}^3$, or with periodic boundary conditions, in which case $x \in \Omega = [0,1]^3$ The variable $\xi \in \mathbb{R}^3$ is the velocity and u is the particle distribution function in the phase space. The collision frequency ν and the scattering kernel k are known functions which are assumed to be independent of x.

The whole analysis is performed in the Hilbert space

$$\mathcal{H} = L_2(\mathbb{R}^3_x, H) = L_2(\mathbb{R}^6, \mathbf{m}^{-1}(\xi)d\xi dx),$$

where $\mathbf{m}$ is the normalized Maxwellian distribution in a given temperature θ

$$\mathbf{m}(\xi) = (2\pi\theta)^{-3/2}\exp(-\xi^2/2\theta). \tag{6.2}$$

Hence, we consider equation (6.1) as an evolution equation in the space $\mathcal{H}$ and, following the results of Section 2, we apply to it the Fourier transform with respect to x to obtain the equivalent problem in $L_2(\mathsf{P}^n, H_\xi)$, given by

$$\begin{aligned}
\partial_t \hat{u}(t,p,\xi) &= ip\partial_x \hat{u}(t,p,\xi) - \nu(\xi)\hat{u}(t,p,\xi) + \int\limits_{\mathbb{R}^3} k(\xi,\xi')\hat{u}(t,p,\xi)d\xi' \\
\hat{u}(p,\xi,0) &= \mathring{\hat{u}}(p,\xi).
\end{aligned} \tag{6.3}$$

As in Section 2, $\mathsf{P} = \mathbb{R}^n$ in the free space case and $\mathsf{P} = \mathbb{Z}^n$ in the periodic case. Thanks to Theorem 2.1 we can analyze problem (6.3) in H, treating p as a parameter.

Also as before we denote by $\mathcal{C}$, $\mathcal{S}$ ..., the collision, streaming, etc. operators, respectively, acting in $\mathcal{H}$ and by C, S ..., their H counterparts.

Let us introduce assumptions necessary to analyze (6.3). The scattering kernel k is usually written in the following form:

$$k(\xi,\xi') = \mathbf{m}(\xi)\phi(\xi,\xi'), \tag{6.4}$$

where ϕ is called collision cross-section. We assume that:

A.1 for almost all $\xi', \xi \in \mathbb{R}^3$ and some constant c_1

$$0 < \phi(\xi, \xi') < c_1.$$

Our analysis does not require ϕ to be a symmetric function. However, in realistic cases the so-called principle of detailed balance [5] asserts that

$$k(\xi, \xi')\mathbf{m}(\xi) = k(\xi', \xi)\mathbf{m}(\xi') \tag{6.5}$$

and that yields

$$\phi(\xi, \xi') = \phi(\xi', \xi). \tag{6.6}$$

Another assumption taken from the transport theory is that the operator C is conservative, that is for every $v \in L_1(\mathbb{R}^n)$ we have

$$\int_{\mathbb{R}^3} Cv d\xi = 0. \tag{6.7}$$

At this stage we introduce another technical assumption, namely we postulate that:

A.2 for every $\xi \in \mathbb{R}^3$ and some constants c_2, c_3

$$0 < c_2 \leq \nu(\xi) \leq c_3.$$

Proposition 6.1 *If the assumptions A.1, A.2, (6.7) and (6.6) are satisfied, then the operator C has property P.1 of Section 2.*
If instead of (6.6) we have, for some constant c_4,

$$\phi(\xi, \xi') \geq c_4 > 0 \quad \xi, \xi' \in \mathbb{R}^3, \tag{6.8}$$

then C has property P.1' Section 5.

The operator S_p formally is given by the formula:

$$S_p u(\xi) = ip\xi u(\xi) = \sum_{k=1}^{3} ip_k \xi_k u(\xi). \tag{6.9}$$

where $\xi = (\xi_1, \xi_2, \xi_3)$, $p = (p_1, p_2, p_3) \in \mathbb{R}^3$. The space D_S, discussed in Section 2, is given by

$$D_S = L_2(\mathbb{R}^3, (1 + |\xi|)\mathbf{m}^{-1}(\xi)d\xi) \tag{6.10}$$

The semigroup generated by S_p is given by the formula:

$$T_{S_p}(t)u = e^{ip\xi t}. \tag{6.11}$$

This semigroup is clearly conservative so that assumption P.2 is satisfied.

Since $D(C) = H$, we have $D(C) \cap D_S = D_S = L_2(\mathbb{R}^3, (1+|\xi|)\mathbf{m}^{-1}(\xi)d\xi)$ which is dense in H, as $C_0^\infty(\mathbb{R}^3) \subset L_2(\mathbb{R}^3, (1+|\xi|)\mathbf{m}^{-1}(\xi)d\xi)$ and $C_0^\infty(\mathbb{R}^3)$ is dense in H. Clearly, $\mathbf{m}$ satisfies all assumptions (2.4), (4.8) and (5.4).

The operator K_p generates a semigroup by the bounded perturbation theorem [11]. However, the Trotter product formula gives a better estimate of its growth. We obtain in particular that T_{K_p} is a contraction semigroup. The bounded perturbation theorem also shows that $D(K_p) = D(S_p)$ so that assumptions P.4 and P.5 are satisfied.

Next we have to turn to assumptions of Section 4. The operator QCQ is defined in a natural way in a subspace $W \subset H$. We denote by $(QCQ)_r$ its restriction to $W \cap H_r$, where

$$H_r := L_2(\mathbb{R}^n, (1+|\xi|)^r \mathbf{m}^{-1}(\xi)d\xi)$$

and $r \in \mathbb{N}_0$. We have

Proposition 6.2 *Under assumptions of this Section, for any $r \in \mathbb{N}_0$ the operator $(QCQ)_r$ generates uniformly continuous semigroup in H_r of (negative) type which is independent of r.*

Moreover, since $\mathbf{m}$ is the Maxwellian defined in (6.2), we see that the assumption (5.4) is satisfied (yielding of course (2.4) and (4.8)). Therefore all assumptions of Theorems 5.1 and 5.2 (or Theorem 5.3) are satisfied and the general theory is applicable.

Before we formulate the main theorem we translate some abstract formulae of the previous sections into the language of this section. We start with the coefficients of the diffusion operator defined in (3.4). From the previous considerations it follows that $PSQ(QCQ)^{-1}QSP$ operates in the space $L_2(\mathbb{R}^n_x) \otimes Lin\{\mathbf{m}\}$ that is it acts on functions of the form $u(x,\xi) = \rho(x)\mathbf{m}(\xi)$ in the following way

$$PSQ(QCQ)^{-1}QSP(\rho\mathbf{m}) = (D\rho)\mathbf{m} \tag{6.12}$$

for some operator D. It can be proved [2] (see also [12]) that D has the properties listed in the proposition below.

Proposition 6.3 *Operator D is a second order elliptic differential operator*

$$(D\rho)(x) = \sum_{k,l=1}^{3} d_{kl}\partial^2_{x_k,x_l}\rho(x), \tag{6.13}$$

with real coefficients d_{kl}, $k = 1,2,3$ given by the formula

$$d_{kl} = \int_{\mathbb{R}^3} d_k(\xi')\xi'_l d\xi', \tag{6.14}$$

where d_k is the unique function in W satisfying

$$(Cd_k)(\xi) = \xi_k\mathbf{m}(\xi), \quad k = 1,2,3. \tag{6.15}$$

If operator C is self-adjoint, then D is formally self-adjoint, that is

$$d_{kl} = d_{lk}.$$

If the collision frequency ϕ satisfies the following assumption: for any linear isometric mapping r of $\mathbb{R}^3$ we have

$$\phi(r\xi, r\xi') = \phi(\xi, \xi'), \quad \xi, \xi' \in \mathbb{R}^3, \tag{6.16}$$

then

$$d_{kl} = d\delta_{kl} \tag{6.17}$$

for some $d > 0$, where δ_{kl} is the Kronecker delta.

Let $\bar{\mathfrak{v}}(t, x, \xi) = \rho(t, x)\mathbf{m}(\xi)$. The initial value for (3.4) is given by

$$\bar{\mathfrak{v}}(0) = \mathring{v} - \epsilon PSQ(QCQ)^{-1}\mathring{w},$$

where $\mathring{v}(x, v) = \mathbf{m}(\xi)\mathring{\varrho}(x) = \mathbf{m}(\xi) \int_{\mathbb{R}^3} \mathring{u}(x, \xi')d\xi'$ and $\mathring{w} = Q\mathring{u} = \mathring{u} - \mathbf{m}\mathring{\varrho}$. Here $\mathring{u}$ is the initial value of the original equation (6.1).

If we define $\mathring{d}$ to be the unique solution in W to

$$(C\mathring{d})(x, \xi) = \mathring{u}(x, \xi) - \mathbf{m}(\xi)\mathring{\varrho}(x)$$

(such a solution exists as the left-hand side belongs to W) then it follows that

$$(PSQ(QCQ)^{-1}\mathring{w})(x, \xi) = \mathbf{m}(\xi) \sum_{k=1}^{3} \int_{\mathbb{R}^3} \xi' \partial_{x_k} \mathring{d}(x, \xi')d\xi'$$

and consequently we obtain the following initial value problem for ρ:

$$\begin{aligned} \partial_t \rho(t, x) &= \sum_{k,l=1}^{3} d_{kl} \partial^2_{x_k, x_l} \rho(t, x) \\ \rho(0) &= \mathring{\varrho} - \epsilon \sum_{k=1}^{3} \int_{\mathbb{R}^3} \xi'_k \partial_{x_k} \mathring{d}(x, \xi')d\xi'. \end{aligned} \tag{6.18}$$

Finally, the initial layer for the hydrodynamic part of the solution is given by

$$\tilde{v}_1\left(\frac{t}{\epsilon}, x, \xi\right) = \mathbf{m}(\xi)\tilde{\rho}\left(\frac{t}{\epsilon}, x\right) = \mathbf{m}(\xi) \sum_{k=1}^{3} \int_{\mathbb{R}^3} \xi'_k \partial_{x_k} \tilde{d}\left(\frac{t}{\epsilon}, x, \xi'\right) d\xi', \tag{6.19}$$

where $\tilde{d}$ is the unique solution in W to

$$(C\tilde{d})\left(\frac{t}{\epsilon}, x, \xi\right) = \tilde{w}_0\left(\frac{t}{\epsilon}, x, \xi\right) = \exp\left(\frac{t}{\epsilon}QCQ\right) \mathring{w}(x, \xi).$$

Here, due to the boundedness of QCQ, $\exp QCQ$ has the meaning of the standard exponent.

Using these results we can write down the approximation $\mathfrak{u}$ to the solution u of (6.1) obtained by the compressed asymptotic method in the following form:

$$\mathfrak{u} = \rho \mathbf{m} + \epsilon \bar{v}_1 + \tilde{w}_0 + \epsilon(\bar{w}_1 + \tilde{w}_1). \tag{6.20}$$

where ρ and the remaining terms of the expansion are defined by (6.18), (6.19) and counterparts of (3.8) and (3.10), respectively.

Alternatively, if we are interested only in the approximation of the hydrodynamic part of the solution u, defined as

$$\varrho(t, x) = \int_{\mathbb{R}^3} u(t, x, \xi') d\xi', \tag{6.21}$$

then the approximation to ϱ is given by

$$\mathfrak{g}(t, x) = \rho(t, x) + \epsilon \tilde{\rho}(t/\epsilon, x) \tag{6.22}$$

where $\tilde{\rho}$ was defined in (6.19). Theorem 5.2, specified to the linear Boltzmann equation, has the following form:

Theorem 6.1 *If assumptions A.1, A2, (6.7) and either (6.6) or (6.8) are satisfied and*

$$\max_{0 \le k \le 3} \left\{ \int_{\mathsf{P}^3} \|p^k \mathring{\hat{u}}\|^2_{H_2} dp \right\} < +\infty \tag{6.23}$$

(or, equivalently $\mathring{u}$ belongs to the Sobolev space $W^3_2(\Omega, H_2)$), then for any T, $0 < T < \infty$, there is a constant C such that for every $t \in [0, T]$ we have

$$\|u(t) - \mathfrak{u}(t)\|_{\mathcal{H}} \le C\epsilon^2 \tag{6.24}$$

and

$$\|\rho(t) - \mathfrak{g}(t)\|_{L_2(\mathbb{R}^3)} \le C\epsilon^2 \tag{6.25}$$

where $\mathfrak{u}$ and $\mathfrak{g}$ were defined in (6.20) and (6.22), respectively.

References

[1] J. Banasiak, Asymptotic analysis of abstract linear kinetic equation, *Math. Meth. in the Appl. Sci.*, to appear.

[2] J. Banasiak, Diffusion approximation and analysis of initial layer for evolution equations of kinetic type, Rapporto n° 7 - Luglio 1994, Universitá degli Studi di Ancona.

[3] J. Banasiak, J. R. Mika, Asymptotic analysis of the Fokker- Planck equation of Brownian motion, *Mathematical Models and Methods in Applied Sciences*, **4** (1) (1994), 17–33.

[4] C. Bardos, R. Santos, R. Sentis, Diffusion approximation and computation of the critical size, *Trans. AMS.*, **284** (2) (1984), 617–649.

[5] C. Cercignani, *Theory and Application of the Boltzmann Equation*, Springer-Verlag, New York, 1988.

[6] T. Kato, *Perturbation theory for linear operators*, Springer-Verlag, Berlin Heidelberg New York, 1966.

[7] E. W. Larsen, J. B. Keller, Asymptotic solution of neutron transport problems for small mean free paths, *J. Math. Phys.*, **15** (1) (1974), 75–81.

[8] J. R. Mika, New asymptotic expansion algorithm for singularly perturbed evolution equations, *Math. Meth. in the Appl. Sci.*, **3** (1981), 172–188.

[9] J. R. Mika, J. Banasiak, Asymptotic analysis of a model kinetic equation, *Mathematical Models and Methods in Applied Sciences*, to appear.

[10] A. Palczewski, Exact and Chapman-Enskog solutions for the Carleman model, *Math. Meth. in the Appl. Sci.*, **6** (1984), 417–432.

[11] A. Pazy, *Semigroup of linear operators and applications to partial differential equations*, Springer-Verlag, Berlin New York, 1983.

[12] F. Poupaud, Diffusion approximation of the linear semiconductor Boltzmann equation: analysis of boundary layers, *Asym. Anal.*, **4** (1991), 293–317.

[13] F. Poupaud, Runaway Phenomena and Fluid Approximation Under High Fields in Semiconductor Kinetic Theory, *Zeitschrift für angewandte Mathematik und Mechanik* **72**(8) (1992), 359– 372.

[14] M. Reed, B. Simon, *Methods of modern mathematical physics*, v. 1, 2, Academic Press, New York London, 1972.

[15] E. Ringeisen, R. Sentis, On the diffusion approximation of a transport process without time scaling, *Asym. Anal.*, **5** (1991), 145–159.

C J K BATTY

The spectral bound of Schrödinger operators

1. Background

A Schrödinger operator on $\mathbf{R}^N$ is a partial differential operator of the form $H_m = H_0 + m$, where $H_0 = \frac{1}{2}\Delta = \frac{1}{2}\sum_{j=1}^N \partial^2/\partial x_j^2$ and $m : \mathbf{R}^N \to [-\infty, \infty]$ is a measurable function. Under suitable assumptions on m, H_m can be defined as an upper-bounded self-adjoint operator on $L^2(\mathbf{R}^N)$ (or on $L^2(Z_m)$ for some subset Z_m of $\mathbf{R}^N$). If σ_m is the supremum of the spectrum of H_m, then $-\sigma_m$ is the *ground state energy*. The Schrödinger semigroup e^{tH_m} generated by H_m represents the flow of heat in the presence of a potential m, the positive part m_+ of m corresponding to excitation and the negative part m_- to absorption. Other physical and mathematical reasons for studying the Schrödinger semigroup, the ground state energy and any eigenfunction of H_m with eigenvalue σ_m are discussed in Section A1 of the survey article by Simon [13], which includes many mathematical results on these topics. In this article, we report on some recent results concerning σ_m.

If m is bounded, there is no difficulty in interpreting H_m as an operator. In this case, m defines a bounded multiplication operator on $L^p(\mathbf{R}^N)$ for each $1 \le p \le \infty$, so H_m is a bounded perturbation of the operator H_0. In fact, it is possible to define H_m under much weaker assumptions on m. This is usually done by means of the associated quadratic form $\mathbf{a}_m$:

$$D(\mathbf{a}_m) = \left\{ u \in W^{1,2}(\mathbf{R}^N) : \int_{\mathbf{R}^N} |mu^2| < \infty \right\},$$

$$\mathbf{a}_m(u) = \int_{\mathbf{R}^N} |\nabla u|^2 - \int_{\mathbf{R}^N} m|u|^2.$$

We shall assume throughout that m_+ is in the Kato class K_N, the definition and properties of which are given in [13, Section A.2] (see [13, Section A.3] and [15, Section 5] for further possibilities). Then $\overline{D(\mathbf{a}_m)} = L^2(Z_m)$ for some measurable subset Z_m of $\mathbf{R}^N$ (see [10] and [2]), $\mathbf{a}_m$ is associated with a lower-bounded self-adjoint operator $-H_m$ on $L^2(Z_m)$, and the semigroup $\left\{e^{tH_m} : t \ge 0\right\}$ interpolates to provide a semigroup $\{S_{m,p}(t) : t \ge 0\}$ of operators on each $L^p(Z_m)$ $(1 \le p \le \infty)$, which is strongly continuous for $1 \leq p < \infty$. Moreover, the semigroups are given by the Feynman-Kac formula:

$$(S_{m,p}(t)f)(x) = \mathbf{E}^x \left[\exp\left(\int_0^t m(B(s))\, ds \right) f(B(t)) \right].$$

Here, $\{B(t) : t \geq 0\}$ is Brownian motion on $\mathbf{R}^N$ and $\mathbf{E}^x$ denotes expectation with respect to Wiener measure $\mathbf{P}^x$ corresponding to motion starting at x. If m_- is locally integrable, then $Z_m = \mathbf{R}^N$. If m_- is locally in Kato's class, then $C_c^\infty(\mathbf{R}^N)$ is a core for the form $\mathbf{a}_m$.

The spectral bound σ_m can be defined from the quadratic form $\mathbf{a}_m$:

$$\sigma_m = -\inf \left\{ \mathbf{a}_m(u) : u \in D(\mathbf{a}_m), \int_{\mathbf{R}^N} |u|^2 = 1 \right\}, \tag{1}$$

so $\|S_{m,2}(t)\|_{\mathcal{B}(L^2)} = e^{t\sigma_m}$. Simon [12, Theorem 1.3], [13, Theorem B.5.1] has shown that σ_m is also the growth bound of $S_{m,p}$ for any $1 \leq p \leq \infty$:

$$\sigma_m = \lim_{t \to \infty} \frac{1}{t} \log \|S_{m,p}(t)\|_{\mathcal{B}(L^p)}.$$

In particular, taking $p = \infty$ and putting $f = 1$ in the Feynman-Kac formula, we obtain an alternative formula for σ_m:

$$\sigma_m = \lim_{t \to \infty} \frac{1}{t} \sup_{x \in \mathbf{R}^N} \log \mathbf{E}^x \left[\exp\left(\int_0^t m(B(s))\, ds \right) \right]. \tag{2}$$

For a given potential m, neither (1) nor (2) is particularly easy to evaluate. However, it is possible in some cases to obtain either estimates or qualitative information about σ_m. For example, Fefferman and Phong [7] and Schechter [11] have given some estimates when $m \geq 0$, and in Corollary 2 below we give a succinct criterion whether $\sigma_m < 0$ when $m \leq 0$. If m takes both positive and negative values, the usual technique is to separate the effects of the positive and negative parts, so results when m has a particular sign are useful even for the general case.

One way to study the spectral bound is to introduce a parameter $\lambda \geq 0$ and to consider the *spectral function* $s_m(\lambda) = \sigma_{\lambda m}$. Although this parameter may have no physical significance, it has mathematical advantages. Moreover, it arises in linearisations of non-linear problems [9] (including some arising in mathematical biology [8]), where the main interest is in finding values of λ such that $s_m(\lambda) = 0$ and a ground state exists.

It is easily seen that s_m is convex, $s_m(0) = 0$ and $s_m(\lambda) \to \infty$ as $\lambda \to \infty$ if $m_+ \neq 0$ (a.e.). We shall see that it is possible to determine the asymptotic behaviour of $s_m(\lambda)$ for large λ if $m \leq 0$, and for small λ if $m \geq 0$.

2. The case when $m \leq 0$

When $m \leq 0$, s_m is a non-increasing convex function. We shall see that the value of $s_m(\infty) := \lim_{\lambda\to\infty} s_m(\lambda)$ can be given in terms of the integral of m over classes of subsets of $\mathbf{R}^N$.

For a Borel subset E of $\mathbf{R}^N$, let σ_E be the spectral bound of $\frac{1}{2}\Delta_E$, where Δ_E is the Dirichlet-Laplacian on E. Thus

$$\begin{aligned}\sigma_E &= -\inf\left\{\int_{\mathbf{R}^N} |\nabla u|^2 : u \in W^{1,2}(\mathbf{R}^N), \widetilde{u} = 0 \text{ q.e. in } \mathbf{R}^N \setminus E, \int_E |u|^2 = 1\right\} \\ &= \lim_{t\to\infty} \frac{1}{t} \sup_{x\in\mathbf{R}^N} \left\{\log \mathbf{P}^x\left[B(s) \in E \text{ for all } s \leq t\right]\right\}.\end{aligned}$$

Here, $\widetilde{u}$ is a quasi-continuous version of u, and "q.e." represents "quasi-everywhere". In the case of an open set Ω, this reduces to the more familiar formula

$$\sigma_\Omega = -\inf\left\{\int_\Omega |\nabla u|^2; u \in C_c^\infty(\Omega), \int_\Omega |u|^2 = 1\right\}.$$

For $\sigma < 0$, let $\mathcal{F}_\sigma$ be the class of all Borel sets such that $\sigma_E > \sigma$, and let $\mathcal{F}$ be the class of all Borel sets E such that $\sigma_E = 0$. Thus an open set Ω belongs to $\mathcal{F}$ if and only if Poincare's inequality does not hold in Ω, i.e. there does not exist a constant $c > 0$ such that $\int_\Omega |\nabla u|^2 \geq c \int_\Omega |u|^2$ for all u in $C_c^\infty(\Omega)$.

Theorem 1 [3, Theorem 4.9]. *Suppose that $m \leq 0$. Then*

$$\lim_{\lambda\to\infty} s_m(\lambda) = \inf\left\{\sigma < 0 : \inf_{E\in\mathcal{F}_\sigma} \int_E |m| > 0\right\}. \tag{3}$$

Corollary 2 [5, Theorem 4.12]. *Suppose that $m \leq 0$. The following are equivalent:*

(1) $\sigma_m < 0$;
(2) $\int_E |m| = \infty$ *for all* E *in* $\mathcal{F}$;
(3) *There exist* $\sigma < 0$ *and* $c > 0$ *such that* $\int_E |m| > c$ *for all* E *in* $\mathcal{F}_\sigma$.

Special cases. The following special cases are worthy of note.

1. Suppose that $N = 1$. Then $E \in \mathcal{F}_\sigma$ if and only if E contains an interval of length $\pi(-2\sigma)^{-1/2}$. Hence if $m \leq 0$,

$$\lim_{\lambda\to\infty} s_m(\lambda) = -\frac{\pi^2}{2d^2},$$

where $d = \inf\{\delta > 0 : \inf_{x\in\mathbf{R}} \int_x^{x+\delta} |m| > 0\}$ [3, Corollary 4.11].

2. Suppose that $m \in L^1_{\text{loc}}(\mathbf{R}^N)$ and $m \leq 0$. Then Theorem 1 and Corollary 2 remain valid if $\mathcal{F}_\sigma$ and $\mathcal{F}$ are replaced by the classes $\mathcal{F}^o_s$ and $\mathcal{F}^o$ of open sets in $\mathcal{F}_\sigma$ and $\mathcal{F}$ respectively [3, Corollary 4.10], [5, Corollary 4.15].

3. Suppose that $m \in L^1(\mathbf{R}^N) + L^\infty(\mathbf{R}^N)$ and $m \leq 0$. Then the conditions (1)–(3) of Corollary 2 are equivalent to:

(4) $\int_E |m| = \infty$ whenever E contains arbitrarily large balls (i.e. whenever, for all $r > 0$, there exists x in $\mathbf{R}^N$ such that $B(x,r) := \{y \in \mathbf{R}^N : \|y-x\| < r\} \subseteq E$).

(5) There exist $r > 0$ and $c > 0$ such that $\int_{B(x,r)} |m| > c$ for all x in $\mathbf{R}^N$.

[1, Theorem 1.2], [5, Proposition 4.19], [6, Theorem 5.1].

The proofs. Various proofs are known that the left-hand side of (3) is at least as great as the right-hand side (or of the corresponding implication (1) $\implies$ (3) in Corollary 2). The proof in [5] uses probabilistic terminology, but only in a weak way (the strong Markov property of Brownian motion is not needed). It is therefore not surprising that it is possible to give analytic proofs, using either semigroup techniques or variational methods. Proofs of each of these types appear in [3].

The only known proof of the reverse inequality in (3) (or the reverse implication in Corollary 2) is strongly stochastic, using the strong Markov property. The sets E which arise in the argument are of the form

$$E = \left\{x \in \mathbf{R}^N : \mathbf{P}^x \left[\int_0^t |m(B(s))|\, ds \leq \alpha\right] \geq \eta\right\}$$

for some $\alpha > 0$ and $\eta > 0$.

In the third special case above, a proof of the equivalence of condition (1) in Corollary 2 with (4) and (5) was given in [1], using variational methods. Another analytic proof was outlined in [5, p.488], and a similar proof has appeared in [6].

Since the statements of Theorem 1 and Corollary 2 have no stochastic content, it must in principle be possible to find an analytic proof. This would be useful, since it would be accessible to many people working in this field who are not familiar with stochastic methods. On the other hand, the evident difficulty of finding such a proof for a known result illustrates the power of the insight which the strong Markov property can provide and the advantages which can accrue from having these techniques at hand.

A supplementary question. Suppose that $m \leq 0$ and $0 > s_m(\infty) > -\infty$. One may ask whether s_m is strictly decreasing (the alternative is that s_m becomes constant for large λ)? This question has been addressed in [3, Section 5], and the following partial answers have been obtained.

1. There exists m in $L^\infty_{\text{loc}}(\mathbf{R}^N)$ with $m \leq 0$ and $0 > s_m(\infty) > -\infty$ such that $s_m(\lambda) = s_m(\infty)$ for all sufficiently large λ.

2. If $N = 1$, $m \in L^\infty(\mathbf{R})$, $m \leq 0$ and $s_m(\lambda) < 0$ for $\lambda > 0$, then s_m is strictly decreasing.

3. If $N \geq 2$ and $m \in L^\infty(\mathbf{R}^N)$ with $m \leq 0$, the question remains open. However, it is equivalent to another question as follows.

For $\sigma < 0$, let

$$\nu(\sigma) = \inf \{ s_{\mathbf{1}_{\Omega^c}}(1) : \Omega \in \mathcal{F}_\sigma, \Omega \text{ bounded open} \},$$

where $\mathbf{1}_{\Omega^c}$ is the characteristic function of $\mathbf{R}^N \backslash \Omega$. It is easily seen that $s_{\mathbf{1}_{\Omega^c}}(1) > \sigma_\Omega$ for each Ω. It is shown in [3] that the following are equivalent:

(1) For all m in $L^\infty(\mathbf{R}^N)$ with $m \leq 0$ and $s_m(\lambda) < 0$ for $\lambda > 0$, s_m is strictly decreasing;

(2) For all $\sigma < 0$, $\nu(\sigma) > \sigma$.

Intuitively, there seem to be two possibilities—either $\nu(\sigma) = \sigma$ or $\nu(\sigma) = s_{\mathbf{1}_{B^c}}(1)$ where B is a ball of such a radius that $\sigma_B = \sigma$.

3. The case when $m \geq 0$

When $0 \leq m \in K_N$, s_m is a non-decreasing convex function, and $s_m(\lambda) \to \infty$ as $\lambda \to \infty$ (unless $m = 0$ a.e.). It is possible to estimate s_m in terms of L^p-averages of m, and to find the value of the right-derivative of s_m at $\lambda = 0$.

Let ω_N be the volume of the unit ball in $\mathbf{R}^N$, and let

$$\|m\|_{p,r} = \sup_{x \in \mathbf{R}^N} \left(\frac{1}{\omega_N r^N} \int_{B(x,r)} m^p \right)^{1/p} \qquad (1 \leq p < \infty, 0 < r < \infty).$$

Let $\psi_t(x) = \int_0^t (2\pi s)^{-N/2} e^{-|x|^2/2s}\, ds$. Note that a non-negative function m belongs to K_N if and only if $\|m * \psi_t\|_\infty \to 0$ as $t \to 0+$ [15, Proposition 5.1].

Fefferman and Phong [7, Theorem 5, p.145] and Schechter [11, Corollary 3.3] used Fourier analysis and variational methods to show that there are positive constants c_1 and C_p $(p > 1)$ (depending only on N and p) such that

$$\sup_{r>0} \left(c_1 \lambda \|m\|_{1,r} - \frac{1}{r^2} \right) \leq s_m(\lambda) \leq \sup_{r>0} \left(C_p \lambda \|m\|_{p,r} - \frac{1}{r^2} \right)$$

for any $p > 1$. The following upper bound involving the L^1-average of m and the formula in Theorem 4 for the right-derivative of s_m at $\lambda = 0$ can be obtained rather easily from (2), Khashmin'skii's Lemma [13, Lemma B.1.2], and Jensen's inequality.

Theorem 3 [4]. *There are positive constants C_1, C_1' and δ such that, for any $0 \leq m \in K_N$,*

(1) $s_m(\lambda) \leq C_1 \lambda \|m\|_{1,r}$ *whenever* $\lambda \|m * \psi_{r^2}\|_\infty \leq 1/2$;
(2) $s_m(\lambda) \leq C_1' \lambda \|m\|_{1,r}$ *whenever* $\lambda \int_0^r \rho \|m\|_{1,\rho}\, d\rho \leq \delta$.

Theorem 4 [4]. *Let $0 \leq m \in K_N$. Then*

$$s_m'(0+) = \lim_{t \to \infty} \frac{1}{t} \|m * \psi_t\|_\infty.$$

Moreover, $s_m'(0+) = 0$ if and only if $\lim_{r \to \infty} \|m\|_{1,r} = 0$.

4. The case when m changes sign

If m takes both positive and negative values (on sets of positive measure), then s_m is convex and $s_m(\lambda) \to \infty$ as $\lambda \to \infty$. Most techniques for estimating s_m depend on separating the positive and negative parts of m and using the inequality

$$s_m(\lambda) \leq \frac{1}{2} \left(s_{m_-}(2\lambda) + s_{m_+}(2\lambda) \right).$$

Hence

$$s_m'(0+) \leq s_{m_-}'(0+) + s_{m_+}'(0+).$$

Thus Theorem 3 and some upper bounds for $s'_{m_-}(0+)$ obtained in the proof of Theorem 1 give a sufficient condition that $s'_m(0+) < 0$ and hence that there exists a unique $\lambda_1 > 0$ such that $s_m(\lambda_1) = 0$. The precise estimates are complicated [3, Remark 4.7], but we can read off the following special case from Corollary 2 and Theorem 4.

Theorem 5 [4, Theorem 3.2]. *Suppose that $m \in L^1_{\text{loc}}(\mathbf{R}^N)$, $0 \neq m_+ \in K_N$, and*

(1) $\int_\Omega m_- = \infty$ *for all open sets Ω in $\mathcal{F}$;*

(2) $\lim_{r\to\infty} \|m_+\|_{1,r} = 0$.

Then there exists a unique $\lambda_1 > 0$ such that $s_m(\lambda_1) = 0$.

5. Some generalisations

1. *Elliptic operators.* It is not important that H_0 is $\frac{1}{2}\Delta$; it is possible for H_0 to be any symmetric, strongly elliptic, operator of the form $\sum_{i,j=1}^N \partial/\partial x_i(a_{ij}(x)\partial/\partial x_j)$. The results are unchanged in nature; for $\sigma < 0$, the class $\mathcal{F}_\sigma$ depends on H_0 ($\mathcal{F}$ does not); in the formula for ψ_t in Section 3, the Gaussian kernel is replaced by the appropriate kernel for H_0. The role of Brownian motion is taken by the appropriate diffusion process. More details may be found in [2, Section 8] and [5, Section 6].

2. *Singular potentials.* In the case of a negative potential (Section 2), it is possible to allow the potential to be very singular (see [14]). Thus m need not be a function, but may be of the form $-\mu$, where μ is a positive (not necessarily σ-finite) measure on the Borel subsets of $\mathbf{R}^N$ such that $\mu(E) = 0$ for all polar sets E. The quadratic form is given by:

$$\mathbf{a}_m(u) = \int_{\mathbf{R}^N} |\nabla u|^2 + \int_{\mathbf{R}^N} |\tilde{u}|^2 \, d\mu.$$

In Theorem 1 and Corollary 2, $\int_E |m|$ is replaced by $\mu(E)$.

If Ω is an open set and $m : \Omega \to [-\infty, 0]$ is measurable, we can put

$$\mu(E) = \begin{cases} \int_E |m| & \text{if } E \setminus \Omega \text{ is polar} \\ \infty & \text{otherwise.} \end{cases}$$

Then $s_m(\lambda)$ is the spectral bound of the Schrödinger operator $\frac{1}{2}\Delta_\Omega + \lambda m$ with Dirichlet boundary conditions on $L^2(\Omega)$. More details may be found in [5, Section 6].

3. *Periodic time-dependent coefficients.* Daners and Koch Medina [6] have extended one case of Corollary 2 to the case of a bounded negative potential $m(x,t)$ which depends smoothly and periodically (with period T) on t, and is uniformly continuous in x. They show that the solutions of

$$\frac{\partial u}{\partial t} = H_0 u + m(x,t)u$$

converge to 0 at an exponential rate if and only if $\int_0^T \int_E |m(x,t)|\, dx\, dt = \infty$ whenever E contains arbitrarily large balls.

References

1. W. Arendt and C.J.K. Batty, *Exponential stability of a diffusion equation with absorption*, Diff. Integral Equations **6** (1993), 1009-1024.
2. W. Arendt and C.J.K. Batty, *Absorption semigroups and Dirichlet boundary conditions*, Math. Ann. **295** (1993), 427–448.
3. W. Arendt and C.J.K. Batty, *The spectral bound of Schrödinger operators*, Potential Anal. (to appear).
4. W. Arendt and C.J.K. Batty, *The spectral function and principal eigenvalues for Schrödinger operators*, preprint.
5. C.J.K. Batty, *Asymptotic stability of Schrödinger semigroups: path integral methods*, Math. Ann. **292** (1992), 457–492.
6. D. Daners and P. Koch Medina, *Exponential stability, change of stability and eigenvalue problems for linear time-periodic parabolic equations on* $\mathbb{R}^N$, Diff. Integral Equations (to appear).
7. C.L. Fefferman, *The uncertainty principle*, Bull. Amer. Math. Soc. **9** (1983), 129–206.
8. W.H. Fleming, *A selection-migration problem in population genetics*, J. Math. Biol. **2** (1975), 219–233.
9. P. Hess and T. Kato, *On some linear and nonlinear eigenvalue problems with an indefinite weight function*, Comm. Partial Diff. Equations **5** (1980), 999–1030.
10. H.P. McKean, *$-\Delta$ plus a bad potential*, J. Math. Phys. **18** (1977), 1277–1279.
11. M. Schechter, *The spectrum of the Schrödinger operator*, Trans. Amer. Math. Soc. **312** (1989), 115–128.
12. B. Simon, *Brownian motion, L^p properties of Schrödinger operators, and the localization of binding*, J. Funct. Anal. **35** (1980), 215–229.
13. B. Simon, *Schrödinger semigroups*, Bull. Amer. Math. Soc. **7** (1982), 447–526.
14. P. Stollmann and J. Voigt, *Perturbation of Dirichlet forms by measures*, preprint (1992).
15. J. Voigt, *Absorption semigroups, their generators, and Schrödinger semigroups*, J. Funct. Anal. **67** (1986), 167–205.

C.J.K. Batty
St. John's College
Oxford OX1 3JP
England

L BERLYAND* AND J XIN†

Renormalization group technique for asymptotic behavior of a thermal diffusive model with critical nonlinearity

1 Introduction

The familiar thermal diffusive model [15], describing a premixed flame from a one-step chemical reaction $A \to B$ reads:

$$\begin{aligned} u_t &= \Delta_x u + v f(u), \\ v_t &= \Lambda^{-1}\Delta_x v - v f(u), \end{aligned} \tag{1.1}$$

where $x \in R^n$, u is the temperature, v is the mass fraction of the reactant A; Λ is the Lewis number, strictly positive; $f(u)$ is an Arrehnius reaction term of the form $e^{-\frac{E}{u}}$, with E being the activation constant.

There have been many studies on system (1.1) for front like L^∞ data. Existence, stability and instability of traveling waves can be found in [5], [15], [16] among others; and related Cauchy problem regarding decay and boundedness in [2], [3], [4], [12] and references therein. In [2], [3], [4] and elsewhere, results on decay of v typically assume strict positivity of initial temperature field at one end of the infinities in case $n = 1$.

In this paper, we are interested in the decay of solutions for system like (1.1) when both initial data u_0 and v_0 are integrable and belong to the space $L^1(R^1) \cap L^2(R^1) \cap L^\infty(R^1)$. Since there is no positive constant lower bound for u, the function v will in general not decay exponentially.

*Dept of Math & Material Research Lab, Penn State Univ, Univ Park, PA 16802.
†Department of Mathematics, University of Arizona, Tucson, AZ 85721.

Notice that $f(u) \leq c_p u^p$, for $u > 0$ and any $p > 0$, and some constant c_p. By maximum principle, solution (u, v) of (1.1) is bounded from above by that of system (c_p normalized to one):

$$\begin{aligned} u_t &= u_{xx} + vu^{p-1}, \\ v_t &= \Lambda^{-1} v_{xx}, \end{aligned} \tag{1.2}$$

with the same inital data. When $p > 3$, by either an integral equation and contraction mapping argument or the recent RG method of Bricmont, Kupiainen, and Lin [7], we know that the solution of system (1.2) decays to zero with rate $O(t^{-\frac{1}{2}})$ if the initial data are small enough. This means that for small initial data the Arrhenius reaction $e^{-\frac{E}{u}}$ or u^{p-1}, $p - 1 > 2$ is too weak so diffusion dominates and leads to decay.

We are motivated to consider the following system instead:

$$u_t = u_{xx} + vu^{p-1}, \quad u(x,0) = u_0(x) \geq 0, \tag{1.3}$$

$$v_t = \Lambda^{-1} v_{xx} - vu^{p-1}, \quad v(x,0) = v_0(x) \geq 0, \tag{1.4}$$

where $p \in (1, 3]$. Existence of globally bounded solutions to system (1.3)-(1.4) and its analogue in several space dimensions have been studied in [1], [10], [14] for bounded domains. Some of their arguments readily extend to R^1 for spatially decaying initial data in L^p, $1 \leq p < \infty$, see [13]. However, whether or not the solutions on R^1 decay to zero is not clear in general.

On the other hand our study was motivated by the scalar semilinear diffusion problem

$$u_t = u_{xx} \pm u^p, \quad u(x,0) = u_0(x) \geq 0. \tag{1.5}$$

It is well known that in case of small initial data $p = 3$ is the critical exponent, see [7], [9], [11] and references therein. This can be shown in various ways, but for our further needs we recall the renormalization group (RG) argument due to Bricmont, Kupiainen and Lin [7]. We pick a number $L > 1$ and observe that the rescaled solution

$$u^{(L)}(x,t) \equiv Lu(Lx, L^2 t) \tag{1.6}$$

satisfies the equation

$$u_t^{(L)} = u_{xx}^{(L)} \pm [u^{(L)}]^p L^{3-p}. \tag{1.7}$$

After n iterations of the rescaling, we obtain the factor $L^{n(3-p)}$ in front of the nonlinearity. Because of (1.7), the nonlinear term has almost no effect for long times (weakly nonlinear behavior) for $p > 3$, see [7] for the RG argument. It was also shown in [7] that if the critical nonlinear term $(-\lambda u^3)$ appears in (1.5), with $\lambda \ll 1$ an additional small parameter, the long time behavior is that of the case $p > 3$ with a logarithmic correction.

For the system (1.3)–(1.4) similar rescaling argument shows that weakly or strongly nonlinear behavior is determined by the critical exponent $p = 3$. For $p = 3$, small initial data, and the nonlinear terms not necessarily small, we obtain large time asymptotics of solutions of (1.3–1.4) with decay exponent for v different from the linear case. This is possible due to interaction between nonlinear terms which have different signs (source and sink). Our consideration is based on RG method analogous to [7]. However, a straightforward attempt of using this method faces a serious difficulty. Namely, the method of [7] was essentially based on a self-similar solution. An easy calculation shows that the system (1.3)-(1.4) has no self similar solutions of the form

$$(u, v) = \left(t^{-\alpha_1} f_1\left(\frac{x}{\sqrt{t}}\right), t^{-\alpha_2} f_2\left(\frac{x}{\sqrt{t}}\right)\right). \tag{1.8}$$

In fact, this is also true for $1 < p < 3$ due to the sign difference. The key observation is that instead of finding exact asymptotics one can evaluate the rate of decay by employing super (sub) self-similar solutions. In brief, our argument goes as follows. First we observe that by maximum principle the solution of (1.1) is always above the solution $\underline{u}$ of the heat problem $u_t = u_{xx}, u(x, 0) = u_0(x)$. Since $\underline{u}$ converges to a self similar solution, as $t \to \infty$ we obtain

$$u \geq a\frac{1}{\sqrt{t}}f_0^\star + h.o.t \equiv \frac{a}{\sqrt{4\pi t}}e^{-\frac{x^2}{4t}} + h.o.t \tag{1.9}$$

for large enough t and some constant a depending only on the initial data. Ignoring the higher order terms for the moment, we see that by maximum principle, v is bounded from above by $\overline{v}$ which solves the equation

$$\overline{v}_t = \Lambda^{-1}\overline{v}_{xx} - \frac{a^2}{t}(f_0^\star)^2\overline{v} \tag{1.10}$$

for $t \geq t_0^\star \gg 1$, and $\overline{v}(x, t_0^\star) = v(x, t_0^\star)$. The equation (1.10) admits self-similar solutions of the form $\overline{v} = t^{-\frac{\alpha}{2}} f_\alpha^\star(\frac{x}{\sqrt{t}})$. The exponent α and the function $f_\alpha^\star$ are principal eigenvalue and eigenfunction of a Sturm Liouville problem. This allows us to find a lower bound for α of the form $\alpha = 1 + g(\Lambda, \epsilon)$ with $g(\Lambda, \epsilon) > 0$, which means that $\overline{v}$ decays faster than $O(t^{-1/2})$, i.e. nonlinearity essentially affects the long time behavior. We further notice that for $t \geq t_0^\star$, u is bounded from above by $\overline{u}$, which satisfies the equation

$$\overline{u}_t = \overline{u}_{xx} + \overline{v}\,\overline{u}^2, t \geq t_0^\star,$$

$$\overline{v}(x, t_0^\star) = u(x, t_0^\star). \tag{1.11}$$

Since $\alpha > 1$, the RG method applies and shows that $\overline{u}$ decays to zero like $O(t^{-1/2})$. Further use of maximum principle and RG method gives lower bounds of decay of similar type. We denote $\|u\| \equiv \sup_{k\in R^1}(1+k^2)(|\hat{u}(k)| + |\frac{d}{dk}\hat{u}(k)|)$, where $\hat{u}(k)$ is the Fourier transform of u. Our main result is

Theorem 1.1 *Consider the system (1.3–1.4) with $p = 3$ and nonnegative initial data such that $u_0, v_0 \in L^1(R)\cap L^2(R)\cap L^\infty(R)$, $u_0 \not\equiv 0, v_0 \not\equiv 0$ and $\|u_0\|+\|v_0\| \leq \epsilon$ for some $\epsilon > 0$. Fix any $\delta \in (0, \frac{1}{2})$. Then there exists an $\epsilon_0 \in (0,1)$ such that if $\epsilon \in (0, \epsilon_0)$ there exist numbers $\alpha = \alpha(\epsilon) > 1, \alpha' = \alpha'(\epsilon) \geq \alpha$, $\alpha'(\epsilon) \to 1$ as $\epsilon \to 0$ and positive numbers A, A', B, B' which depend on ϵ and initial data (u_0, v_0) so that*

$$A'\frac{1}{t^{1/2}}f_0^\star\left(\frac{x}{\sqrt{t}}\right) + O\left(\frac{1}{t^{(1-\delta)}}\right) \leq u(x,t) \leq A\frac{1}{t^{1/2}}f_0^\star\left(\frac{x}{\sqrt{t}}\right) + O\left(\frac{1}{t^{\frac{\alpha}{2}}}\right) \tag{1.12}$$

$$B'\frac{1}{t^{\alpha'/2}}f_{\alpha'}^\star\left(\frac{x}{\sqrt{t}}\right) + O\left(\frac{1}{t^{\alpha'/2+\alpha/2-1/2}}\right) \leq v(x,t) \leq B\frac{1}{t^{\alpha/2}}f_\alpha^\star\left(\frac{x}{\sqrt{t}}\right) + O\left(\frac{1}{t^{\alpha/2+\frac{1}{2}-\delta}}\right) \tag{1.13}$$

for all $x \in R^1$, as $t \to \infty$. The functions $f_0^\star(\cdot)$ and $f_\alpha^\star(\cdot)$ are strictly positive and exponentially decay at infinities.

The rest of the paper is organized as follows. In section 2, we derive the spectral problem which determines the self-similar supersolution $\overline{v}$, and show upper and lower bounds on the new exponent α. In section 3, we briefly describe the ideas of the renormalization method and outline how it is applied to our problem. We refer to [7] for more details of the method and [6] for the complete proof of the main theorem.

We remark that if $p \in (1,3)$, the equation (1.10), with $p-1$ in place of exponent 2, does not admit self-similar solutions of the form $t^{-\alpha/2}f(x/\sqrt{t})$. So one has to come up with a different approach to investigate the long time behavior.

2 Self-Similar Supersolutions

In this section, we study the self-similar solutions of the equation (1.10).As it was mentioned in the Introduction there are no self-similar solutions to system (1.3) and (1.4) of the form

(1.8) with positive f_1 and f_2 required by physical nature of the problem (see [6] for details). We look for a self-similar solution of the form $\frac{1}{t^{\frac{\alpha}{2}}}f(\xi)$, $\xi = \frac{x}{\sqrt{t}}$, to equation (1.10),where $\xi = \frac{x}{\sqrt{t}}$. Upon substitution, we have:

$$-\frac{\alpha}{2}t^{-1-\frac{\alpha}{2}}f - \frac{1}{2}t^{-1-\frac{\alpha}{2}}\frac{x}{t^{1/2}}f' = \Lambda^{-1}t^{-1-\frac{\alpha}{2}}f'' - \frac{a^2}{4\pi t}e^{-\frac{x^2}{2t}}t^{-\frac{\alpha}{2}}f, \tag{2.1}$$

or

$$-\Lambda^{-1}f'' - \frac{1}{2}\xi f' + \frac{a^2}{4\pi}e^{-\xi^2/2}f = \frac{\alpha}{2}f, \quad f \in L^2(R^1),\ f > 0. \tag{2.2}$$

We see that α and f are principal eigenvalue and eigenfunction of the elliptic operator in (2.2). We also observe that if we had u^{p-1}, $p \neq 3$, in (1.3) and (1.4), the powers of t would not cancel in equation (2.1). Suppose that (2.2) has desired solutions, integration over ξ shows:

$$\frac{1}{2}\int_{R1} f d\xi < \frac{1}{2}\int_{R^1} f d\xi + \frac{a^2}{4\pi}\int_{R^1} e^{-\xi^2/2} f d\xi = \frac{\alpha}{2}\int_{R^1} f d\xi,$$

or

$$\alpha > 1. \tag{2.3}$$

Let us analyze the problem (2.2). Making the change of variable, $f = e^{-\xi^2\Lambda/8}\psi$, we get rid of the first derivative term:

$$-\Lambda^{-1}\psi'' + \left(\frac{1}{16}\Lambda\xi^2 + \frac{a^2}{4\pi}e^{-\frac{\xi^2}{2}} + \frac{1}{4}\right)\psi = \frac{\alpha}{2}\psi \tag{2.4}$$

and we require $\psi \in L^2(R^1)$. This is a standard eigenvalue problem and there exists a positive principal eigenfunction whose corresponding eigenvalue gives α. Using smallness of a and perturbation method, we have the bound:

$$\frac{\alpha}{2} \geq \frac{1}{2} + \frac{a^2}{4\pi}\frac{1}{\sqrt{2\Lambda^{-1}+1}} - \frac{a^4}{4\pi}\left[\frac{1}{\sqrt{4\Lambda^{-1}+1}} - \frac{1}{2\Lambda^{-1}+1}\right] + O(a^6). \tag{2.5}$$

An upper bound for α can be obtained by variational method for any a, and it reads:

$$\frac{\alpha}{2} \leq \frac{1}{2} + \frac{a^2}{4\pi\sqrt{2\Lambda^{-1}+1}}. \tag{2.6}$$

We summarize our analysis into:

Proposition 2.1 *Equation (1.10) admits a unique self-similar solution of the form*

$$v = \frac{1}{t^{\frac{\alpha}{2}}} f_\alpha^\star\left(\frac{x}{\sqrt{t}}\right)$$

such that $\frac{\alpha}{2} > \frac{1}{2}$ *satisfies (2.5), (2.6);* $f_\alpha^\star(\xi)$ *is positive and decays faster then* $O(e^{-\Lambda\xi^2/8})$ *at* ξ *infinities. Uniqueness is up to a constant multiple of* $f_\alpha^\star$.

3 RG Method and Large Time Asymptotics

In this section, we use the self-similar supersolutions of section 2, the RG method [7], and the maximum principle to study the long time decay of solutions of system (1.3) and (1.4). This will allow us to show decay of solutions for a larger class of initial data than those by employing only the maximum principle. Recall the norm:

$$||u|| \equiv \sup_{k \in R^1} (1+k^2)(|\hat{u}(k)| + |\hat{u}'(k)|), \tag{3.1}$$

where $\hat{u}$ is the Fourier transform of u, and prime denotes the derivative in k. Define the Banach space B of functions f with $\hat{f} \in C^1(R^1)$ equipped with the norm (3.1). It is straightforward to check that B is continuously imbedded into $L^1(R^1) \cap L^2(R^2) \cap L^\infty(R^1)$.

Now we review the basic ideas and procedures of the RG method on a formal level(see [7] for details). For this, we consider the scalar equation:

$$\begin{aligned} u_t &= u_{xx} + |u|^p, \quad t \geq 1 \\ u|_{t=1} &= u_0(x). \end{aligned} \tag{3.2}$$

Since we expect u to behave in a self-similar way as $t \to \infty$, scale u with constant $L > 1$ and consider:

$$u_L(x,t) \equiv Lu(Lx, L^2 t).$$

The function u_L satisfies the equation:

$$\begin{aligned} u_{L,t} &= u_{L,xx} + L^{3-p}|u_L|^p \\ u_L|_{t=1} &= Lu(Lx, L^2) \equiv R_L u_0(x), \end{aligned} \tag{3.3}$$

where R_L is called the renormalization group map. After n times of such rescaling, the function u_{L^n} satisfies:

$$\begin{aligned} u_{L^n,t} &= u_{L^n,xx} + (L^n)^{3-p}|u_{L^n}|^p, \\ u_{L^n}|_{t=1} &= R_{L^n} \circ R_{L^{n-1}} \circ \cdots \circ R_L u_0(x), \end{aligned} \tag{3.4}$$

where R_{L^n} denotes the n-th RG map. Note that each time the nonlinear term changes, so R_{L^n} depends also on the rescaled nonlinearity. We skip this dependence here for ease of presentation. If $p > 3$, and $n \to \infty$, the rescaled nonlinearities go to zero, and R_{L^n} is approximately the RG map of the heat equation for large n. The product

$$R_{L^n} \circ R_{L^{n-1}} \cdots \circ R_L u_0(x) \to f_0^*(x),$$

as $n \to \infty$, where $f_0^\star$ is a fixed point of the RG map of the heat equation. It follows that:

$$L^n u(L^n x, L^{2n}) \to f_0^\star(x).$$

Letting $t = L^{2n}$, we have:

$$t^{\frac{1}{2}} u(t^{\frac{1}{2}} x, t) - f_0^\star(x) \to 0, \tag{3.5}$$

as $t \to \infty$, uniformly in x since the limit will be justified in our norm $||\cdot||$ that imbeds into L^∞. Replacing $t^{\frac{1}{2}} x$ by x in (3.5), we get:

$$u(x,t) - t^{-\frac{1}{2}} f_0^\star(\frac{x}{\sqrt{t}}) = o(t^{-\frac{1}{2}}),$$

as $t \to \infty$. We will show below that up to a multiplicative constant

$$f_0^\star(x) = \frac{1}{\sqrt{4\pi}} \exp\{-\frac{x^2}{4}\},$$

so what we just showed is simply the convergence of solutions to the fundamental solution of the heat equation for large time.

To recover the functional form of $f_0^\star$ for the heat equation $u_t = u_{xx}$, we start with: $R_0 u_0(x) = Lu(Lx, L^2)$, or $\widehat{R_0 u_0} = \hat{u}(\frac{k}{L}, L^2)$. Recall that:

$$\hat{u}(\frac{k}{L}, L^2) = \exp\{-(\frac{k}{L})^2 (L^2 - 1)\} \hat{u}_0(\frac{k}{L}) = \exp\{-k^2(1 - L^{-2})\} \hat{u}_0(\frac{k}{L}).$$

The fixed point $f_0^\star$ satisfies:

$$\hat{f_0^\star}(k) = \exp\{-k^2(1 - L^{-2})\} \hat{f_0^\star}(\frac{k}{L}),$$

which implies when passing to the limit $L \to \infty$ that: $\hat{f_0^\star}(k) = e^{-k^2} \hat{f_0^\star}(0)$. Normalizing $\hat{f_0^\star}(0)$ to one and taking the inverse Fourier transform then gives the result. It is easy to check that if $\hat{g}(0) = 0$,then

$$||R_0 g|| \leq C L^{-1} ||g||,$$

for constant C independent of L,which provides the contractive property of R_0 if we choose $L > C$.To show convergence of solutions of heat equations to $f_0^\star$, we decompose initial data as:

$$u_0(x) = c_1 f_0^\star + g_0(x),$$

such that $c_1 = \hat{u}_0(0)$, $\hat{g}_0(0) = 0$. It follows that

$$R_0^n u_0 = c_1 f_0^\star(x) + R_0^n g_0(x),$$

where

$$||R_0^n g_0(x)|| \leq (CL^{-1})^n ||g_0|| \to 0,$$

as $n \to \infty$. In the presence of nonlinearities, one has to analyze the product $R_{L^n} \circ \cdots \circ R_L$ by iteration with more estimates, and update constant c_1 at each step of iteration. Eventually, the c_1's converge and the second term in the decomposition decays to zero.

Applying the RG method to our system, we analyze:

$$\begin{aligned} \overline{v}_t &= \Lambda^{-1}\overline{v}_{xx} - \overline{v}\,(\underline{u}^\star)^2, \quad t > t_0^\star, \\ \overline{v}|_{t=t_0^\star} &= v(x, t_0^\star) \equiv v_0^\star(x). \end{aligned} \tag{3.6}$$

where $\underline{u}^\star \equiv at^{-\frac{1}{2}} f_0^\star(\frac{x}{\sqrt{t}}) - bt^{\delta-1} f_1^\star(\frac{x}{\sqrt{t}})$, a and b two positive constants, $\delta \in (0, \frac{1}{2})$; $f_1^\star$ is a smooth spatially decaying function; $t_0^\star$ is roughly the time for the solution of the heat equation to approach its fundamental solution up to a higher order correction. Also we study the equation:

$$\begin{aligned} \overline{u}_t &= \overline{u}_{xx} + \overline{v}\,\overline{u}^2, \quad t > t_0^\star, \\ \overline{u}|_{t=t_0^\star} &= u(x, t_0^\star) \equiv u_0^\star(x). \end{aligned} \tag{3.7}$$

The functions $\overline{u}$ and $\overline{v}$ bound u and v from above respectively. Based on our knowledge of self-similar solutions, we define the RG map:

$$(u_L(x,1), v_L(x,1)) \equiv (Lu(Lx, L^2), L^\alpha v(Lx, L^2)).$$

We have:

Proposition 3.1 *Let $(\overline{u}, \overline{v})$ be the supersolution defined by (3.6), (3.7). Then*

$$\overline{u}(x,t) \leq At^{-1/2} f_0^\star(\frac{x}{\sqrt{t}}) + O(t^{-\alpha/2}) \tag{3.8}$$

$$\overline{v}(x,t) \leq Bt^{-\frac{\alpha}{2}} f_\alpha^\star(\frac{x}{\sqrt{t}}) + O(t^{-(\frac{1}{2}-\delta)-\frac{\alpha}{2}}) \tag{3.9}$$

for some numbers $A = A(u_0^\star), B = B(v_0^\star)$ and $t \geq t_0^\star$. Here $0 < \delta < \frac{1}{2}$.

By maximum principle, $v \geq \underline{v}$, and $\underline{v}$ solves:

$$\begin{aligned} \underline{v}_t &= \Lambda^{-1}\underline{v}_{xx} - \underline{v}\overline{u}_\star^2, \\ \underline{v}|_{t=t_0^\star} &= v(x, t_0^\star), \end{aligned} \tag{3.10}$$

where $\overline{u}_\star$ denotes the right hand side of (3.8). Applying the same RG analysis shows that there exist constants $B' = B'(\epsilon)$ and $\alpha'(\epsilon) > 1$, $f^\star_{\alpha'}$ strictly positive such that

$$v(x,t) \geq \underline{v} = B't^{-\frac{\alpha'}{2}} f^\star_{\alpha'}(\frac{x}{\sqrt{t}}) + O(t^{\frac{1}{2}-\frac{\alpha+\alpha'}{2}})$$

as $t \to \infty$ with $\alpha \leq \alpha'$. Combining these upper and lower bounds for the solutions (u,v) completes the proof of Theorem 1.1.

Acknowledgements.
We wish to thank J.Avrin, P.Fife, J. Goldstein, M. Parrott, M. Pierre, B. Sleeman and G.Wayne for kindly providing references and their interest. The work of J. Xin was partially supported by NSF grant DMS-9302830 and the Foreign Travel Grant Program of the University of Arizona.

References

[1] N. Alikakos, *L^p bounds of solutions of the reaction-diffusion equations*, Comm. PDE 4(1979), pp 827-868.

[2] J. D. Avrin, *Qualitative theory for a model of laminar flames with arbitrary nonnegative initial data*, J. D. E, 84(1990), pp 290-308.

[3] J. D. Avrin, *Decay and boundedness results for a model of laminar flames with complex chemistry*, Proc. AMS, Vol 110, No. 4, 1990, pp 989-995.

[4] J. D. Avrin, *Behavior at $\pm\infty$ for a model of laminar flames with applications to questions of flame propagation versus extinction*, Proc. Royal Soc. of Edinburgh, 117A, 103-108, 1991.

[5] H. Berestycki, B. Nicolaenko, and B. Sheurer, *Traveling wave solutions to combustion models and their singular limits*, SIAM J. Math. Anal., 16(1985), pp 1207-1242.

[6] L. Berlyand, J. Xin, *Large Time Asymptotics of Solutions to a Model Combustion System with Critical Nonlinearity*, Penn. State U. preprint, No. AM 144, 1994 (submitted for publication).

[7] J. Bricmont, A. Kupiainen, and G. Lin, *Renormalization Group and Asymptotics of Solutions of Nonlinear Parabolic Equations*, Comm. Pure and Applied Math, to appear.

[8] J. Bricmont, A. Kupiainen, *Renormalization Group and the Ginzburg-Landau Equation*, Comm. Math. Phys., 150, pp 193-208, 1992.

[9] H. Fujita, *On the blowing up of solutions of the Cauchy problem for* $u_t = \Delta u + u^{1+\alpha}$, J. Fac. Sci. Univ. Tokyo, (I) 13 (1966), 109-124.

[10] S. Hollis, R. Martin, M. Pierre, *Global existence and boundedness in reaction-diffusion systems*, SIAM J. Math Anal., Vol 18, No. 3, 1987, pp. 744-761.

[11] Levine, H. A. *The role of critical exponents in blowup thoerems*, SIAM Review, 32, pp 262-288(1990).

[12] M. Marion, *Qualitative properties of a nonlinear system for laminar flames without ignition temperature*, Nonlinear Analysis, TMA, 9(1985), pp 1269-1292.

[13] R.H.Martin, and M. Pierre, *Nonlinear reaction-diffusion systems*, in Nonlinear Equations in the Applied Sciences, W.F.Ames and C. Rogers ed., Academic Press, Boston, 1992.

[14] K. Masuda, *On the global existence and asymptotic behavior of solutions of reaction-diffusion equations*, Hokkaido math J., 12(1983), pp. 360-370.

[15] Matkowsky, B. J., and Sivashinsky, G. I., *An asymptotic derivation of two models in flame theory associated with the constant density appoximation*, SIAM J. Appl. Math., 37(1979), p 686.

[16] D. Terman, *Stability of planar wave solutions to a combustion model,* SIAM J. Math. Anal., 21 (1990), pp 1139-1171.

[17] F. B. Weissler, *Existence and nonexistence of global solutions for a semilinear heat equation.*, Isreal J. Math. 38 (1981), 29-40.

I CIORANESCU AND G LUMER

On $K(t)$-convoluted semigroups

For problems of the type $u' = Au, u(0) = x$, in a Banach space X, we consider the regularized problems $v' = Av + K(t)x, v(0) = 0$ (K being a scalar kernel) and study the evolution operators $S_K(t)$ giving the (local mild) solutions; we obtain generation results generalizing and improving earlier Hille-Yosida type results and give an application to multiplication operators in L^p−spaces.

Let X be a Banach space, $0 < \tau \leq \infty$, and A a closed linear operator with domain $D(A) \subset X$. We consider the problem

$$u' = Au, \ u(0) = x, \ 0 \leq t < \tau \leq \infty, \ x \in X \tag{1}$$

to which we associate the K-regularized equation on $[0, \tau)$ which is

$$v' = Av + K(t)x, \ v(0) = 0, \ 0 \leq t < \tau \leq \infty, \ x \in X \tag{2}$$

where K is a scalar function on $[0, \ \infty)$ with $K(0) = 0$.

Definition. *If there exists a strongly continuous operator family $S_K = \{S_K(t)\}_{0 \leq t < \tau}$ such that $\int_0^t S_K(s)xds \in D(A)$ and $S_K(t)x = A\int_0^t S_K(s)xds + K(t)x$ for all $x \in X$ and $0 \leq t < \tau$, then we call S_K a $K(t)$−convoluted semigroup if A satisfies the uniqueness condition, respectively a $K(t)$−convoluted semigroup in the extended sense, if no uniqueness condition is assumed. We briefly write K−c.s.g, respectively K−c.s.g.e. and we always say that A is the (respectively a) generator of S_K.*

It is not difficult to prove that the problem (1) is well-posed if and only if A is the generator of a K−c.s.g. We also note that if $K(t) = t^k/k!$ or $K(t) = E(t)$ where $E(t)$ is the fundamental solution of an ultradifferential operator of Gevrey type we reobtain the concepts of local k−times integrated semigroups [1], [2], [12], [13] and local E−convoluted semigroups [7] respectively.

Remark 1. Suppose that $K(t)$ is C^1 on $[0, \ \infty)$; then
i) one can prove that the following functional equation is satisfied on $[0, \ \tau)$

$$S_K(s)S_K(t) = \int_s^{t+s} K'(t+s-r)S_K(r)dr - \int_0^t K'(t+s-r)S_K(r)dr. \tag{3}$$

It follows in particular that

$$S_K(s)S_K(t) = S_K(t)S_K(s), \ s, t \in [0, \ \tau).$$

ii) one can extend $S_{K'*K} = K' * S_K$ from $[0, \ \tau)$ to $[0, \ 2\tau)$ by the formula inspired by (3)

$$S_{K'*K}(t) = S_K(t)S_K(t-s) + (K'_{-s} * S_K)(t-s) + (K'_{-(t-s)} * S_K)(s) \tag{4}$$

for $s \in [0,\ \tau')$, $0 < \tau' < \tau$ and $t \in [\tau',\ 2\tau')$. The result is also true for S_{K*K} so that, in particular, if A generates a $K-$c.s.g. on $[0,\ \tau)$then it generates a $K * K-$c.s.g. on $[0,\ 2\tau)$. We shall assume further that $K(t)$ satisfies the following properties:

$K(0) = 0$, $K(t)$ is C^1 on $[0, \infty)$, $|K'(t)| \leq ce^{\omega t}$ for some $c > 0$, $\omega \geq 0$ and that $\tilde{K}(t) = \int_0^\infty e^{-tz} K(t)dt \neq 0$, for $Re z > \omega$.

In what follows we are interested in resolvent characterizations and generation results, i.e. in Hille-Yosida type theorems.

We define the finite Laplace transform of S_K as $L_K(z,t) = \int_0^t e^{-sz} S_K(s)ds$, for $z \in \mathbb{C}$, $0 \leq t < \tau$, and the approximate resolvent of A

$$R_K(z,t) = \tilde{K'}^{-1}(z) L_K(z,t),\ \ Re z > \omega,\ 0 \leq t < \tau.$$

Lemma. *For all $x \in X$, $R_K(z,t)x \in D(A)$ and*

$$(z - A)R_K(z,t) = I - B_K(z,t),\ \ 0 \leq t < \tau,\ Re z > \omega \tag{5}$$

where

$$B_K(z,t) = \tilde{K'}^{-1}(z) \left[e^{-tz} S_K(t) + \int_t^\infty e^{-sz} K'(s)ds \right].$$

Proof. We can write

$$L_K(z,t) = e^{-tz} \int_0^t S_K(r)dr + z \int_0^t e^{-sz} \left(\int_0^s S_K(r)dr \right) ds.$$

It follows that $L_K(z,t) \in D(A)$ for every $x \in X$ and

$$\begin{aligned}
(z - A)L_K(z,t) &= zL_K(z,t) - e^{-tz}(S_K(t) - K(t)) - z \int_0^t e^{-sz}(S_K(s) - K(s))ds \\
&= -e^{-tz} S_K(t) + e^{-tz} K(t) + z \int_0^t e^{-sz} K(s)ds \\
&= -e^{-tz} S_K(t) + \int_0^t e^{-sz} K'(s)ds \\
&= -e^{-tz} S_K(t) + \tilde{K'}(z) - \int_t^\infty e^{-sz} K'(s)ds
\end{aligned}$$

We obtain (5) by multiplication with $\tilde{K'}^{-1}(z)$.

□

Let Φ be a real valued positive function on $[r_0,\ \infty)$, $r_0 \geq 0$, of class C^1 with $\Phi' > 0$ and such that $\lim_{r \to \infty} \Phi(r) = \infty$. We introduce

$$\chi(\Phi) = \chi = \limsup_{r \to \infty} \frac{\Phi(r)}{r}$$

$$\sigma(\Phi) = \sigma = \limsup_{r \to \infty} \frac{\ln r}{\Phi(r)}$$

and

$$\mu(\Phi) = \mu = \liminf_{r\to\infty} \frac{\ln r}{\Phi(r)}$$

For $\alpha,\ \beta > 0$ we denote

$$\Gamma_{\alpha\beta}(\Phi) = \Gamma_{\alpha\beta} = \{z \in \mathbb{C};\ Re z \geq \beta,\ Re z \geq \alpha\Phi(|z|)\}.$$

Theorem I. *Suppose that A is the generator of a $K-$c.s.g.e. on $[0,\tau)$ and that $|\tilde{K}^{-1}(z)| = O(e^{L\Phi(|z|)})$, for some $L > 0$. Then $\forall 0 < (L-\mu)\tau^{-1} < \alpha < \chi^{-1}$, there exists β with $[\beta,\ \infty) \subset \Gamma_{\alpha\beta} \subset \rho(A)$ and $M > 0$ such that*

$$\|R(z,A)\| \leq \frac{M}{|z\tilde{K}(z)|} = O\left(\frac{e^{L\Phi(|z|)}}{|z|}\right), \quad \textit{for} \quad z \in \Gamma_{\alpha\beta}.$$

Proof. Let $0 < (L-\mu)\tau^{-1} < \alpha < \chi^{-1}$;there is $r_1 > \max(\omega, r_0)$ such that $\dfrac{\Phi(r)}{r} < \dfrac{1}{\alpha}$ for $r > r_1$ and consequently, $[r_1,\ \infty) \subset \Gamma_{\alpha r_1}$. Consider now $z \in \Gamma_{\alpha r_1}$, $r = |z|$, $0 < t < \tau$ and estimate $B_K(z,t)$ of the above lemma; we obtain:

$$\text{(6)} \qquad \|B_K(z,t)\| \leq const.\frac{e^{L\Phi(|z|)}}{|z|}\left(e^{-t_0 Re z}\|S_K(t)\| + C\frac{e^{(\omega - Re z)t}}{Re z - \omega}\right)$$

$$\leq const.e^{(L-\alpha t)\Phi(r)-\ln r}.$$

Since $L < \alpha\tau + \mu$, there are $0 < t_0 < \tau$ and $\mu_0 < \mu$ such that

$$\text{(7)} \qquad L < \alpha t_0 + \mu_0.$$

Moreover, there is $\overline{r} > r_1$ with

$$\text{(8)} \qquad \mu_0\Phi(r) - \ln r < 0 \quad \text{for} \quad r > \overline{r}.$$

Then (6) and (8) yield

$$\begin{aligned} \|B_K(z,t_0)\| &\leq const.e^{(L-\alpha t_0-\mu_0)\Phi(r)+\mu_0\Phi(r)-\ln r} \\ &\leq const.e^{(L-\alpha t_0-\mu_0)\Phi(r)}, \quad z \in \Gamma_{\alpha\overline{r}}. \end{aligned}$$

Since $\lim_{r\to\infty} \Phi(r) = \infty$, we can now choose $\beta \geq \overline{r}$ such that by (7)

$$\|B_K(z,t_0)\| \leq \frac{1}{2} \quad \text{for} \quad z \in \Gamma_{\alpha\beta}(\subset \Gamma_{\alpha\overline{r}} \subset \Gamma_{\alpha r_1}).$$

Thus we can invert $I - B_K(z,t_0)$ and $\|(I - B_K(z,t_0))^{-1}\| \leq 1$. Consequently, $\Gamma_{\alpha\beta} \subset \rho(A)$ and by formula (5)

$$\begin{aligned} \|R(z,A)\| &= \|R_K(z,t_0)(I - B_K(z,t_0))^{-1}\| \\ &\leq \frac{\|L_K(z,t_0)\|}{|\tilde{K}'(z)|} \leq \frac{M}{|z\tilde{K}(z)|} = O\left(\frac{e^{L\Phi(|z|)}}{|z|}\right) \quad \text{for} \quad z \in \Gamma_{\alpha\beta}. \end{aligned}$$

where $M = \left(\sup_{0\le s\le t_0} \|S_K(s)\|\right) t_0$.

□

We now get via Ljubich's uniqueness result the following

Corollary. *If in the context of Theorem I we have*

$$\liminf_{r\to\infty} \frac{\ln|\tilde{K}(r)|}{r} \ge 0$$

then the uniqueness property holds for A and S_K is indeed a $K-$c.s.g.

We also have the following generation result

Theorem II. *If for $0<\alpha<\chi^{-1}$, $\beta>0$ and $-1<\gamma<l-\sigma$ with $l>0$ one has $\Gamma_{\alpha\beta}\subset\rho(A)$ and*

$$|\tilde{K}(z)| = O\left(e^{-l\Phi(|z|)}\right), \quad \text{and} \quad \|R(z,A)\| = O\left(e^{\gamma\Phi(|z|)}\right), \quad z\in\Gamma_{\alpha\beta}, \tag{9}$$

then A is a generator of a K_1-c.s.g.e. on $[0,\ \tau)$ with $\tau=(l-\gamma-\sigma)\alpha^{-1}$ and a K_1-c.s.g. if $\chi=0$ where $K_1=D^{-1}K=1*K$.

Proof. Since $\alpha<\chi^{-1}$ we can choose $\beta>\max(\omega, r_0)$ such that $[\beta,\ \infty)\subset\Gamma_{\alpha\beta}$ (see the beginning of the proof of Theorem I) and we can now assume without loss of generality that this is the β of our above statement. Consequently $\Gamma_{\alpha\beta}$ is a nonvoid region with C^1 boundary $\partial\Gamma_{\alpha\beta}$ where for $z=re^{i\theta}$ we have $|dz|\le const.dr$ (for $Rez>\beta$).

For $t\ge0$ we define

$$S(t) = \frac{1}{2\pi i}\int_{\partial\Gamma_{\alpha\beta}} e^{tz}\tilde{K}(z)R(z,A)dz. \tag{10}$$

We have for $z\in\partial\Gamma_{\alpha\beta}$

$$\|e^{tz}\tilde{K}(z)R(z,A)\| \le const.e^{(t\alpha+\gamma-l)\Phi(r)}. \tag{11}$$

Let $0<\tau'<\tau=(l-\gamma-\sigma)\alpha^{-1}$; then $\tau'\alpha+\gamma-l+\sigma<\tau\alpha+\gamma-l+\sigma=0$. We choose now a $\overline{\sigma}$ such that $\sigma<\overline{\sigma}<\sigma+(\tau-\tau')\alpha$; there are $\varepsilon>0$ and $r'>\beta$ such that $1<1+\varepsilon<\dfrac{\sigma+(\tau-\tau')\alpha}{\overline{\sigma}}$ and $\dfrac{\ln r}{\Phi(r)}<\overline{\sigma}$.

Then for $t\le\tau'$ and $r\ge r'$ we obtain

$$\begin{aligned}(t\alpha+\gamma-l)\Phi(r) &= (\tau\alpha+\gamma-l+\sigma)\Phi(r)-(\sigma+(\tau-t)\alpha)\Phi(r)\\ &\le -\frac{\sigma+(\tau-\tau')\alpha}{\overline{\sigma}}\ln r \le -(1+\varepsilon)\ln r.\end{aligned}$$

It follows from (11) that

$$\|e^{tz}\tilde{K}(z)R(z,A)\| \le \frac{const}{r^{1+\varepsilon}}, \qquad z\in\partial\Gamma_{\alpha\beta},\ |z|>r'$$

uniformly for $0 < t \leq \tau' < \tau$, so that the integral in (10) exists.

We further have

$$\begin{aligned}\int_0^t S(s)ds &= \int_0^t ds \left[\frac{1}{2\pi i} \int_{\partial\Gamma_{\alpha\beta}} e^{sz} \tilde{K}(z) R(z, A) dz \right] \\ &= \frac{1}{2\pi i} \int_{\partial\Gamma_{\alpha\beta}} \frac{e^{tz} - 1}{z} \tilde{K}(z) R(z, A) dz\end{aligned}$$

and

$$\begin{aligned}A \int_0^t S(s)ds &= \frac{1}{2\pi i} \int_{\partial\Gamma_{\alpha\beta}} \frac{e^{tz} - 1}{z} \tilde{K}(z)(zR(z, A) - I)dz \\ &= \frac{1}{2\pi i} \int_{\partial\Gamma_{\alpha\beta}} e^{tz} \tilde{K}(z) R(z, A) dz - \frac{1}{2\pi i} \int_{\partial\Gamma_{\alpha\beta}} e^{tz} \frac{\tilde{K}(z)}{z} dz \\ &= S(t) - K_1(t).\end{aligned}$$

(By Cauchy's contour theorem, $\frac{1}{2\pi i} \int_{\partial\Gamma_{\alpha\beta}} \frac{\tilde{K}(z)}{z} dz = 0$ and $\frac{1}{2\pi i} \int_{\partial\Gamma_{\alpha\beta}} \frac{\tilde{K}(z)}{z} zR(z, A)dz = 0$ and by the Laplace inversion formula, $K_1(t) = \frac{1}{2\pi i} \int_{-i\infty}^{i\infty} e^{itz} \tilde{K_1}(z)dz = \frac{1}{2\pi i} \int_{\partial\Gamma_{\alpha\beta}} e^{tz} \frac{\tilde{K}(z)}{z} dz$).

It is now clear that $\{S(t)\}_{0 \leq t < \tau}$ is a K_1–c.s.g.e. generated by A. Finally, we note that for every $\varepsilon > 0$ and $r \geq r(\varepsilon)$

$$\frac{\ln \|R(r, A)\|}{r} \leq \frac{\gamma\Phi(r) + const}{r} \leq \frac{\gamma(\chi + \varepsilon)}{r} + \frac{const}{r}$$

so that $\limsup_{r \to \infty} \frac{\ln \|R(r, A)\|}{r} \leq \chi\gamma$.

It follows that if $\chi = 0$, Ljubich's uniqueness condition works.

□

Remark 2. Our results permit not only to unify earlier generation results but also to improve them.

Indeed, in order to reobtain the results on local k–times integrated semigroups, we take $\Phi(r) = \ln r$ $(r \geq r_0 > 1)$; then $\chi = 0$ and $\sigma = \mu = 1$. With $L = k + 1$, $K(t) = \frac{t^k}{k!}$ and α becoming α^{-1}, Theorem I essentially gives Theorem 2.1 of [2]. If in Theorem II we take $\alpha, \beta > 0$, $-1 < \gamma = k < p - 1$ and $l = p$ where $p \in \mathbb{N}$ then $\tau = (l - \gamma - \sigma)\alpha^{-1} = (p - (k+1))\alpha^{-1}$, and $\|R(z, A)\| = O(|z|^k)$. Hence we reobtain Theorem 2.2 from [2]. Similarly, one also finds earlier results of [7] on E–convoluted semigroups; we take in this case essentially $\Phi(r) = r^a$, $0 < a < 1$. Then $\chi = \sigma = \mu = 0$, $K(t) = E(t)$ the fundamental solution of some ultradifferential operator.

If we take in Theorem II $\alpha, \beta, \gamma > 0$, l being now l^a, $\Phi(r) = r^a$, we reobtain Theorem 2.5 of [7]. With L^a for L, $\Phi(r) = r^a + \frac{1}{L} \ln r (\sim r^a$ for large $r)$ we reobtain Theorem 2.2 of [7] from Theorem I above.

We shall further show that the generation Theorem II contains and extends Chazarain's results on abstract Cauchy problems well-posed in the sense of ultradistributions [5], [6], as well as further generalizations due to Beals [3], [4], Emamirad [10] and Cioranescu-Zsidó [9].

Let M_k, $k = 0, 1, 2, ...$ be a sequence of positive numbers satisfying the conditions

$$M_0 = 1, \ M_k^2 \le M_{k-1}M_{k+1} \text{ and } \sum_{k=1}^{\infty} \frac{M_{k-1}}{M_k} < \infty, \tag{11}$$

and define the associated function $M(r) = \sup_k \ln \frac{r^k}{M_k}$. Then $\lim_{r\to\infty} M(r) = \infty$, M is of class C^1 and $M'(r) > 0$ for large r and $\int_1^\infty \frac{M(r)}{r} dr < \infty$. Moreover $\chi = \sigma = 0$ (see [11]).

Let $m_k = \frac{M_k}{M_{k-1}}$, $k \in \mathbb{N}$; then $\sum_{k=1}^{\infty} \frac{1}{m_k} < \infty$ and we can define the following entire function $P(z) = \prod_{k=1}^{\infty}(1 + \frac{z}{m_k})$.

For $Re z > 0$ we have $|P(z)| \ge \sup_k \prod_{j=1}^{k} |1 + \frac{z}{m_j}| \ge \sup_k \prod_{j=1}^{k} \frac{|z|}{m_j} = \sup_k \frac{|z|^k}{M_k}$ so that $|P(z)| \ge e^{M(|z|)}$ for $Re z > 0$. Then we directly obtain from Theorem II the

Proposition. *Suppose there are* α, β, $\gamma > 0$ *with* $\Gamma_{\alpha\beta} \subset \rho(A)$ *and* $\|R(z, A)\| = O(e^{\gamma M(|z|)})$, $z \in \Gamma_{\alpha\beta}(M)$; *then* A *generates a* K_1*−c.s.g. on* $[0, \tau)$ *with* $\tau = (l - \gamma)\alpha^{-1}$, l *being an integer* $> \gamma$ *and* K *defined by* $\tilde{K}^{-1}(z) = \prod_{k=1}^{\infty}(1 + \frac{z}{m_k})^l$ *for* $Re z > 0$.

Under the conditions of this proposition but assuming A densely defined, J. Chazarain [5], [6] proved that A generates an ultradistribution semigroup of class (M_k). More generally

Corollary. *Let* $\Psi : [0, \infty) \longrightarrow [0, \infty)$ *be increasing such that* $\int_1^\infty \frac{\Psi(r)}{r^2} dr < \infty$ *and suppose that there are* α, β, $\gamma > 0$ *with* $\Gamma_{\alpha\beta}(\Psi) \subset \rho(A)$ *and* $\|R(z, A)\| = O\left(e^{\gamma\Psi(|z|)}\right)$, $z \in \Gamma_{\alpha\beta}$. *Then there is a kernel* K *and* $\tau > 0$ *such that* A *generates a* K_1*−c.s.g. on* $[0, \tau)$.

For the proof, we note that under the conditions on the function Ψ there is a sequence (M_k) satisfying (11) such that $\Psi(r) \le const.M(r)$, $r > 0$ (see Theorem 1.6 in [9]).

Thus we can apply the above proposition.

We note that the classes of densely defined operators such that $\Gamma_{\alpha\beta}(\Psi) \subset \rho(A)$ but with polynomial growth for $\|R(z, A\|$ were considered by Beals [3], Cioranescu-Zsidó [9] and Emamirad [10].

An application. Let Ω be a σ−finite measure space, $m : \Omega \longrightarrow \mathbb{C}$ a measurable function and define in $X = L^p(\Omega)$, $1 \le p \le \infty$ the operator A by $Af = mf$, with the usual domain.

Let K and Φ be as in Theorem II and suppose that $\Gamma_{\alpha\beta}(\Phi) \subset \rho(A)$. We define

$$(S_K(t))\, f(\zeta) = \left(\int_0^t K(t-s)e^{m(\zeta)s}ds\right) f(\zeta), \ 0 \le t < \tau, \ f \in L^p(\Omega).$$

Then one can prove that S_K is a K−c.s.g.e. on $[0, \tau)$ for $\tau\alpha \le l$ whose generator is A.

Finally although we do not at all develop this matter here (it is announced also in [8] and will be included in a later joint paper) we briefly indicate that (and how) $K-$convoluted semigroups are a particular case of $K-$evolution operators $\mathcal{S}_K(t)$ ($K-$e.o., whose theory was developed earlier for $K(t) = t^n/n!$, $n \geq 0$, in [12], [13]). $K()$ can now have (bounded) linear operator values (i.e. in B(X)). While for general $\mathcal{S}_K(t)$ we do not have generation results of Hille-Yosida type, other useful results-some mentioned above like the functional equations (3), (4), and the extension result from $[0,\ \tau)$ to $[0,\ 2\tau)$-admit generalizations to the $\mathcal{S}_K(t)$ context. In that context $K:\ [0,\tau) \longrightarrow B(X);\ t \mapsto K(t)$ is C^1 (strongly), the $K()$ commute and commute with A on $D(A)$ (A having the uniqueness property),

$$Z_K = Z_K(\tau) = \{x \in X;\ \exists \text{ a } \ C^1 \ \text{ solution } \ v_K = v_K(t,x) \text{ of the equation } \ (2) \text{ on } \ [0,\tau)\}$$

$\mathcal{S}_K(t)$ is defined on Z_K by $\mathcal{S}_K(t)x = v'_K(t,x)$. $K-$c.s.g. correspond to $Z_K = X$. Set $K(t) = C + K_0(t)$ where $C = K(0)$ (this setup includes $C-$semigroups for $K_0() = 0$). As said many results are still true in this context. For instance, (3) still holds in some form on $Z_K(\tau)$ if $C = 0$, but also a generalized form holds when $C \neq 0$; also approximate resolvents R_K can be treated and the analogue of the lemma above is true (see also Section 4, Part I of [13]). There are indeed many situations (for instance for certain multiplication operators on $L^p(\Omega)$, normal operators on Hilbert spaces,...) where A does not generate a $K-$c.s.g. but does generate a $K-$e.o. (i.e. a family $\{\mathcal{S}_K(t)\}_{0\leq t<\tau}$ of $K-$evolution operators $\mathcal{S}_K(t)$) with dense Z_K spaces.

References

[1] Arendt W., *Vector-valued Laplace transforms and Cauchy problems,* Israel. J. Math. **59** (1987), 327-352.

[2] Arendt W., El-Mennaoui O. and Keyantuo V., *Local integrated semigroups: evolution with jumps of regularity,* to appear in J. Math. Anal. Appl.

[3] Beals R., *Semigroups and abstract Gevrey spaces, J. Funct. Anal.* **10**(1972), 281-299.

[4] Beals R., *On the abstract Cauchy problem, J. Funct. Anal.* **10**(1972), 300-308.

[5] Chazarain J., Problèmes de Cauchy abstraits et applications à quelques problèmes mixtes, *J. Funct. Anal.* **7**(1971), 386-445.

[6] Chazarain J., *Problèmes de Cauchy dans les espaces d'ultradistributions,* C. R. Acad. *Sci.* **310** *Série A, (1968), 564-566.*

[7] Cioranescu I., *Local Convoluted Semigroups,* to appear.

[8] Cioranescu I. and Lumer G., *Problèmes d'évolution régularisés par un noyau général $K(t)$. Formule de Duhamel, prolongement, théorèmes de génération,* to appear in C. R. *Acad. Sci.*

[9] Cioranescu I. and Zsidó L., *ω−ultradistributions and their applications to Operator Theory, Banach Center Publications, vol. 8, Spectral Theory, Warsaw, 1982, pp. 77-220.*

[10] Emamirad H., *Systèmes pseudodifférentiels d'évolution bien posés au sens des distributions de Beurling, Bolletino U.M.I., Analisi Funzionale e Appl., Serie VI, vol. I, 1982, p.303-322.*

[11] Komatsu H., *Ultradistributions, I J. Fac. Sci. Univ. Tokyo, Sec IA,* **20***(1973), 25-105.*

[12] Lumer G., *Solutions généralisées et semi-groupes intégrés, C. R. Acad. Sci.* **310** *Série I, (1990) 577-582*

[13] Lumer G., *Evolution equations. Solutions for irregular evolution problems via generalized solutions and generalized initial values. Application to periodic shock models, Annales Saraviensis, vol. 5, No 1, (1994), 1-102.*

Ioana Cioranescu:
Department of Mathematics and Computer Science, University of Puerto Rico,
Box 23355 Rio Piedras, San Juan, Puerto Rico 00931

Gunter Lumer:
Institut de Mathématiques et Informatique, Université de Mons,
B-7000 Mons, Belgium.

J COOPER AND H KOCH

Remarks on the spectrum of a linear wave operator with time periodic boundary condition

In a system of ordinary differential equations with periodic coefficients the evolution operator through one time period has eigenvalues and the corresponding eigenvectors can be used to construct Floquet type solutions of the form

$$e^{i\rho t}v(t)$$

where $v(t)$ is periodic. In some cases this can also be done for partial differential equations. A parabolic equation with a time periodic potential term was treated in [1] and a hyperbolic equation with a time periodic potential was discussed in [2]. In both cases it was found that Floquet type solutions existed.

In this note we summarize some of the important points of [3] which treats the following example of a hyperbolic partial differential equation with a time periodic boundary condition. We have found a situation where the usual Floquet theory is not possible because the evolution operator has no eigenvalues. The boundary value problem is for the linear wave equation in one space dimension. Let $s(t)$ be a smooth function of period T such that $s(t) > 0$ and $|s'(t)| < 1$ for all t. Let Q be the region in (x,t) space

$$Q = \{(x,t) : 0 < x < s(t), \quad t \in R\}.$$

We assume that $s(0) = 1$. The boundary value problem is

$$u_{tt} - u_{xx} = 0 \qquad \text{in } Q \tag{1}$$

$$u(0,t) = u(s(t),t) = 0 \qquad \text{for all } t \in R \tag{2}$$

$$u(x,0) = u_0(x), \qquad u_t(x,0) = u_1(x) \quad \text{for } 0 < x < 1. \tag{3}$$

The hypothesis $|s'(t)| < 1$ implies that the initial boundary value problem is well posed in the following sense: If $u_0 \in H_0^1(0,1)$ and $u_1 \in L^2(0,1)$, then there exists a unique weak solution $u(x,t)$ of (1), (2), (3), such that $t \to u(.,t)$ is continuous with values in $H^1(R) \times L^2(R)$ when u is extended by zero outside Q. One can also demonstrate well posedness for initial data in closed subspaces $X_m \subset H^m(0,1) \times H^{m-1}(0,1)$ where X_m incorporates appropriate compatibilty conditions of the initial data with the boundary conditions.

We want to study the spectrum of the evolution operator

$$U_m(0,T) : X_m \to X_m$$

which takes the initial data (u_0, u_1) at time zero into $(u(.,T), u_t(.,T))$. The spectral properties of $U_m(0,T)$ are determined by the manner in which characteristics are affected by reflection at the moving boundary. Instead of considering the two characteristics through each point (x,t) we construct a problem in a symmetrized domain $\hat{Q}$ where we need deal with only one characteristic through each point. Specifically, let

$$\hat{Q} = \{(x,t) : -s(t) < x < s(t), \quad t \in R\}.$$

$\hat{Q}$ is an unfolding of the two sheeted covering over Q. Characteristics with slope $+1$ are left in the right half of $\hat{Q}$, while characteristics with slope -1 become characteristics with slope $+1$ in the left half of $\hat{Q}$. With a solution of (1), (2), (3), we wish to associate a function $\hat{u}(x,t)$ in $\hat{Q}$ that solves

$$\hat{u}_t + \hat{u}_x = 0 \qquad \text{in } \hat{Q} \tag{4}$$

by writing $u(x,t) = \hat{u}(x,t) - \hat{u}(-x,t)$. Then the equations

$$\begin{aligned} u_0(x) &= u(x,0) = \hat{u}(x,0) - \hat{u}(-x,0) \\ u_1(x) &= u_t(x,0) = \hat{u}_t(x,0) - \hat{u}_t(-x,0) \\ &= -\hat{u}_x(x,0) + \hat{u}_x(-x,0). \end{aligned}$$

determine $\hat{u}(x,0)$, and hence $\hat{u}(x,t)$ up to a constant. The boundary condition for $\hat{u}$ is

$$\hat{u}(-s(t),t) = \hat{u}(s(t),t). \tag{5}$$

$\hat{u}$ is uniquely determined by u modulo constants. Let H_P^m be the quotient space of functions with period 2 on R modulo constants with norm

$$\|f\|_{H_P^m} = \|f'\|_{H^{m-1}(-1,1)}.$$

Then the mapping $(u_0, u_1) \leftrightarrow \hat{u}(x,0) = f$ is one to one and onto from X_m to H_P^m. Note that the energy of u at $t = 0$

$$\frac{1}{2}\int_0^1 [(u_0')^2 + u_1^2]dx = \int_{-1}^1 |\hat{u}(x,0)|^2 dx.$$

Then with the operator $U_m(0,T)$ we associate the operator

$$A_m : H_P^m \to H_P^m$$

which takes $\hat{u}(x,0)$ into $\hat{u}(x,T)$.

Next we define the mapping of the interval $[-1,1]$ onto itself associated with the characteristic flow. If $x_0 \in [-1,1)$ we follow the characteristic $x = x_0 + t$ until it intersects the line $[-1,1] \times \{T\}$ at the point (y_0, T), or meets the right boundary of $\hat{Q}$ at the point $(s(\tau), \tau)$. In the former case we set $\varphi(x_0) = y_0$. In the latter case we follow the characteristic $x = -s(\tau) + (t - \tau)$ and repeat the process. We call the characteristic follwed in this manner a 'broken' characteristic. In this way we define a continuous, piecewise smooth, mapping $\varphi : [-1,1] \to [-1,1]$ which is one to one and onto. Because $\hat{u}$ solves (4), $\hat{u}$ is constant on the 'broken' characteristics. Consequently

$$(A_m f)(x) = f \circ \varphi^{-1}. \tag{6}$$

Now extend φ to all of R by the rule

$$\varphi(x+2) = \varphi(x).$$

We can think of φ as a mapping of S onto S where S is the circle of circumference 2. In the proper coordinates on S, φ is a diffeomorphism of S. Let $\tilde{H}^m(S)$ be $H^m(S)$ modulo constants. Then A_m can be thought of as a mapping of $\tilde{H}^m(S)$ onto $\tilde{H}^m(S)$ and the spectrum of $U_m(T,0)$ is the same as the spectrum of A_m. We have reduced our original problem to that of determining the spectrum of a mapping given by (6). In [4] Lopes made a similar reduction for a wave equation with time periodic coefficients and obtained results similar to our, but less complete.

With a diffeomorphism $\varphi : S \to S$ we can associate a rotation number and we state our results in two cases, depending on whether the rotation number is rational or irrational.

Theorem 1 If the rotation number of φ is irrational, the spectrum

$$\sigma(A_m) = \{\lambda : |\lambda| = 1\} \quad \text{for all } m \geq 1.$$

Next suppose that the rotation number of φ is rational in which case φ has periodic points. We replace φ by a sufficiently high iterate φ^j such that φ^j has only fixed points. For simplicity of exposition we assume that φ has a finite number of fixed points $p_1, \ldots, p_n$ and we set $\Xi = \{\varphi'(p_i), i = 1, \ldots, n\}$. Let

$$\mu_- = \max \Xi \qquad \text{and } \mu_+ = \min \Xi.$$

Clearly $\mu_+ \leq 1 \leq \mu_-$.

Theorem 2 If φ has only a finite number of fixed points

a) $A_m - \lambda I$ is one to one for all $\lambda \in C$.

b) The closure of the range has infinite codimension if and only if $\mu_-^{1/2-m} < |\lambda| < \mu_+^{1/2-m}$ or $|\lambda| = 1$.

c) The range of $A_m - \lambda I$ is closed if and only if $|\lambda| \notin \Xi$.

d) The range of $A_m - \lambda I$ is dense if $|\lambda| \neq 1$ and $|\lambda| \leq \mu_-^{1/2-m}$ or $|\lambda| \geq \mu_+^{1/2-m}$.

The fact that A_m has no eigenvalues precludes the existence of any Floquet type solutions.

One of the key ideas here is that oscillations in the initial data f become compressed around the attracting fixed points of φ. We give an indication of the methods used in the proof of Theorem 2. Consider the case $m = 1$, and suppose that $\varphi : [0,1] \to [0,1]$ with the only fixed points being $x = 0$ and $x = 1$, and that $\varphi'(0) = a < 1$ and $\varphi'(1) = b > 1$. Then

$$\lim_{k\to\infty} \|A^k f\|_{\tilde{H}^1(0,1)}^{1/k} = a^{-1/2} \tag{7}$$

for all nontrivial initial data f. First we can make a change of coordinates so that $\varphi(x) = ax$ for $0 \leq x \leq 1/3$ and $\varphi(x) = 1 + b(x-1)$ for $2/3 \leq x \leq 1$. Now from (6) we see that

$$A^k f = f \circ \varphi^{-k}$$

so that

$$\|A^k f\|_{\tilde{H}^1(0,1)}^2 = \int_0^1 |f'(\varphi^{-k}(x))|^2 |(\varphi^{-k})'(x)|^2 dx = \int_0^1 |f'(y)|^2 \frac{dy}{(\varphi^k)'(y)}.$$

Furthermore

$$(\varphi^k)'(y) = \varphi'(\varphi^{k-1}(y))\varphi'(\varphi^{k-2}(y)) \cdots \varphi'(y).$$

Hence there is a constant $C > 0$ such that

$$(\varphi^k)'(y) \geq C^{-1} a^k$$

for all $y, 0 \leq y \leq 1$ and all k sufficiently large. Therefore

$$\|A^k f\|_{\tilde{H}^1(0,1)}^2 \leq C a^{-k} \|f\|_{\tilde{H}^1(0,1)}^2.$$

On the other hand, if f is not constant, then f' must differ from zero on the interval $[0, \varphi^{-k_0}(1/3)]$ for some $k_0 > 0$. Then

$$\|A^{k+k_0} f\|_{\tilde{H}(0,1)}^2 = \int_0^1 |(f \circ \varphi^{-k_0})'|^2 \frac{dy}{(\varphi^k)'}$$

$$\geq a^{-k}\int_0^{1/3}|(f\circ\varphi^{-k_0})'|^2dy = a^{-k}\int_0^{\varphi^{-k_0}(1/3)}|f'(z)|^2\frac{dz}{(\varphi^{k_0})'}\geq C_1a^{-k}$$

which establishes (7).

We illustrate these results with a simple example. Suppose $s(t) = 1+\varepsilon\sin(\pi t)$, $\varepsilon\pi < 1$, so that $T = 2$. The fixed points of φ are $x = 0$ and $x = \pm 1$. Then

$$\mu_+ = a = \varphi'(0) = \frac{1-\varepsilon\pi}{1+\varepsilon\pi} < 1,$$

$$\mu_- = \varphi'(\pm 1) = \frac{1+\varepsilon\pi}{1-\varepsilon\pi} = \frac{1}{a}.$$

The spectrum of A_m is $\sigma(A_m) = \{a^{m-1/2}\leq|\lambda|\leq a^{1/2-m}\}$. The circles $|\lambda| = a^{m-1/2}$ and $|\lambda| = a^{1/2-m}$ are continuous spectrum and for λ between the two circles, $A_m - \lambda I$ has closed range with infinite codimension.

As a further application of our results consider the damped equation

$$u_{tt} - u_{xx} + 2du_t + d^2u = 0$$

with the same boundary conditions. The function $v = \exp(dt)u$ satisfies $v_{tt} - v_{xx} = 0$ in Q and we have

$$\|v(t)\|_{H^m} + \|v_t(t)\|_{H^{m-1}} \approx a^{t(1/2-m)/2}$$

so that

$$\|u(t)\|_{H^m} + \|u_t(t)\|_{H^{m-1}} \approx [e^d a^{(1/2-m)/2}]^t.$$

Thus if

$$e^d a^{(1/2-m)/2} > 1,$$

then $u(t)$ grows in the H^m norm, but if

$$e^d a^{(1/2-m)/2} < 1,$$

then $u(t)$ decays in the H^m norm.

References

[1] S.N.Chow, K.Lu, and J. Mallet-Paret. Floquet theory for parabolic equations: the time periodic case, *J.of Differential Equations*, to appear, 1994.

[2] J. Cooper, G. Perla-Menzala and W. Strauss. On the scattering frequencies of time-dependent potentials. *Math. Meth. in the Appl. Sci.*, 8:576-584, 1986

[3] J. Cooper and H. Koch. The spectrum of a hyperbolic evolution operator. *to appear*.

[4] O. Lopes. On the structure of the spectrum of a linear time periodic wave equation. *J. d'Analyse Mathematique*, 47:55-68, 1986.

Jeffery Cooper
Department of Mathematics
University of Maryland
College Park, MD 20742, USA

Herbert Koch
Institut für Angewandte Mathematik
Universität Heidelberg
D - 69120 Heidelberg, Germany

R DE LAUBENFELS

Entire vectors and entire existence families

I. INTRODUCTION AND SOME GENERAL THEORY. Most of this section, and Example 2.1, are in [4, chapters VII and VIII]. Example 2.3 will appear in [6]. Example 2.2 is new.

I will give at most outlines of proofs.

I will be discussing the many physical problems that may be modelled as an *abstract Cauchy problem*

$$\frac{d}{dt}u(t,x) = A(u(t,x))\ (t \geq 0),\ \ u(0,x) = x. \tag{1.1}$$

By a *solution* I will mean a strong solution, that is, $t \mapsto u(t,x) \in C([0,\infty),[\mathcal{D}(A)]) \cap C^1([0,\infty),X)$, and u satisfies (1.1).

Throughout, A is a closed linear operator on a Banach space X, with domain $\mathcal{D}(A)$, spectrum $\sigma(A)$, resolvent set $\rho(A)$. I will write $B(X)$ for the space of bounded linear operators from X to itself.

I will attack the abstract Cauchy problem with what is known as an entire vector.

Definition 1.2. I will write $C^\infty(A)$ for $\cap_{k=0}^{\infty}\mathcal{D}(A^k)$. By an *entire vector* for A I mean $x \in C^\infty(A)$ such that

$$\sum_{k=0}^{\infty}\frac{s^k}{k!}\|A^k x\| < \infty,\ \ \forall s > 0. \tag{1.3}$$

I will write $\mathcal{E}(A)$ for the set of all entire vectors for A.

This is a very old idea (see [10]). An analytic vector is one where the series in (1.3) converges for some $s > 0$.

One may write down simple characterizations of generators of groups and sufficient conditions for generating a semigroup, in terms of analytic or entire vectors (see [10], [2], [8], [3] and their references).

Example 1.4. Let A be $-\frac{d}{ds}$ on $X \equiv \{f \in C[0,\infty) \mid f(0) = 0 = \lim_{s\to\infty} f(s)\}$, the generator of right translation,

$$(e^{tA}f)(s) \equiv f(s-t)\ (s,t \geq 0),$$

where $f(s) \equiv 0$, when $s < 0$.

Then it is not hard to see that A has no nontrivial analytic vectors, although A generates a bounded strongly continuous semigroup.

Remarks 1.5. It may be shown that a generator of a strongly continuous *group* has a core of entire vectors (see [3]; compare this with Example 1.4).

In [1] it is shown that, if $-A$ generates a strongly continuous holomorphic semigroup and u is a solution of (1.1), then u has an entire extension. In other words, the set of all x for which (1.1) has a solution equals $\mathcal{E}(A)$.

In order that $\mathcal{E}(A) = X$, A must be bounded.

Definitions 1.6. It is clear that the definition of an entire vector is exactly what we need to define, for any complex z, $x \in \mathcal{E}(A)$,

$$e^{zA}x \equiv \sum_{k=0}^{\infty} \frac{z^k}{k!} A^k x. \tag{1.7}$$

When $A \in B(X)$, this is the way we define the group $\{e^{zA}\}_{z\in\mathbf{C}}$ generated by A.

The map $t \mapsto e^{tA}x$, from (1.7), is a solution of the abstract Cauchy problem. In fact, $\mathcal{E}(A)$ consists precisely of those initial data x for which (1.1) has an entire solution, and that solution will then be given by (1.7) (see [1, Theorem 1]).

Note that I have not mentioned uniqueness of the solution. Although we may not have uniqueness, it is the case that (1.7) will be the unique *analytic* solution of (1.1), when $x \in \mathcal{E}(A)$.

We may make $\mathcal{E}(A)$ into a Frechet space, using e^{zA}, as follows (this is a slight variation of the seminorms in [9]). For any nonnegative integer n, $x \in \mathcal{E}(A)$, define

$$\|x\|_n \equiv \sup_{|z|\leq n} \|e^{zA}x\|. \tag{1.8}$$

We topologize $\mathcal{E}(A)$ with the seminorms $\{\| \ \|_n \mid n = 0, 1, 2, \dots\}$. Convergence of a sequence $\{x_k\}_k$, with respect to this topology, is uniform convergence of the functions $\{z \mapsto e^{zA}x_k\}_k$, on compact subsets of the complex plane.

Let me summarize the properties of $\mathcal{E}(A)$ and e^{zA}.

Proposition 1.9.

(1) *$\mathcal{E}(A)$ is a Frechet space.*

(2) *Both A and e^{zA}, for any complex z, map $\mathcal{E}(A)$ to itself, and are bounded on $\mathcal{E}(A)$.*

(3) *$\{e^{zA}\}_{z\in\mathbf{C}}$ is an entire group generated by A.*

I would like to discuss estimating $\mathcal{E}(A)$ by finding bounded operators C such that $Im(C) \subseteq \mathcal{E}(A)$.

Throughout, $C \in B(X)$, and I will write $[Im(C)]$ for the Banach space $Im(C)$, with the norm

$$\|y\|_{[Im(C)]} \equiv \inf\{\|x\| \mid Cx = y\}. \tag{1.10}$$

Definition 1.11. An *entire C-existence family for A* is a family $\{W(z)\}_{z \in \mathbf{C}} \subseteq B(X)$ such that, for any $y \in X$, the map $t \mapsto W(t)y$ is an entire solution of (1.1), with $x = Cy$.

Note that $z \mapsto W(z)$ is an entire map into $B(X)$.

When C is injective and $W(z)$ commutes with $W(w)$, for all complex z and w, an algebraic definition is possible; it may be shown that the entire family of bounded operators $\{W(z)\}_{z \in \mathbf{C}}$ is an entire C-existence family for some closed operator if and only if $W(z)W(w) = CW(z + w)$, for all complex z, w, and $W(0) = C$. This is an *(entire) C-regularized semigroup.*

When A has an entire C-existence family, then we have the following analogue of well-posedness. When $y_n \to y$ in X, as $n \to \infty$, then $u(z, Cy_n) \equiv W(z)y_n \to u(z, Cy) \equiv W(z)y$, uniformly for z in compact subsets of the complex plane.

When C is injective and commutes with A, this may be expressed as follows. If $\{x_n\}_{n=0}^{\infty}$ is in the image of C, and $C^{-1}x_n \to C^{-1}x_0$, as $n \to \infty$, then $u(z, x_n) \to u(z, x_0)$, as $n \to \infty$, uniformly for z in compact subsets of the complex plane.

This is continuous dependence of the solutions on the initial data, except that we have different topologies on the initial data than on the solutions. We are putting a stronger topology on the initial data.

The choice of C measures how far from bonafide well-posedness we are; we'd like the image of C to be as large as possible.

Similarly, a C-existence family for A, and a C-regularized semigroup generated by A, have been defined. For basic material on the subject, including references, and the precise relationships between these concepts and (1.1), see [4] and [5].

In the following, I am writing "$\hookrightarrow$" to mean "is continuously embedded in."

Proposition 1.12. *The following are equivalent.*

(a) $Im(C) \subseteq \mathcal{E}(A)$.

(b) $Im(C) \subseteq C^{\infty}(A)$ *and*

$$\sum_{k=0}^{\infty} \frac{s^k}{k!} \|A^k C\| < \infty, \ \forall s > 0.$$

(c) *There exists an entire C-existence family for A, $\{W(z)\}_{z\in\mathbf{C}}$.*

(d) $[Im(C)] \hookrightarrow \mathcal{E}(A)$.

Then

$$W(z) = e^{zA}C = \sum_{k=0}^{\infty} \frac{z^k}{k!} A^k C,$$

for all complex z.

Thus I give you a choice. One can have something analogous to well-posedness, on the original space, or one can go down to the continuously embedded subspace $\mathcal{E}(A)$, where A generates a strongly continuous group; this is a common definition of (1.1) being well-posed. The disadvantage in the second choice is that $\mathcal{E}(A)$ is a Frechet space, rather than a Banach space.

II. EXAMPLES. I would like to give three examples of choices of C, whose image can be placed inside $\mathcal{E}(A)$, in ill-posed problems. In all these examples, $Im(C)$ is dense, thus we are obtaining entire solutions of (1.1), for all initial data x in a dense set.

Example 2.1: Reversibility of parabolic problems. By a "parabolic problem," I mean (1.1) with A generating a strongly continuous holomorphic semigroup; for example, the heat equation. By "reversibility" I mean letting t assume negative values in (1.1); for example, the backwards heat equation; we are letting time run backwards. An equivalent way to run time backwards in (1.1) is to have $-A$, rather than A, generate a strongly continuous holomorphic semigroup.

Proposition 2.1(1). *Suppose $-A$ generates a strongly continuous holomorphic semigroup. Then there exists $\alpha > 1$ and real ω such that*

$$Im(e^{-(A+\omega)^{\alpha}}) \subseteq \mathcal{E}(A).$$

Let me make it clear what I mean by e^B. This means that B generates a strongly continuous semigroup $\{e^{tB}\}_{t\geq 0}$, and e^B is the member of that semigroup when $t = 1$.

Outline of Proof: There exists real ω and positive $\theta \leq \frac{\pi}{2}$ so that $-(A+\omega)$ generates an exponentially decaying strongly continuous holomorphic semigroup $\{e^{-t(A+\omega)}\}_{t\geq 0}$ of angle θ. For ϕ between $\frac{\pi}{2}$ and $\frac{\pi}{2} - \theta$, this semigroup may be represented with an unbounded analogue of the Riesz-Dunford functional calculus,

$$e^{-t(A+\omega)} = \int_{\Gamma_\phi} e^{-tz}(z-(A+\omega))^{-1} \frac{dz}{2\pi i},$$

where Γ_ϕ is defined to be the boundary of $\{re^{i\psi} \mid |\psi| < \phi\}$, oriented counter-clockwise.

This representation may also be used for fractional powers. For any positive α such that $\alpha(\frac{\pi}{2} - \theta) < \frac{\pi}{2}$, $-(A+\omega)^\alpha$ generates a strongly continuous holomorphic semigroup $\{e^{-t(A+\omega)^\alpha}\}_{t\geq 0}$ given by

$$e^{-t(A+\omega)^\alpha} = \int_{\Gamma_\phi} e^{-tz^\alpha}(z-(A+\omega))^{-1}\,\frac{dz}{2\pi i}, \qquad (2.1(2))$$

where ϕ is now chosen so that $\phi > \frac{\pi}{2} - \theta$ and $\alpha\phi < \frac{\pi}{2}$.

It may be shown that, for any nonnegative integer k,

$$A^k e^{-(A+\omega)^\alpha} = \int_{\Gamma_\phi} (z-\omega)^k e^{-z^\alpha}(z-(A+\omega))^{-1}\,\frac{dz}{2\pi i}.$$

The intuition here is that, as with the Cauchy integral formula, just replace z by $A+\omega$, everywhere in the integrand besides the $(z-(A+\omega))^{-1}$. Thus, for any $s > 0$, letting $C \equiv e^{-(A+\omega)^\alpha}$,

$$\begin{aligned}\sum_{k=0}^{\infty} \frac{s^k}{k!}\|A^k C\| &\leq \int_{\Gamma_\phi} \sum_{k=0}^{\infty} \frac{s^k}{k!} |(z-\omega)^k e^{-z^\alpha}|\,\|(z-(A+\omega))^{-1}\|\,\frac{dz}{2\pi} \\ &\leq \int_{\Gamma_\phi} e^{s|(z-\omega)|}|e^{-z^\alpha}|\,\|(z-(A+\omega))^{-1}\|\,\frac{dz}{2\pi}.\end{aligned}$$

By choosing $\alpha > 1$, with $\alpha\phi < \frac{\pi}{2}$, as in (2.1(2)), this integral will converge for all $s > 0$, as desired. ∎

Example 2.1(3): Backwards heat equation. Suppose Ω is a bounded open set in $\mathbf{R}^n$, with smooth boundary. Let $w(\vec{s}, t)$ be the temperature at time t and position $\vec{s}$, and consider

$$\begin{aligned}&\frac{\partial w}{\partial t}(\vec{s},t) + \Delta w(\vec{s},t) = 0 \quad (\vec{s} \in \Omega,\ t \geq 0) \\ &w(\vec{s},t) = 0 \quad (\vec{s} \in \partial\Omega,\ t \geq 0) \\ &w(\vec{s},0) = f(\vec{s}) \quad (\vec{s} \in \Omega),\end{aligned}$$

where Δ is the Laplacian in $\mathbf{R}^n$.

We may apply Proposition 2.2(1), since, if $1 \leq p < \infty$, and $A \equiv -\Delta$, $\mathcal{D}(A) \equiv W^{2,p}(\Omega) \cap W_0^{1,p}(\Omega)$, then $-A$ generates a strongly continuous holomorphic semigroup on $X \equiv L^p(\Omega)$. Thus the map $t \mapsto w(\cdot, t)$, from $[0,\infty)$ into X, extends to an entire function, for f in a dense subspace of X.

This is saying that the heat equation is reversible on a dense set.

Example 2.2: An ill-posed heat equation. Time will run forward here, but we will still be ill-posed. Take the usual heat equation in one dimension; we let $w(t,s)$ be the temperature of something long and skinny, at time t and position s; but instead of having initial data, that is, the temperature at $t=0$, we will specify the temperature and heat flux at one end.

$$\frac{\partial}{\partial t}w(t,s) = (\frac{\partial}{\partial s})^2 w(t,s)\ (s,t \geq 0),\ w(t,0) = g_1(t),\ \frac{\partial}{\partial s}w(t,0) = g_2(t)\ (t \geq 0). \tag{2.2(1)}$$

So we know everything at one end ($s=0$), and we want to predict what's happening everywhere else.

To write (2.2(1)) as an abstract Cauchy problem, we must interchange the role of s and t, so that we will have t replaced by s in (1.1). This may cause confusion, but it's unavoidable, given the axiom that time must be represented by t.

So define, for any $s \geq 0$, the function $v(s)$ by

$$[v(s)](t) \equiv w(t,s)\ (t \geq 0).$$

Letting $B \equiv \frac{d}{dt}$, we may then write (2.2(1)) as

$$(\frac{d}{ds})^2 v(s) = B(v(s))\ (s \geq 0),\ v(0) = g_1,\ \frac{d}{ds}v(0) = g_2. \tag{2.2(2)}$$

More precisely, let's choose $v(s) \in C_0([0,\infty)) \equiv \{f \in C[0,\infty) \mid \lim_{t\to\infty} f(t) = 0\}$, and B will be the generator of left translation on $C_0([0,\infty))$,

$$(e^{sB}f)(t) \equiv f(s+t)\ (s,t \geq 0).$$

(2.2(2)) is a second order abstract Cauchy problem; we do the usual matrix reduction to a first order problem, by defining A on $X \equiv (C_0([0,\infty)))^2$ to be

$$A \equiv \begin{bmatrix} 0 & I \\ B & 0 \end{bmatrix},\ \mathcal{D}(A) \equiv \mathcal{D}(B) \times C_0([0,\infty)),$$

$$u(s,\vec{g}) \equiv \begin{bmatrix} v(s) \\ \frac{d}{ds}v(s) \end{bmatrix}\ (s \geq 0).$$

Then (2.2(1)) becomes

$$\frac{d}{ds}u(s,\vec{g}) = A(u(s,\vec{g}))\ (s \geq 0),\ u(0,\vec{g}) = \vec{g}. \tag{2.2(3)}$$

Formally, the solution of (2.2(3)) is given by

$$[u(s,\vec{g})] = e^{sA}\vec{g};$$

but this is only formal, because A does not generate a strongly continuous semigroup. However, we can choose C, with dense image, such that $Im(C) \subseteq \mathcal{E}(A)$.

Let's write I_2 for the 2×2 identity matrix.

Proposition 2.2(4). *For $\frac{1}{2} < \gamma < 1$, $Im(e^{-((-B)^{\frac{1}{2}}+I)^{2\gamma}} I_2) \subseteq \mathcal{E}(A)$.*

Outline of Proof: This follows fairly quickly from the previous example, when one notes that

$$A^2 = BI_2,$$

thus, letting $C \equiv e^{-((-B)^{\frac{1}{2}}+I)^{2\gamma}} I_2$,

$$\sum_{k=0}^{\infty} \frac{s^k}{k!}\|A^k C\| = \sum_{n=0}^{\infty} \frac{s^{2n}}{(2n)!}\|((-B)^{\frac{1}{2}})^{2n} C\| + \sum_{n=0}^{\infty} \frac{s^{2n+1}}{(2n+1)!}\|((-B)^{\frac{1}{2}})^{2n} AC\|,$$

so that, since $-(-B)^{\frac{1}{2}}$ generates a bounded strongly continuous holomorphic semigroup of angle $\frac{\pi}{4}$, Proposition 2.1(1) implies that this series converges. ∎

See [7] for more results about this sort of ill-posed heat equation.

It is clear from the proof of Proposition 2.2(4) that the only relevant property of B was that it generated a bounded strongly continuous semigroup. More generally, to deal with (2.2(2)) as we did in Proposition 2.2(4), all we need is to have $-(\alpha B)^{\gamma}$ defined, and generating a strongly continuous holomorphic semigroup, for some $\gamma > \frac{1}{2}$, complex α. For this, it is sufficient to have αB be what is sometimes called a *positive operator*. This means that $(-\infty, 0) \subseteq \rho(\alpha B)$ and there exists a constant M so that

$$\|r(r+\alpha B)^{-1}\| \leq M, \ \forall r > 0.$$

Thus Proposition 2.2(4) could also be applied to the Cauchy problem for the Laplace equation; see [4, chapter IX], for details.

Example 2.3: Adjoints of symmetric operators. For this example only, I will have X equal to a Hilbert space. Note that, if B is symmetric and densely defined, then $\overline{B}B^*$ is positive self-adjoint, thus $-\overline{B}B^*$ generates a bounded strongly continuous holomorphic semigroup $\{e^{-t\overline{B}B^*}\}_{t\geq 0}$.

In the following, I must emphasize that B may not have a self-adjoint extension on X, thus the spectral theorem is not available. We could, for example, choose $B \equiv i\frac{d}{ds}$, on $L^2([0,\infty))$.

Proposition 2.3(1). *If B is symmetric and densely defined, then*

$$Im(e^{-\overline{B}B^*}) \subseteq \mathcal{E}(B^*).$$

Outline of Proof: Without loss of generality, we may assume B is closed. We use two facts. First, since $B \subseteq B^*$, it follows that $C^\infty(BB^*) \subseteq C^\infty(B^*)$. Second, since $\{e^{-tBB^*}\}_{t\geq 0}$ is a bounded strongly continuous holomorphic semigroup, there exists a constant M, so that, for any $x \in X$, $t > 0$, $e^{-tBB^*}x \in C^\infty(BB^*)$, with

$$\|BB^*e^{-tBB^*}x\| \leq \frac{M}{t}\|x\|.$$

Thus, for any nonnegative integer n,

$$\|(B^*)^{2n}e^{-BB^*}\| = \|(BB^*)^n e^{-BB^*}\| = \|(BB^*e^{-\frac{1}{n}BB^*})^n\| \leq (Mn)^n.$$

A little more calculation with the inner product gives us

$$\|(B^*)^{2n+1}e^{-BB^*}\| \leq M^{n+1}(n+1)^{n+1},$$

for any nonnegative n. This implies that

$$\sum_{k=0}^{\infty} \frac{s^k}{k!}\|(B^*)^k e^{-BB^*}\| < \infty, \ \forall s > 0,$$

as desired. ∎

Example 2.3(2). Symmetric differential operators, especially if they are defined only on a bounded set, generally have boundary conditions. Obtaining a self-adjoint extension, when this is possible, often involves a very careful choice of boundary conditions.

Passing to the adjoint of a symmetric operator, as in Proposition 2.3(1), causes us to remove all or some of the boundary conditions. Proposition 2.3(1) asserts that we are still guaranteed entire solutions of the corresponding abstract Cauchy problem, after this removal.

In the following example, we will put strong boundary conditions on B, so that B^* will have no boundary conditions.

On $X \equiv L^2(\mathbf{R}^n)$, define

$$(Bf)(\vec{s}) \equiv \sum_{i,j=1}^{n} \frac{\partial}{\partial s_i}(a_{i,j}(\vec{s})\frac{\partial f}{\partial s_j}(\vec{s})) \ (\vec{s} \in \mathbf{R}^n),$$

with $\mathcal{D}(B) \equiv C_c^\infty(\mathbf{R}^n)$, where, for $1 \leq i, j \leq n$, $a_{i,j}$ and its first order partial derivatives are real-valued functions in $L_{loc}^\infty(\mathbf{R}^n)$, and for each $\vec{s} \in \mathbf{R}^n$, $a_{i,j}(\vec{s}) = a_{j,i}(\vec{s})$.

Then B is symmetric, and B^* is also given by

$$(B^* f)(\vec{s}) = \sum_{i,j=1}^{n} \frac{\partial}{\partial s_i}(a_{i,j}(\vec{s}) \frac{\partial f}{\partial s_j}(\vec{s})) \ (\vec{s} \in \mathbf{R}^n),$$

but now $\mathcal{D}(B^*)$ equals $\{f \in L^2(\mathbf{R}^n) \,|\, B^* f \in L^2(\mathbf{R}^n)\}$.

Thus Proposition 2.3(1) asserts that $Im(e^{-\overline{B}B^*}) \subseteq \mathcal{E}(B^*)$.

We should comment that vectors in the deficiency subspaces of B^*, when B is symmetric, are clearly contained in $\mathcal{E}(B^*)$, since they are eigenvectors for B^*. However, the span of these subspaces, unlike $Im(e^{-\overline{B}B^*})$, may not be dense.

REFERENCES

[1] L. Autret, *Entire vectors and time reversible Cauchy problems,* Semigroup Forum 46 (1993), 347–351.

[2] P. R. Chernoff, *Some remarks on quasi-analytic vectors,* Trans. Amer. Math. Soc. 167 (1972), 105–113.

[3] E. B. Davies, "One-Parameter Semigroups," Academic Press, London, 1980.

[4] R. deLaubenfels, "Existence Families, Functional Calculi and Evolution Equations," Springer-Verlag Lecture Notes in Math. 1570 (1994).

[5] R. deLaubenfels, S. Guozheng and S. Wang, *Regularized semigroups, existence families and the abstract Cauchy problem,* submitted.

[6] R. deLaubenfels and F. Yao, *Entire solutions of the abstract Cauchy problem in a Hilbert space,* Proc. Amer. Math. Soc. (to appear).

[7] J. R. Dorroh, *Continuous dependence of nonnegative solutions of the heat equation on noncharacteristic Cauchy data,* preprint.

[8] J.A. Goldstein, "Semigroups of Operators and Applications," Oxford, New York, 1985.

[9] R. Goodman, *Complex Fourier analysis on a nilpotent Lie group,* Trans. Amer. Math. Soc. 160 (1971), 373–391.

[10] E. Nelson, *Analytic vectors,* Ann. of Math. 70 (1959), 572–615.

J R DORROH

Continuous dependence of nonnegative solutions of the heat equation on noncharacteristic Cauchy data

The sideways Cauchy problem for the one-dimensional heat equation , in which the solution and its first-order spatial derivative are specified on an interval of the time axis, is well known to be ill-posed. Nevertheless, we establish continuous dependence on the Cauchy data of nonnegative solutions satisfying the extra smoothness requirement that the time derivative is continuous on the initial manifold. This is done by first establishing explicit bounds for such solutions and certain of their derivatives and then applying a result of J. Cannon. Most of the estimates, which may be of interest in themselves, do not depend on the extra smoothness assumption. In spite of the fact that the admissible Cauchy data is highly non-arbitrary, our results include the fact that the set of admissible Cauchy data is closed under uniform convergence of the data and one first-order derivative of the data.

Let $a, T > 0$, and let $f, g \in C(0,T)$. We consider the following Cauchy problem for the one-dimensional heat equation.

(PDE)	$u_t(x,t) = u_{xx}(x,t)$,	$0 < x < a,\ 0 < t < T$,
(IC)	$u(0+,t) = f(t)$,	$0 < t < T$,
(IC)	$u_x(0+,t) = g(t)$,	$0 < t < T$.

This is the problem one must consider if one boundary is all that is accessible. We require that the solution u and its derivative u_x be continuous on $[0,a) \times (0,T)$. We denote this problem by $\mathcal{P}(a,T,f,g)$. For $f, g \in C[0,T]$, we denote by $\mathcal{P}^*(a,T,f,g)$ the problem $\mathcal{P}(a,T,f,g)$ with the added restriction that u and u_x be continuous on $[0,a] \times [0,T]$. This is only a technical distinction; any solution of $\mathcal{P}(a,T,f,g)$ is also a solution of $\mathcal{P}^*(a^*,T^*,f,g)$ for $0 < a^* < a$, $0 < T^* < T$ after a change of variable $t \to t - \delta$, $0 < \delta < T - T^*$. If $f \in C^{(1)}(0,T)$, then we denote by $\mathcal{P}'(a,T,f,g)$ the problem $\mathcal{P}(a,T,f,g)$ with the additional restriction the u_t is continuous on $[0,a) \times (0,T)$. This amounts to an extra smoothness assumption for the solution.

This problem is well known to be ill-posed. By taking $f(t) = r^{-3}\cos(2r^2t)$ and $g(t) = r^{-2}[\cos(2r^2t) - \sin(2r^2t)]$, where $r > 0$, one easily sees that solutions do not depend continuously on the Cauchy data, for the solution is given by $u(x,t) = r^{-3}\exp(rx)\cos(rx + 2r^2t)$. Since solutions typically represent temperature or density of a diffusing substance, it is quite reasonable to consider nonnegative solutions, and we are able to give some some fairly simple explicit estimates and establish continuous dependence on Cauchy data for *nonnegative solutions* of $\mathcal{P}(a,T,f,g)$. Even here, the situation is delicate; consider the example $f(t) = \exp(-rb + r^2(t-T))$, $g(t) = -r\exp(-rb + r^2(t-T))$, where $r > 0$ and $0 < b < a$. The solution u of $\mathcal{P}(a,T,f,g)$ is then given by $u(x,t) = \exp(r(x-b) + r^2(t-T))$. This example will also suffice to show that our estimates cannot be easily strengthened. The following is our main result; at present we are unable to eliminate the extra smoothness assumption, but it may be possible to do this.

Theorem. *Let $a, T > 0$. Suppose that $f_n \in C^{(1)}(0,T)$, $g_n \in C(0,T)$, and $\mathcal{P}'(a,T,f_n,g_n)$ has a nonnegative solution u_n for each n. Suppose further that $\{f_n\}$, $\{f_n'\}$, $\{g_n\}$ converge uniformly on each closed subinterval of $(0,T)$ to f, f', g. Then $\mathcal{P}(a,T,f,g)$ has a nonnegative solution u, and $\{u_n\}$ converges to u uniformly on each set $[0,b] \times [T_1,T_2]$ with $0 < b < a$ and $0 < T_1 < T_2 < T$.*

The first example shows that nothing of the kind is true if one omits the assumption of existence of nonnegative solutions. The result is especially surprising when one considers the highly non-arbitrary nature of those f, g for which $\mathcal{P}(a,T,f,g)$ has any kind of solution whatsoever, much less a nonnegative one. For example, in order that $\mathcal{P}^*(a,T,f,g)$ have a solution, it is necessary and sufficient that the function ζ defined by

$$\zeta(t) = f(t) + \frac{1}{\sqrt{\pi}} \int_0^t \frac{g(\tau)}{\sqrt{t-\tau}}\, d\tau$$

be infinitely differentiable on $(0,T)$, with derivatives satisfying

$$|\zeta^{(n)}(t)| \leq M \frac{(2n)!}{R^{(2n)}}$$

on each closed subinterval of $(0,T)$ for some $M, R > 0$. This may be seen as follows. Let

$$v(x,t) = -2 \int_0^t g(\tau) k(x, t-\tau)\, d\tau,$$

where k is the fundamental solution of the heat equation, and let $w = u - v$. Then $w_x(0+,t) = 0$, [7, Thm. 7.3, p 72], and w_x has a continuous extension to $[0,a] \times [0,T]$. Therefore, by the Schwarz reflection principle, [7, Thm. 7, p. 115], the odd extension of w_x satisfies the heat equation on $(-a,a) \times (0,T)$. Therefore, the even extension of w satisfies the heat equation on $(-a,a) \times (0,T)$. Since $\zeta(t) = w(0,t)$, the assertion follows; see [7, Thm. 13, p. 84].

Conditions equivalent to this were established by Holmgren in [4]. Sufficient conditions on f and g in order that the solution of $\mathcal{P}$ be nonnegative or satisfy a given bound are much more elusive. In spite of this subtlety, it is true that the set of all (f,g) in $C^{(1)}(0,1) \times C(0,1)$ such that $\mathcal{P}(a,T,f,g)$ has a nonnegative solution is a closed set in the natural metric of this space; namely uniform convergence of f, f', g on closed subintervals of $(0,T)$.

For earlier results on nonnegative solutions, see Pucci [7]. For similar results for solutions satisfying a prescribed bound, see Cannon ([1], [2]), F. John [5], and Lavrent'ev [6].

Even though we establish continuous dependence on the Cauchy data, this does not solve the problem of producing an approximate solution from approximate Cauchy data, because approximately known functions f, g will almost certainly lack the property that $\mathcal{P}(a,T,f,g)$ has a nonnegative solution, and it is not at all clear how to find nearby data that does have this property. Cannon does address this problem for solutions satisfying a prescribed bound, and we show that his results can be applied to our problem; see our concluding remark.

The proof of the main theorem requires two lemmas. These are of some interest and potential further use in themselves, especially the second lemma, which gives estimates for a nonnegative solution of $\mathcal{P}^*(a,T,f,g)$ and some of its derivatives.

Lemma 1. *Suppose φ is a nonnegative function in $C^1[0,a] \cap C^2(0,a)$, that $\lambda > 0$, and that $|\varphi''(x)| \leq \lambda^2 \varphi(x)$ for $0 < x < a$. Then*

$$\varphi(x) \leq \varphi(0)\cosh\lambda x + \varphi'(0)\frac{\sinh\lambda x}{\lambda},$$

and

$$|\varphi'(x)| \leq \varphi(0)\left(\lambda\sinh\lambda x\right) + |\varphi'(0)|\cosh\lambda x$$

for $0 \leq x \leq a$.

In order to state the second lemma, we need some notation. Let k denote the source solution of the heat equation, h the derived source solution, and erfc the complimentary error function; that is,

$$k(x,t) = \frac{1}{\sqrt{4\pi t}}\exp\left(\frac{-x^2}{4t}\right),\ h(x,t) = \frac{x}{\sqrt{4\pi t^3}}\exp\left(\frac{-x^2}{4t}\right),\ \operatorname{erfc}(z) = \frac{2}{\sqrt{\pi}}\int_z^\infty e^{-y^2}\,dy.$$

Also, $\|\cdot\|$ will denote the supremum norm on $[0,T]$, and $\|\cdot\|_1$ will denote the L^1 norm on $[0,T]$. The following theorem gives estimates for nonnegative solutions that are needed for the proof of Theorem 1 and are of interest and potential further use themselves.

Lemma 2. *Let $a, T > 0$, and $f, g \in C[0,T]$. If $\mathcal{P}^*(a,T,f,g)$ has a nonnegative solution u, then $u = u^{(1)} + u^{(2)} + u^{(3)}$, where $u^{(1)}(x,t) =$*

$$\int_0^t \left([h(x,t-\tau) - h(2a-x,t-\tau)]\,\frac{f(\tau)}{2} - [k(x,t-\tau) - k(2a-x,t-\tau)]\,g(\tau)\right)d\tau$$

for all $(x,t) \in (0,a] \times [0,T]$, and $u^{(2)}$ and $u^{(3)}$ are nonnegative. Furthermore,

$$u^{(2)}(x,t) \leq \left(\|f\| + 2\sqrt{\frac{T}{\pi}}\|g\|\right)\left(\cosh\lambda x + \left(\frac{a}{2t_0} + \frac{1}{a}\right)\frac{\sinh\lambda x}{\lambda}\right),$$

$$|u_x^{(2)}(x,t)| \leq \left(\|f\| + 2\sqrt{\frac{T}{\pi}}\|g\|\right)\left(\lambda\sinh\lambda x + \left(\frac{a}{2t_0} + \frac{1}{a}\right)\cosh\lambda x\right),$$

and

$$|u_{xx}^{(2)}(x,t)| \leq \lambda^2 u^{(2)}(x,t)$$

for $0 < t_0 \leq t \leq T$, $0 \leq x \leq a$, where $\lambda = [(a^2/4t_0^2) + (3/2t_0)]^{1/2}$. Furthermore,

$$|u^{(3)}(x,t)| \leq \left(\|f\|_1 + 2\sqrt{\frac{T}{\pi}}\|g\|_1\right)\frac{1}{\operatorname{erfc}\left(a/\sqrt{4(T-t)}\right)}\frac{3\sqrt{6/\pi e^3}}{(a-x)^2},$$

$$|u_x^{(3)}(x,t)| \leq \left(\|f\|_1 + 2\sqrt{\frac{T}{\pi}}\|g\|_1\right)\frac{1}{\operatorname{erfc}\left(a/\sqrt{4(T-t)}\right)}\frac{3\sqrt{6/\pi e^3} + 25\sqrt{10/\pi e^5}}{(a-x)^3},$$

and

$$|u_{xx}^{(3)}(x,t)| \leq \left(\|f\|_1 + 2\sqrt{\frac{T}{\pi}}\|g\|_1\right)\frac{1}{\operatorname{erfc}\left(a/\sqrt{4(T-t)}\right)}\frac{75\sqrt{10/\pi e^5} + 7^3\sqrt{14/\pi e^7}}{(a-x)^4}$$

for $0 \leq x < a$ *and* $0 < t < T$*. Of course, we have the trivial estimate*

$$|u^{(1)}(x,t)| \leq \|f\| + 2\sqrt{\frac{T}{\pi}}\|g\|,$$

and this estimate holds on all of $[0,a] \times [0,T]$*, whether or not* $\mathcal{P}(a,T,f,g)$ *has a nonnegative solution, or any solution at all, for that matter. Furtermore, if* $f \in C^{(1)}[0,T]$*, then* $u^{(1)}$ *can be rewritten as*

$$\begin{aligned} u^{(1)}(x,t) &= f(0)\operatorname{erfc}(x/\sqrt{4t}) \\ &+ \int_0^t \left[\operatorname{erfc}\left(x/\sqrt{4(t-\tau)}\right) - \operatorname{erfc}\left((2a-x)/\sqrt{4(t-\tau)}\right)\right] \frac{f'(\tau)}{2}\,d\tau \\ &- \int_0^t [k(x,t-\tau) - k(2a-x,t-\tau)]\,g(\tau)\,d\tau. \end{aligned}$$

This gives

$$\begin{aligned} u_x^{(1)}(x,t) = -2f(0)k(x,t) &- \int_0^t [k(x,t-\tau) + k(2a-x,t-\tau)]\,f'(\tau)\,d\tau \\ &+ \frac{1}{2}\int_0^t [h(x,t-\tau) + h(2a-x,t-\tau)]\,g(\tau)\,d\tau. \end{aligned}$$

This yields the estimate

$$|u_x^{(1)}(x,t)| \leq \frac{1}{\pi t_0} f(0) + 2\sqrt{\frac{T}{\pi}}\|f'\| + \|g\|$$

for $t_0 \leq t \leq T$, $0 \leq x \leq a$.

The proofs will appear in [3].

References

[1] J. R. Cannon, "The One-Dimensional Heat Equation," The Encyclopedia of Mathematics, vol. 23, Addison-Wesley, 1984.

[2] J. R. Cannon, A Cauchy problem for the heat equation, *Annali di Mat. Pura ed Appl.* (Serie IV) **66** (1964), 155-165.

[3] J. R. Dorroh, Continuous dependence of nonnegative solutions of the heat equation on non-characteristic Caucy data, *Applicable Analysis*, to appear.

[4] E. Holmgren, Om Cauchys problem vid de lineära partielle differentialek vationerna of 2:dra ordningen, *Arkiv för Mat., Astr. och Fys.* **2** (1906), No. 24, 1 - 13.

[5] F. John, Continuous dependence on data for solutions of partial differential equations with a prescribed bound, *Communications Pure and Applied Math.* **8** (1960), 551-585.

[6] M. M. Lavrent'ev, V. G. Romonov, and S. P. Shishatskii, "Ill-posed Problems of Mathematical Physics and Analysis," Translations of Mathematical Monographs, vol. 64, American Math. Soc., 1986.

[7] Carlo Pucci, Alcune limitazioni per le soluzioni di equazioni paraboliche, *Annali di Mat. Pura ed Appl.* (Serie IV)**48** (1959), 161 - 172.

[8] D. V. Widder, "The Heat Equation," Pure and Applied Mathematics, vol. 67, Academic Press, 1975.

J R DORROH AND J W NEUBERGER

A theory of strongly continuous semigroups in terms of Lie generators

Let X denote a complete separable metric space, and let $\mathcal{C}(X)$ denote the linear space of all bounded continuous real-valued functions on X. The Lie generator of a strongly continuous semigroup T of continuous transformations in X is the linear operator in $\mathcal{C}(X)$ consisting of all ordered pairs (f, g) such that $f, g \in \mathcal{C}(X)$, and for each $x \in X$, $g(x)$ is the derivative at 0 of $f(T(\cdot)x)$. We completely characterize such Lie generators and establish the canonical exponential formula for the original semigroup in terms of powers of resolvents of its Lie generator. The only topological notions needed in the characterization are two notions of sequential convergence, pointwise and strict. A sequence in $\mathcal{C}(X)$ converges strictly if the sequence is uniformly bounded in the supremum norm and converges uniformly on compact subsets of X. Our sufficient conditions do not involve powers of the resolvent higher than the first power.

Let $\mathcal{F}(X)$ denote the collection of all continuous transformations from X into X. A *strongly continuous semigroup of continuous transformations in* X is a function T from $[0, \infty)$ into $\mathcal{F}(X)$ such that $T(0)$ is the identity transformation on X, $T(t)T(s) = T(t+s)$ for $s, t \geq 0$, and for each x in X, the function $T(\cdot)x$ is continuous from $[0, \infty)$ to X. The semigroup T is commonly denoted by $\{T(t)\}_{t \geq 0}$. Denote the collection of all strongly continuous semigroups of continuous transformations in X by $\mathcal{S}(X)$, and let $\mathcal{C}(X)$ denote the linear space of all bounded continuous real-valued functions on X. We need to mention that if $T \in \mathcal{S}(X)$, then the transformation $(t, x) \to T(t)x$ is jointly continuous from $[0, \infty) \times X$ into X; see [2, Theorem 4]. If $T \in \mathcal{S}(X)$, then the *Lie generator* of T is the linear operator A in $\mathcal{C}(X)$ consisting of all ordered pairs (f, g) such that $f, g \in \mathcal{C}(X)$ and

$$g(x) = \lim_{t \to 0} \frac{1}{t} [f(T(t)x) - f(x)]$$

for all $x \in X$. A linear operator A in $\mathcal{C}(X)$ with domain $\mathcal{D}(A)$ is said to be a *derivation* if $f, g \in \mathcal{D}(A)$ implies that $fg \in \mathcal{D}(A)$ and

$$A(fg) = fAg + gAf.$$

It is easy to see that the Lie generator of a semigroup $T \in \mathcal{S}(X)$ is a derivation. A sequence $\{f_n\}_{n=1}^{\infty}$ in $\mathcal{C}(X)$ is said to converge *strictly* to a function $f \in \mathcal{C}(X)$, and we say that f is the *strict limit* of $\{f_n\}_{n=1}^{\infty}$, if $\{f_n\}_{n=1}^{\infty}$ is uniformly bounded in the supremum norm and $\{f_n\}_{n=1}^{\infty}$ converges to f uniformly on compact subsets of X. We will see later that strict convergence is convergence in a topology on $\mathcal{C}(X)$. This topology is called the *strict topology*; see [11]. We say that a linear operator Q from $\mathcal{C}(X)$ into $\mathcal{C}(X)$ is *strictly sequentially continuous* if Q transforms strictly convergent sequences to strictly convergent sequences, and that a collection $\mathcal{Q}$ of such operators is *strictly sequentially equicontinuous* if whenever $\{f_n\}_{n=1}^{\infty} \subset \mathcal{C}(X)$ converges strictly to $f \in \mathcal{C}(X)$, then the collection $\{Qf_n : n \in \mathbb{N},\ Q \in \mathcal{Q}\}$ is bounded in the supremum norm, $\{[Qf_n](x)\}_{n=1}^{\infty}$ converges to $[Qf](x)$ for each $Q \in \mathcal{Q}$ and $x \in X$, and this convergence is uniform for $Q \in \mathcal{Q}$ and x

in compact subsets of X. A subset F of $\mathcal{C}(X)$ is said to be ***strictly sequentially dense*** in $\mathcal{C}(X)$ if each $f \in \mathcal{C}(X)$ is the strict limit of a sequence of functions belonging to F. We can now state our main theorem.

Theorem. *Let A be a linear operator in $\mathcal{C}(X)$, that is, with domain and range contained in $\mathcal{C}(X)$. Then A is the Lie generator of a semigroup $T \in \mathcal{S}(X)$ if and only if*

(i) A is a derivation,

(ii) the domain of A is strictly sequentially dense in $\mathcal{C}(X)$,

(iii) for each $\lambda > 0$, $I - \lambda A$ has a norm nonexpansive and strictly sequentially continuous inverse defined on all of $\mathcal{C}(X)$ (I denotes the identity transformation in $\mathcal{C}(X)$), and

(iv) if $\eta > 0$, then the collection $\{(I - \lambda A)^{-1} : 0 < \lambda \leq \eta\}$ is strictly sequentially equicontinuous.

Furthermore, if A is the Lie generator of $T \in \mathcal{S}(X)$, then

$$f \circ T(t) = \lim_{n\to\infty} (I - (t/n)A)^{-n} f$$

for $t > 0$ and $f \in \mathcal{C}(X)$, where the limit is the strict limit.

The main theorem characterizes Lie generators and establishes the canonical exponential formula. [6] characterized Lie generators, but the characterization here is much simpler in that it only involves pointwise and strict sequential convergence, whereas the characterization in [6] involved a locally convex topology on $\mathcal{C}(X)$. Also, the sufficient conditions in [6] included an equicontinuity assumption on $(I - \lambda A)^{-n}$ for $n > 1$, whereas this result does not. The proofs, which will appear elsewhere, are more nearly "self-contained" in that they do not appeal to any theory of strongly continuous semigroups in topological vector spaces, whereas [6] did. Furthermore, [6] did not establish the exponential formula. A theory like that in [6] is given in [4] in the case X is a locally compact Hausdorff space, not necessarily metric. The paper [5] contains a relevant result on the strict topology. In [9], it was proved that if A is the Lie generator of $T \in \mathcal{S}(X)$, then

$$f(T(t)x) = \lim_{n\to\infty} \left[(I - (t/n)A)^{-n} f\right](x)$$

for all $t \geq 0$, $f \in \mathcal{C}(X)$, and $x \in X$.

References

[1] C. J. K. Batty, Derivations on the line and flows along orbits, Pacific J. Math. **126** (1987), 209 - 225.

[2] P. Chernoff and J. E. Marsden, On continuity and smoothness of group actions, Bull. American Math. Society **76** (1970), 1044 - 1049.

[3] M. G. Crandall and T. M. Liggett, Generation of semigroups of nonlinear transformations on general Banach spaces, American J. Math. **93** (1971), 265 - 298.

[4] J. R. Dorroh, Semigroups of maps in a locally compact space, Canadian J. Math. **19** (1967), 688 - 696.

[5] ______________, The localization of the strict topology via bounded sets, Proc. American Math. Soc. **20** (1969), 413 - 414.

[6] J. R. Dorroh and J. W. Neuberger, Lie generators for semigroups of transformations on a Polish space, Electronic J. Diff. Eq. **1993/01** (1993).

[7] N. Dunford and J. Schwartz, *Linear Operators, Part I*, Interscience Publishers, 1958.

[8] J. Goldstein, *Semigroups of Linear Operators and Applications*, Oxford Univ. Press, 1985.

[9] J. W. Neuberger, Lie generators for one parameter semigroups of transformations, J. reine angew. Math. **258** (1973), 315-318.

[10] A. P. Robertson and W. J. Robertson, *Topological Vector Spaces*, Cambridge University Press, 1964.

[11] F. Dennis Sentilles, Bounded continuous functions on a completely regular space, Trans. American Math. Society **168** (1972), 311 - 336.

[12] H. H. Schaefer, *Topological Vector Spaces*, Springer-Verlag, 1970.

G FERREYRA AND O HIJAB

Smooth fit for some Bellman equations

ABSTRACT. Two linear-convex deterministic singular control problems in dimensions one and two are considered in this paper. They are solved using the dynamic programming method. The interest here is the explicitness of the results and the relation between the geometry of the drift along the free boundary of these problems and the principle of smooth fit.

0. INTRODUCTION

The method of dynamic programming reduces the study of an optimal control problem to the study of a nonlinear partial differential equation, the Hamilton-Jacobi-Bellman equation (see [6]). The value function for the optimal control problem is a solution of this equation. Singular optimal control problems generally give rise to a free boundary problem for said p.d.e. A basic step in solving the p.d.e. then is that of finding the free boundary. In the problems we consider here, convexity leads to the value function being $C^{1,1}$ (Lipschitz first partial derivatives) across the free bounday. Then a central issue is to determine whether or not the value function is C^2 across the free boundary. The C^2 case is referred to as satisfying the *smooth fit principle.* In stochastic control problems, smooth fit occurs in the case of the linear-quadratic-Gaussian problem and it is also known to occur in other examples with nondegenerate diffusion (see [8] and its references). In fact, the property of smooth fit was instrumental in solving the celebrated monotone follower problem [1]. On the other hand [8] gives an example where smooth fit does not occur.

In this paper we present problems in one and two dimensions having no diffusion at all; i.e., deterministic problems. These problems have linear dynamics and a nonnegative control. Other work dealing with problems with nonnegative control appear in [2]-[5],[7],[9],[10]. Because of the singular nature of our variational problems, we expect the optimal control to be extreme or to be singular. Since our controls are nonnegative this implies that we expect optimal controls to equal zero, infinity, or to be singular. The free boundary separates the null region (where the

---

1991 *Mathematics Subject Classification.* 35F30; 49L20; 49L25.

Key words and phrases. Singular Control, Free Boundary Problem, Bellman Equation.

The second author is supported by a National Science Foundation grant.

optimal control is zero) and the jump region (where the optimal control is impulsive). We find that in the one dimensional case the free boundary is just one point and smooth fit is a property that depends on the parameters of the problem. In the two dimensional case we find that smooth fit and non C^2-fit coexist (on different pieces of the free boundary). Our examples lead us to conjecture that generally smooth fit occurs if and only if in the optimal synthesis the zero control flow along the free boundary leads away from the null region (as opposed to leading into the null region).

1. A ONE DIMENSIONAL PROBLEM.

We consider the scalar control system

$$\dot{x} = b(x) + u, \qquad x(0) = x_0 \in \mathbf{R} \tag{1.1}$$

where u is a nonnegative measurable function of time. We define

$$v^u(x) = \int_0^\infty e^{-t}[f(x(t)) + u(t)]dt, \tag{1.2}$$

and we set

$$v(x) = \inf\{v^u(x): u(\cdot) \geq 0\}. \tag{1.3}$$

Our main assumption is linearity of b and convexity of f. These assumptions imply the convexity of v, which enables us to present a complete analysis of the control problem (1.1), (1.2), (1.3).

We assume the following:

(1) f is C^2 and $f(x) \geq 0$,
(2) $|f'(x)| \leq C_1(1 + f(x))$,
(3) $0 < \mu \leq f''(x) \leq C_2(1 + f(x))$,
(4) $b(x)$ is linear and $b'(x) < 0$.

Theorem 1. *The function v is a classical C^1 solution of the Bellman equation*

$$\max(v - bv' - f, -v' - 1) = 0, \qquad -\infty < x < \infty, \tag{1.4}$$

and there is a point $a \in \mathbf{R}$ such that $v - bv' - f = 0$ on $N = [a, \infty)$ and $-v' - 1 = 0$ on $J = (-\infty, a]$. Moreover $v \in C^2(\mathbf{R} \setminus \{a\})$ and $v \in C^2(\mathbf{R})$ iff $b(a)$ points strictly outward from N, i.e. $b(a) < 0$. The quantity a can be computed in terms of the data of the problem. Even if f is C^∞, v is never C^3 at a.

We note that assumption (3) implies f is strictly convex and assumptions (2),(3) hold for example when f'' is a strictly positive polynomial. Although $b'(x)$ is a constant, we find it helpful not to show this in the notation. Let $f^*(x)$ denote the maximum of f over the line segment joining x and the origin. Since f is convex $f^*(x) = \max(f(x), f(0))$.

Lemma. *The value function v is convex, C^1, and satisfies (1.4). Moreover v'' exists almost everywhere and*

(1) $0 \le v(x) \le f^*(x)$,
(2) $|v'(x)| \le C_1(1+f^*(x))$,
(3) $0 \le v''(x) \le C_2(1+f^*(x))$ a.e.

Proof. Clearly $v(x) \ge 0$. Let u_i be controls satisfying $v^{u_i}(x_0^i) \le v(x_0^i) + \epsilon$, $i = 0, 1$. For $s \in [0,1]$ let $x_0 = (1-s)x_0^0 + sx_0^1$, $u = (1-s)u_0 + su_1$. Then the corresponding solutions of (1.1) satisfy $x(t) = (1-s)x_0(t) + sx_1(t)$, $t \ge 0$, and the convexity of f implies

$$v(x_0) \le v^u(x_0) \le (1-s)v^{u_0}(x_0^0) + sv^{u_1}(x_0^1) \le (1-s)v(x_0^0) + sv(x_0^1) + \epsilon.$$

This shows v is convex.

Since $b' < 0$, $x(t)$ lies on the line segment joining x to 0 when $u \equiv 0$; this implies $v(x) \le v^0(x) \le f^*(x)$. Hence we need only consider controls u in (1.3) satisfying $v^u(x) \le f^*(x)$.

Now

$$|\nabla v^u(x)| \le \int_0^\infty |\nabla f(x(t))| dt \le C_1(1 + v^u(x)) \le C_1(1 + f^*(x))$$

and similarly

$$\nabla^2 v^u(x) \le C_2(1 + f^*(x)).$$

Since the right side of this last inequality is bounded on every compact interval, we conclude for each $a < b$ there is a $k(a,b) > 0$ such that $k(a,b)x^2 - v^u(x)$ is convex on $[a,b]$. Taking the supremum over all u we obtain $k(a,b)x^2 - v(x)$ is convex on $[a,b]$. Thus v is semi-concave; since v is also convex then v is C^1 and v'' exists almost everywhere. Finally the estimates on v', v'' follow from the above estimates for ∇v^u, $\nabla^2 v^u$. The fact that v solves the Bellman equation (1.4) is a standard consequence of dynamic programming (see, for example, [4] or [6]). □

Let $w(x) = v^0(x)$ be the cost corresponding to the zero control. Then by differentiation under the integral sign it follows that w is C^2, strictly convex, and

(1) $0 \le w(x) \le f^*(x)$,
(2) $|w'(x)| \le C_1(1 + f^*(x))$,
(3) $0 < \mu \le w''(x) \le C_2(1 + f^*(x))$,
(4) $w - bw' - f = 0, x \in \mathbf{R}$.

Since $w' : \mathbf{R} \to \mathbf{R}$ is increasing and onto we can define α by $w'(\alpha) = -1$. Similarly since $f' - b'$ is increasing and onto we can define β by $f'(\beta) - b'(\beta) = -1$.

We turn to the definition of the free boundary. Let $a = \inf\{x : -v'(x) - 1 < 0\}$. If $-v' - 1 = 0$ on $\mathbf{R}$ then v is affine hence not bounded below; thus $a < \infty$. If $a = -\infty$ then $v - bv' - f = 0$ on $\mathbf{R}$. Differentiating $e^{-t}v(x(t))$ with $\dot{x} = b(x)$ yields

$$v(x_0) = e^{-T}v(x(T)) + \int_0^T e^{-t} f(x(t)) dt.$$

Letting $T \to \infty$ yields $v(x_0) = w(x_0)$ for all x_0 hence $v'(\alpha) = w'(\alpha) = -1$ hence $\alpha \leq a$ contradicting $a = -\infty$. Hence a is finite. By definition of a and (1.4) $v - bv' - f = 0$ for $x \geq a$. By convexity and (1.4) $-v'(x) - 1 = 0$ for $x \leq a$. In particular at a both equalities hold and we obtain $v(a) = f(a) - b(a)$.

Proposition 1. *$a = \min(\alpha, \beta)$ and $v \in C^2(\mathbf{R} \setminus \{a\})$. Moreover $v \in C^2(\mathbf{R})$ iff $b(a) < 0$ iff $a = \beta < \alpha$.*

Proof. Assume first $b(a) \geq 0$. Then for the control $u(\cdot) \equiv 0$ and initial $x_0 \geq a$, we have $x(t) \geq a$ for all $t \geq 0$; let $\epsilon(T) = \int_T^\infty e^{-t} f(x(t))dt$; then $\epsilon(T) \to 0$ as $T \to \infty$. Differentiating $e^{-t}v(x(t))$ yields

$$
\begin{aligned}
v^u(x_0) &= -e^{-T}v(x(T)) + v(x_0) \\
&\quad + \int_0^T e^{-t}[(f(x(t)) + b(x(t))v'(x(t)) - v(x(t))]dt + \epsilon(T) \\
&= -e^{-T}v(x(T)) + v(x_0) + \epsilon(T).
\end{aligned}
$$

Letting $T \to \infty$ shows that $u \equiv 0$ is optimal at x_0 for all $x_0 \geq a$ which yields $v = w$ for $x \geq a$. This implies $-1 = v'(a) = w'(a)$ hence $a = \alpha$. Also since w is strictly convex and v is affine to the left of a, v is in $C^2(R \setminus \{a\})$ but not in $C^2(\mathbf{R})$. Now differentiating $w - bw' - f = 0$ and inserting $x = \alpha = a$ yields $f'(\alpha) - b'(\alpha) = -1 - b(\alpha)w''(\alpha) \leq -1 = f'(\beta) - b'(\beta)$. Thus $\beta \geq \alpha$. This completes the proof if $b(a) \geq 0$.

Assume now $b(a) < 0$. Then $u(t) \equiv -b(a)$, $t \geq 0$, is a control that is optimal at a since we know $v(a) = f(a) - b(a)$. But this optimal control is in the interior of the control set $[0, \infty)$ and hence we can perform a first variation. Specifically for any bounded control u let $u_\epsilon(t) = -b(a) + \epsilon u(t)$, $t \geq 0$, and let $v^\epsilon(a)$ be the corresponding cost starting from a. Then $v^\epsilon(a) \geq v^0(a)$ for all small real ϵ and hence $(d/d\epsilon)v^\epsilon(a) = 0$ at $\epsilon = 0$. The result of this computation is

$$
0 = \int_0^\infty e^{-t} \left(f'(a) \int_0^t e^{-b'(a)(t-s)} u(s)ds + u(t) \right) dt
$$

which implies

$$
0 = \left(\frac{f'(a)}{1 - b'(a)} + 1 \right) \int_0^\infty e^{-t} u(t)dt;
$$

thus $f'(a) - b'(a) = -1$, $a = \beta$, and $b(\beta) < 0$. Now differentiating $w - bw' - f = 0$ and inserting $x = \beta = a$ yields $(1 + w'(\beta))(1 - b'(\beta)) = w''(\beta)b(\beta) < 0$. Thus $\beta < \alpha$. Now differentiating $v - bv' - f = 0$ at $x = a+$ yields $v''(a+) = 0$. Since $v''(a-) = 0$ we obtain v in $C^2(\mathbf{R})$. □

2. A TWO DIMENSIONAL PROBLEM.

In the previous section we have seen that in the one dimensional situation the property of *smooth fit* can be characterized by the geometry of the drift along the free boundary. In fact, we have shown that v is C^2 at the free boundary iff in the optimal synthesis the integral curves of the zero control flow along the free boundary lead away from the null region (and this depends on the parameters of the problem).

In this section we describe the solution of a prototype problem in two dimensions (for more details, see [5]). We find that in this case both smooth fit and non-C^2 fit coexist (along different pieces of the free boundary). Moreover, the geometric characterization previously described concerning smooth fit along the free boundary is still valid.

Consider the two dimensional linear system in canonical controllable form

$$\dot{x} = y, \qquad x(0) = x \tag{2.1}$$

$$\dot{y} = u, \qquad y(0) = y \tag{2.2}$$

where the control u is a nonnegative measurable function of time. We define

$$v^u(x,y) = \int_0^\infty e^{-t}[u(t) + f(x(t), y(t))]dt, \tag{2.3}$$

where $f(x,y) = x^2 + y^2 + y$. We set

$$v(x,y) = \inf\{v^u(x,y) : u(\cdot) \geq 0\}. \tag{2.4}$$

It turns out that v, the *value function*, is a classical solution of the *free boundary problem*

$$\max(V - yV_x - f, -V_y - 1) = 0, \qquad (x,y) \in \mathbf{R}^2. \tag{2.5}$$

Because of the singular nature of the above variational problem, one expects the optimal control $u(\cdot)$ to be extreme , i.e., to equal zero or infinity, or to be singular. The free boundary – the curve where both terms in (2.5) equal zero – is a connected union of two half lines S_0 and S_1. This free boundary $S_0 \cup S_1$ is the switching curve of the optimal synthesis and it divides the plane into two regions. Below the free boundary lays the jump region, i.e., where the optimal control is impulsive, and above it lays the zero-control region. The jump region is the open set where $v_y = -1$, and the zero-control region is the set where $v = yv_x + f$. In the jump region the optimal control causes an instantaneous jump of the state in the direction of the vector (0,1) sending a state (x,y) to the point of the free boundary directly above it. The optimal control is zero on S_0 and it is singular ($u(x,y) = x + y + 1$) on S_1. In the optimal synthesis the zero-flow along S_0 leads into the null region while along S_1 it leads away from the null region. The curve S_1 has the additional property of being an optimal singular trajectory. The value function turns out to be a classical $C^{1,1}$ solution of the Bellman equation. It is C^2 along S_1 and it fails to be C^2 along S_0. To present the optimal synthesis, we introduce the following subsets of $\mathbf{R}^2$. Let

$$\begin{aligned}
S_0 &= \{(x, -(x+1)/3) : x \leq -1\}, \\
S_1 &= \{(x, \lambda(x+1)) : x > -1\}, \\
N_0 &= \{(x,y) : y \geq 0, y \geq -(x+1)/3\}, \\
N_1 &= \{(x,y) : \lambda(x+1) < y < 0\}, \\
J_0 &= \{(x,y) : x \leq -1, y < -(x+1)/3\}, \\
J_1 &= \{(x,y) : x \geq -1, y < \lambda(x+1)\}, \\
N &= N_0 \cup N_1, \\
J &= J_0 \cup J_1,
\end{aligned}$$

where $\lambda = (1-\sqrt{5})/2$. Then the optimal feedback control and the value function are given as follows. Let $U(x,y) = 0$ on N, $U(x,y) = x+y+1$ on S_1, and $U(x,y) = a(x,y)\delta_0$ on J, where δ_0 is the Dirac impulse at time zero and $a(x,y)$ is the direction and intensity of the impulse with $a(x,y) = (0, -(x+1)/3 - y)$ in J_0 and $a(x,y) = (0, \lambda(x+1) - y)$ in J_1. The value function is $v = v^U$, the cost for the control U. It is given by

$$\begin{aligned}
v(x,y) &= x^2 + 2xy + 3y^2 + y && \text{on } N_0, \\
v(x,y) &= (x^2 + 3y^2 + 2yx + y)(1 - e^{-T}) - [2yT(x+y) + y^2T^2 \\
&\quad + y^2(1+\lambda^2)/(\lambda^2(2\lambda - 1)) + y(2+\lambda)/\lambda - 1]e^{-T}, \\
&\qquad \text{where} \quad T = 1/\lambda - 1/y - x/y, && \text{on } N_1, \\
v(x,y) &= 2(x^2 - x)/3 - y - 1/3 && \text{on } J_0, \\
v(x,y) &= 1 - y - 2(x+1) - (x+1)^2(1+\lambda^2)/(2\lambda - 1) && \text{on } J_1.
\end{aligned}$$

It is simple, although tedious, to check that v is $C^{1,1}$ in $\mathbf{R}^2$, that it is not C^2 along S_0, and that it satisfies the Bellman equation (2.5) in the classical sense. Once this is done, a verification theorem implies that the optimal synthesis has been found.

REFERENCES

1. V. E. Benes, L. A. Shep & H. S. Witsenhausen, *Some Solvable Stochastic Control Problems*, Stochastics **4** (1980) 39-83.
2. J. R. Dorroh & G. Ferreyra, *Optimal Advertising in Exponentially Decaying Markets*, Journal of Optimization Theory and Applications, **79**, 2 (1993) 219-236.
3. ______, *A Multi-State, Multi-Control Problem With Unbounded Controls*, SIAM Journal on Control and Optimization, **32**, 5, (1994) 1322-1331.
4. G. Ferreyra & O. Hijab, *A Simple Free Boundary Problem in R^d*, SIAM Journal on Control and Optimization **32**,2 (1994) 501-515.
5. ______, *Linear-Convex Singular Control in Two Dimensions*, Proc. 33rd. Conference on Decision and Control, Orlando, Florida (1994) to appear.
6. W. H. Fleming & H. M. Soner, *Controlled Markov Processes and Viscosity Solutions*, Applications of Mathematics No 25, Springer-Verlag, 1993.
7. D. H. Jacobson, D. H. Martin, M. Pachter, & T. Geveci, *Extensions of Linear-Quadratic Control Theory*, Lect. Notes in Control Info. Sci. No 27, Springer-Verlag, 1980.
8. J. P. Lehoczky & S. E. Shreve, *Absolutely Continuous and Singular Stochastic Control*, Stochastics **17** (1986) 91-109.
9. M. Pachter, *The Linear-Quadratic Optimal Control Problem with Positive Controllers*, Int. J. Control **32** (1980), 589-608.
10. S. P. Sethi, *Optimal Control of the Vidale-Wolfe Advertising Model*, Operations Research **21** (1973), 998-1013.

DEPARTMENT OF MATHEMATICS, LOUISIANA STATE UNIVERSITY, BATON ROUGE, LA 70803.
E-mail address: mmferr@lsuvax.sncc.lsu.edu

DEPARTMENT OF MATHEMATICS, TEMPLE UNIVERSITY, PHILADELPHIA, PA 19122.
E-mail address: hijab@euclid.math.temple.edu

W E FITZGIBBON AND M E PARROTT

Approximation of strongly damped string equations by strongly damped beam equations

1 Introduction

In this paper we discuss the beginning of a project which examines the validity of approximating the strongly damped string equation

$$\frac{\partial^2 u}{\partial t^2} - \alpha \frac{\partial^3 u}{\partial x^2 \partial t} = \left(\beta + k \int_0^\ell (\partial u / \partial y)^2 \, dy \right) \frac{\partial^2 u}{\partial x^2}, \quad t > 0 \tag{1.1a}$$

$$u(0,t) = u(\ell,t) = 0, \quad t > 0 \tag{1.1b}$$

$$u(x,0) = u_1(x), \quad u_t(x,0) = u_2(x), \quad x \in (0,\ell) \tag{1.1c}$$

by the strongly damped beam equation

$$\frac{\partial^2 u}{\partial t^2} + \epsilon \frac{\partial^4 u}{\partial x^4} - \alpha \frac{\partial^3 u}{\partial x^2 \partial t} = \left(\beta + k \int_0^\ell (\partial u / \partial y)^2 \, dy \right) \frac{\partial^2 u}{\partial x^2}, \quad t > 0 \tag{1.2a}$$

$$u(0,t) = u(\ell,t) = u_{xx}(0,t) = u_{xx}(\ell,t), \quad t > 0 \tag{1.2b}$$

$$u(x,0) = u_1(x), u_t(x,0) = u_2(x), x \in (0,\ell). \tag{1.2c}$$

Equation (1.2a), without the damping term $-\alpha \partial^3 u / \partial x^2 \partial t$, is the model introduced by Woinowsky–Krieger [33] to describe the transverse deflection $u(x,t)$ of an elastic, or extensible, beam of reference, stress-free length ℓ, whose ends are then fixed at $x = 0$ and $x = \ell + \Delta$. Since the beam is constrained to lie on the x-axis, an axial force and bending moment are induced. The coefficients ϵ, β and k have the following physical definitions: $\epsilon = EI/\rho$, $\beta = H/\rho$, $k = EA/2\rho\ell$, where $H = EA\Delta/\ell$, E is Young's modulus, I is the cross-sectional second moment of area, ρ is the density, and A is the cross-sectional area. The constants ϵ, k, and α (the damping coefficient) are positive. It can be seen that if the beam is very thin, then the coefficient ϵ will be small. The constant β is, in general, unrestricted in sign (since H has that property); if β (i.e. H) is positive, it represents a tensile force. The nonlinear term in (1.2a) (i.e. the right hand side of the equation) represents the change in tension of the beam due to its extensibility.

The term $-\alpha\partial^3 u/\partial x^2 \partial t$ represents a strong, structural (internal) type of damping. Another type of damping which is often included in beam and string equations is a weak aerodynamic (external) type of damping, modeled by the term $\delta \partial u/\partial t$ (where δ is usually, but not always, positive). Various other types of viscoelastic and structural damping terms have also been studied in beam and string equations.

The boundary conditions (1.2b) represent hinged, or simply supported, conditions. Another common type of boundary conditions are the clamped conditions

$$u(0,t) = u(\ell,t) = u_x(0,t) = u_x(\ell,t) = 0, \quad t > 0.$$

Other types of boundary conditions, for example, cantilevered boundary conditions, can be important in applications, but present more mathematical difficulties.

The basic questions of existence and uniqueness of solutions of damped or undamped beam equations, and stability of equilibrium solutions of damped beam equations, with various boundary and initial conditions, have been considered by many authors; see, for example, Ball [1], [2], Eisley [11], Reiss and Matkowsky [28], Dickey [7], [8], Holmes and Marsden [20], Pereira [27], De Brito [5], Biler [3], Fitzgibbon [12], and references therein. Most of these stability results for damped beam equations are obtained under the assumption that $\beta \geq -\lambda_1$, where λ_1 is the least positive eigenvalue of the problem

$$\epsilon u_j'''' + \lambda_j u_j'' = 0, \text{ where } \left\| u_j' \right\|^2_{L^2(0,\ell)} = -(\beta + \lambda_j)/k.$$

If $\beta \geq -\lambda_1$, then (cf. Ball [2]) the only equilibrium position is the trivial one. (If $\beta < -\lambda_1$, there exist nontrivial equilibrium positions, the so-called buckled states.) The long-term, global behavior of solutions of damped beam equations for a general β has been considered by Hale [17], Taboada and You [31], and Ševčovič [29], [30].

If $\epsilon = 0$ in equation (1.2a), then the resulting equation (1.1a) is widely considered to model the deflection of a vibrating string. Here, the constants of α, β and k are positive, and have the same physical meaning as described above for the beam equation.

Global existence and uniqueness for the strongly damped string equation has been obtained by Nishihara [23] and Matos and Pereira [22]. In contrast to the beam equation, existence results for the string equation are more sensitive to the presence and type of damping terms included in the equation. For the string equation with weak damping term $\delta u_t\,(\delta > 0)$, it is known that if the initial conditions are sufficiently smooth <u>and</u> sufficiently small, then there exists a unique, global, classical, exponentially decaying solution (cf. [6], [3], [24] and references therein). For the string equation without damping it is known only that if the initial conditions are sufficiently smooth, then there exists a unique local classical solution (cf. [9], [10], [25]).

In the next section we show that solutions of the strongly damped beam equation (1.2a-c) (with $\epsilon,\ \alpha,\ \beta,\ k > 0$) converge on finite time intervals, as $\epsilon \to 0^+$, to solutions of the strongly damped string equation (1.1a-c).

2 Convergence of Solutions

We first formulate equations (1.1a-c) and (1.2a-c) as abstract Cauchy initial value problems in an appropriate function space.

Let $H = L_2(0,\ell)$, with norm $\|\cdot\|$. We define $A : D(A) \subset H \to H$ pointwise for a.e. $x \in (0,\ell)$ by

$$(Au)(x) = -u''(x) \tag{2.1a}$$

with

$$D(A) = \left\{ u \,\middle|\, u \in H^2(0,\ell) \cap H_0^1(0,\ell) \right\}. \tag{2.1b}$$

Here, $H_0^1(0,\ell)$ and $H^2(0,\ell)$ are the usual Hilbert Sobolev spaces. It is well-known that A so defined is a strictly positive definite, self-adjoint operator on H, and positive powers of A, A^γ for $\gamma > 0$, may be computed. By imposing a graph norm, $D(A^\gamma)$ can be made into a Hilbert space H_{A^γ}, and

$$\|u\|_{A^\gamma} = \|A^\gamma u\| \text{ for } u \in D(A^\gamma). \tag{2.2}$$

The damped beam equation (1.2a) can now be written as the abstract second-order evolution equation

$$\ddot{u}(t) + \epsilon A^2 u(t) + \alpha A \dot{u}(t) = -\beta A u(t) - k\left\|A^{1/2}u(t)\right\|^2 A u(t), \quad t > 0 \tag{2.3a}$$

$$u(0) = u_1, \quad \dot{u}(0) = u_2. \tag{2.3b}$$

To convert (2.3a-b) to a first-order evolution equation, we let $X = D(A) \times H$, and define an operator matrix $\widehat{A}_\epsilon$ by

$$\hat{A}_\epsilon = \begin{pmatrix} 0 & -I \\ \epsilon A^2 & \alpha A \end{pmatrix} \tag{2.4a}$$

$$D\left(\hat{A}_\epsilon\right) = D\left(A^2\right) \times D(A) := D_1. \tag{2.4b}$$

Proposition 2.1 *$-\widehat{A}_\epsilon$ is the infinitesimal generator of an analytic semigroup $\left\{\widehat{T}_\epsilon(t) \,|t \geq 0\right\} \in X$. $\left\{\widehat{T}_\epsilon(t)\right\}$ is an analytic semigroup of contractions in $X_\epsilon = H_{\sqrt{\epsilon A}} \times H$.*

Proof. The proof of the first statement follows from results of Chen and Triggiani [4]. A direct calculation shows that $\widehat{A}_\epsilon$ is m-accretive in X_ϵ and hence $-\widehat{A}_\epsilon$ generates a semigroup of contractions in X_ϵ.

If we define a nonlinear operator $F : X \to X$ by

$$F(U_\epsilon) = \begin{pmatrix} 0 \\ -\beta A u - k\left\|A^{1/2}u\right\|^2 A u \end{pmatrix} \tag{2.5}$$

for $U_\epsilon = (u, u_t)^T \in X$, then (2.3a-b) can be written as

$$\frac{dU_\epsilon}{dt} = -\widehat{A}_\epsilon U_\epsilon + F\left(U_\epsilon\right), t > 0 \tag{2.6a}$$

$$U_\epsilon\left(0\right) = U_0 = \begin{pmatrix} u_1 \\ u_2 \end{pmatrix}. \tag{2.6b}$$

The proof of the following theorem follows from the existence results of the various authors listed in the Introduction; the regularity follows, in particular, from [20, Proposition 2.6].

Theorem 2.2 *If* $U_0 = (u_1, u_2)^T \in D_1$ *then there exists a strong, continuously differentiable solution to* (2.6a-b) *on* $[0, \infty)$ *which has variation of parameters representation*

$$U_\epsilon\left(t\right) = \widehat{T}_\epsilon\left(t\right) U_0 + \int_0^t \widehat{T}_\epsilon\left(t-s\right) F\left(U_\epsilon\left(s\right)\right) ds.$$

Moreover, if $n \in Z^+$ *and* $U_0 \in D\left(\widehat{A}_\epsilon^n\right)$, *then* $U_\epsilon\left(t\right) \in D\left(\widehat{A}_\epsilon^n\right)$.

We note that, while $\left\{\widehat{T}_\epsilon\left(t\right) \middle| t \geq 0\right\}$ is an analytic semigroup in X (cf. Proposition 2.1), we can no longer claim that it is a contraction semigroup in X. In fact, due to the singularity imposed by the factor ϵ, one would expect the norm of $\widehat{T}_\epsilon\left(t\right)$ in X to blow up as $\epsilon \downarrow 0$. For this reason, we choose to convert the second-order evolution equation (2.3a-b) into a first-order evolution equation in an alternate manner. We define an operator matrix $\widetilde{A}$ by

$$\widetilde{A} = \begin{pmatrix} 0 & -I \\ \beta A & \alpha A \end{pmatrix} \tag{2.7a}$$

$$D\left(\widetilde{A}\right) = D\left(A\right) \times D\left(A\right). \tag{2.7b}$$

Using results of Webb [32, Prop. 2.2], one can show that $-\widetilde{A}$ is the infinitesimal generator of an analytic semigroup of contractions $\{T\left(t\right) | \, t \geq 0\}$ on X. By defining a new nonlinearity F_ϵ by

$$F_\epsilon \begin{pmatrix} u \\ u_t \end{pmatrix} = \begin{pmatrix} 0 \\ -k\left\|A^{1/2}u\right\|^2 Au - \epsilon A^2 u \end{pmatrix}, \tag{2.8}$$

we can merely regroup terms to write (2.3a-b) as:

$$\frac{dU_\epsilon(t)}{dt} = -\widetilde{A}U_\epsilon(t) + F_\epsilon(U_\epsilon(t)), t > 0 \tag{2.9a}$$

$$U_\epsilon(0) = U_0 = \begin{pmatrix} u_1 \\ u_2 \end{pmatrix} \tag{2.9b}$$

for $U_\epsilon = (u, u_t)^T \in X$. By applying the theory of inhomogeneous Cauchy initial value problems (cf. [26]), we have the following:

Proposition 2.3 *If* $U_0 \in D_1$ *then the strong solution of* (2.6a-b) *may be represented as*

$$U_\epsilon(t) = T(t)U_0 + \int_0^t T(t-s)F_\epsilon(U_\epsilon(s))\,ds.$$

With the operator A as defined by (2.1a-b), the damped string equation (1.1a-c) may be written as the second-order evolution equation

$$\ddot{u}(t) + \alpha A\dot{u}(t) = -\beta Au(t) - k\left\|A^{1/2}u(t)\right\|^2 Au(t), \quad t > 0 \tag{2.10a}$$

$$u(0) = u_1, \quad \dot{u}(0) = u_2. \tag{2.10b}$$

With the operator matrix $\widetilde{A}$ defined by (2.7a-b), and a nonlinear operator F_0 defined by

$$F_0\begin{pmatrix} u \\ u_t \end{pmatrix} = \begin{pmatrix} 0 \\ -k\left\|A^{1/2}u\right\|^2 Au \end{pmatrix}, \tag{2.11}$$

(2.10a-b) may be written as the first-order evolution equation

$$\frac{dU(t)}{dt} = -\widetilde{A}U(t) + F_0(U(t)), \quad t > 0 \tag{2.12a}$$

$$U(0) = U_0 = \begin{pmatrix} u_1 \\ u_2 \end{pmatrix}. \tag{2.12b}$$

By the results of Matos and Pereira [22], we have

Proposition 2.4 *If $U_0 = (u_1, u_2)^T \in D\left(A^{1/2}\right) \times H$, then the strong solution to* (1.1a-c) *(respectively* (2.12a-b)) *exists on* $[0, \infty)$*, and has abstract variation of parameters representation*

$$U(t) = T(t) U_0 + \int_0^t T(t-s) F(U(s)) \, ds.$$

The key to showing convergence of solutions of (1.2a-c) to solutions of (1.1a-c) lies in establishing a priori bounds for solutions of (1.2a-c). We state the following two propositions, whose proofs are given in [13].

Proposition 2.5 *If $(u_1, u_2)^T \in D_1$ then there exists a positive constant M_1, which does not depend on ϵ, so that*

$$\sup_{t>0} \left\{ \|\dot{u}(t)\|^2, \left\|A^{1/2} u(t)\right\|^2, \epsilon \|Au(t)\|^2 \right\} \le M_1.$$

Proposition 2.6 *If $n \in Z^+$, $T > 0$ and $(u_1, u_2)^T \in D(A^{2n}) \times D(A^n)$, then there exists a positive constant $M_n(T)$, which does not depend on ϵ, so that*

$$\sup_{t \in [0,T]} \left\{ \left\|A^{n/2} \dot{u}(t)\right\|^2, \left\|A^{(n+1)/2} u(t)\right\|^2, \epsilon \left\|A^{(n+2)/2} u(t)\right\|^2 \right\} \le M_n(T).$$

Using an Arzela-Ascoli argument, along with the bounds of Propositions 2.5 and 2.6 and results of Kato [21], we obtain our main convergence result, whose proof is given in [13]. In the theorem below, $\|\cdot\|_\infty$ denotes the supremum norm.

Theorem 2.7 *If $U_0 = (u_1, u_2)^T \in D(A^4) \times D(A^2)$ and $T > 0$, then*

$$\lim_{\epsilon \to o^+} \left(\sup_{t \in [0,T]} \|u_\epsilon(\cdot, t) - u(\cdot, t)\|_\infty \right) = 0,$$

where u_ϵ and u are strong solutions of (1.2a-c) *and* (1.1a-c) *respectively.*

3 Future Research

The results contained herein may be viewed as the initiation of a larger project which examines the relationship between the dynamics of solutions of (1.2a-c) and (1.1a-c). We shall apply techniques of Taboada and You [31], and [14], to establish the existence of global attractors A_ϵ

and A corresponding to both (1.2a-c) and (1.1a-c) respectively. The next question becomes: Does A_ϵ converge (in some appropriate sense) to A as $\epsilon \downarrow 0$? If so, then the validity of viewing the strongly damped beam equation as an approximation of the strongly damped string equation is further verified. In seeking to answer this question, we are motivated by results of Hale and Raugel [18], [19], and our recent work on global attractors for singularly perturbed Hodgkin-Huxley equations [15], [16].

Another project of interest involves the convergence of solutions of weakly damped beam equations to weakly damped string equations; that is, the term $\delta\, \partial u/\partial t$, $\delta > 0$, replaces the term $-\alpha\, \partial^3 u/\partial x^2 \partial t$ in equations (1.2a) and (1.1a). A result of this type could possibly improve known existence results for weakly damped string equations.

References

[1] J. Ball, Initial-boundary value problems for an extensible beam, J. Math. Anal. Appl. 42 (1973), 61-90.

[2] J. Ball, Stability theory for an extensible beam, J. Diff. Equations 14 (1973), 399-418.

[3] P. Biler, Remark on the decay for damped string and beam equations, Nonlinear Anal., TMA 10 (1986), 839-842.

[4] S. Chen and R. Triggiani, Proof of extensions of two conjectures on structural damping for elastic systems, Pacific J. Math. 136 (1989), 15-55.

[5] E. De Brito, Decay estimates for the generalized damped extensible string and beam equations, Nonlinear Anal., TMA 8 (1984), 1489-1496.

[6] E. De Brito, The damped elastic stretched string equation generalized: Existence, uniqueness, regularity and stability, Applic. Anal. 13 (1982), 219-233.

[7] R. Dickey, Free vibrations and dynamic buckling of the extensible beam, J. Math. Anal. Appl. 29 (1970), 443-454.

[8] R. Dickey, Dynamic stability of equilibrium states of the extensible beam, Proc. Amer. Math. Soc. 41 (1973), 94-102.

[9] R. Dickey, Infinite systems of nonlinear oscillations related to the string, Proc. Amer. Math. Soc. 23 (1969), 459-468.

[10] J. Dix and R. Torrejón, A quasilinear integrodifferential equation of hyperbolic type, Diff. and Int. Eqns. 6 (1993), 431-447.

[11] J. Eisley, Nonlinear vibrations of beams and rectangular plates, Z. Angew. Math. Phys. 15 (1964), 167-175.

[12] W. Fitzgibbon, Strongly damped quasilinear evolution equations, J. Math. Anal. Appl. 79 (1981), 536-550.

[13] W. Fitzgibbon and M. Parrott, Convergence of singular perturbations of strongly damped nonlinear wave equations, submitted.

[14] W. Fitzgibbon, M. Parrott and Y. You, Global dynamics of coupled system modeling nonplanar beam motion, in Evolution Equations, Marcel Dekker, New York, 1994, to appear.

[15] W. Fitzgibbon, M. Parrott and Y. You, Global dynamics of singularly perturbed Hodgkin-Huxley equations, in Semigroups of Linear and Nonlinear Operations and Applications, G. Goldstein and J. Goldstein (eds.), Kluwer Academic Publ., 1993, 154-176.

[16] W. Fitzgibbon, M. Parrott and Y. You, Finite dimensionality and upper semicontinuity of the attractor of singularly perturbed Hodgkin-Huxley systems, preprint.

[17] J. Hale, Asymptotic Behavior of Dissipative Systems, Amer. Math. Soc., Providence, RI, 1988.

[18] J. Hale and G. Raugel, Upper semicontinuity of the attractor for a singularly perturbed hyperbolic equation, J. Diff. Equations 73 (1988), 197-214.

[19] J. Hale and G. Raugel, Lower semicontinuity of the attractor for a singularly perturbed hyperbolic equation, J. Dynamics and Diff. Eqns. 2 (1990), 19-67.

[20] P. Holmes and J. Marsden, Bifurcation to divergence and flutter in flow-induced oscillations: An infinite dimensional analysis, Automatica 14 (1978), 367-384.

[21] T. Kato, Accretive operators and nonlinear evolution equations, in Proc. Symposium in Pure Mathematics (ed. F. Browder), Amer. Math. Soc., Providence, RI, 1970, 162-169.

[22] M. Matos and D. Pereira, On a hyperbolic equation with strong damping, Funkcial. Ekvac. 34 (1991), 303-311.

[23] K. Nishihara, Degenerate quasilinear hyperbolic equation with strong damping, Funkcial. Ekvac. 27 (1984), 125-145.

[24] K. Nishihara and Y. Yamada, On global solutions of degenerate quasilinear hyperbolic equations with dissipative terms, Funkcial. Ekvac. 33 (1990), 151-159.

[25] D. Oplinger, Frequency response of a nonlinear stretched string, J. Acoust. Soc. Amer. 32 (1960), 1529-1538.

[26] A. Pazy, Semigroups of Linear Operators, Applied Math. Sc. Series 44, Springer-Verlag, Berlin, 1983.

[27] D. Pereira, Existence, uniqueness and asymptotic behavior for solutions of the nonlinear beam equation, Nonlinear Anal., TMA 14 (1990), 613-623.

[28] E. Reiss and B. Matkowsky, Nonlinear dynamic buckling of a compressed elastic column, Quart. Appl. Math. 29 (1971), 245-260.

[29] D. Ševčovič, Existence and limiting behavior for damped nonlinear evolution equations with nonlocal terms, Comment. Math. Univ. Carolinae 31 (1990). 283-293.

[30] D. Ševčovič, Limiting behavior of global attractors for singularly perturbed beam equations with strong damping, Comment. Math. Univ. Carolinae 32 (1991), 45-60.

[31] M. Taboada and Y. You, Global attractor and inertial manifolds of nonlinear damped beam equations, Comm. Partial Diff. Eqns., to appear.

[32] G. Webb, Existence and asymptotic behavior for a strongly damped nonlinear wave equation, Can. J. Math. XXXII (1980), 631-643.

[33] S. Woinowsky-Krieger, The effect of axial force on the vibration of hinged bars, J. Appl. Mech. 17 (1950), 35-36.

W.E. Fitzgibbon
University of Houston
Houston, TX 77204

M.E. Parrott
University of South Florida
Tampa, FL 33620

G R GOLDSTEIN, J A GOLDSTEIN AND S OHARU

The Favard class for a nonlinear parabolic problem

ABSTRACT

The parabolic partial differential equation

$$\partial u/\partial t = \varphi(x,\ \partial u/\partial x)\partial^2 u/\partial x^2$$

is considered for $0 \leq x \leq 1$ and $t \geq 0$. A variety of boundary conditions $x = 0, 1$ are allowed. The function $\varphi = \varphi(x, \xi)$ is allowed to vanish at $x = 0, 1$. The problem (under some assumptions on φ and the boundary conditions) is governed by an m-dissipative operator A and a corresponding contraction semigroup $\{T(t) : t \geq 0\}$ on $C[0, 1]$. The Favard class $\widehat{\mathfrak{D}}(A)$ is explicitly calculated. It follows that if the initial data for u at $t = 0$ is in $\mathfrak{D}(A)$ (or $\widehat{\mathfrak{D}}(A)$), then $u(t)$ is in $\widehat{\mathfrak{D}}(A)$ for all $t > 0$. This implies some spatial regularity since $\widehat{\mathfrak{D}}(A) \subset W^{2,1}(0, 1)$.

§1. Introduction

We are interested in questions of spatial regularity. Consider an autonomous partial differential equation which can be written in the form

$$du(t)/dt = A(u(t)) \tag{1.1}$$

in some Banach space X; A maps its domain $\mathfrak{D}(A) \subset X$ to X. In many situations, (1.1) is governed by a strongly continuous semigroup $T = \{T(t) : t \geq 0\}$ acting on $D = \overline{\mathfrak{D}\ (A)}$. Thus $u(t) = T(t)u_0$ is the unique (mild) solution corresponding to the initial condition $u(0) = u_0$. To illustrate the following ideas, let us first consider the case $X = L^p(\Omega)$ (where $\Omega \subset \mathbb{R}^n$) and $D = X$.

An *invariant set* is a set $D_1 \subset D$ such that $T(t)(D_1) \subset D_1$ for all $t \geq 0$. (Maybe $D_1 = \mathfrak{D}(A)$, maybe not.) We wish to focus on the fact that D_1 contains information on spatial regularity. For example if we could show $W_0^{k,p}(\Omega) \subset D_1 \subset W^{k,p}(\Omega)$, then the statement that D_1 is an invariant set implies that the solution $u(t)$ has spatial derivatives up to order k in $L^p(\Omega)$, for all $t \geq 0$, whenever the initial data $u_0 \in$

D_1. (A precise description of D_1 requires consideration of the boundary conditions associated with A.)

For simplicity suppose that A (which may be multivalued) is m-dissipative on X. Thus for all $\lambda > 0$ and $f \in X$, there is a $u \in \mathfrak{D}(A)$ such that $u - \lambda Au \ni f$. Moreover, if $u_i - \lambda Au_i \ni f_i$, $i = 1, 2$, then $\|u_1 - u_2\| \leq \|f_1 - f_2\|$. Phrased differently, the statement that A is m-dissipative is equivalent to the statements

$$\mathscr{R}(I - \lambda A) = X \text{ and } \|(I - \lambda A)^{-1}\|_{\mathrm{Lip}} \leq 1, \text{ for all } \lambda > 0.$$

Then (by the Crandall-Liggett theorem [7], [1], [2], [13]), A determines semigroup T by the formula

$$T(t)f = \lim_{n\to\infty}(I - \frac{t}{n}A)^{-n}f \quad t \geq 0, \quad f \in D = \overline{\mathfrak{D}(A)}.$$

This semigroup gives the unique mild solution of (1.1) provided that $u(0) = f$ is specified and $u(t)$ is defined to be $u(t) = T(t)f$.

A natural candidate for an invariant set D_1 is $\mathfrak{D}\ (A)$. In the linear case, where A is single valued and $\mathfrak{D}\ (A)$ is dense, $\mathfrak{D}\ (A)$ is always invariant (see e.g. [4], [12]). But this is false in general in the nonlinear case (see e.g. [7]). (It *is* true when X is reflexive.)

An important example of a nonlinear operator on a nonreflexive space is a (single) conservation law

$$\partial u/\partial t + \partial(\varphi(u))/\partial x = 0 \tag{1.2}$$

where $\varphi \in C^2(\mathbb{R})$, $\varphi'' > 0$. This is governed by an m-dissipative operator on $L^1(\mathbb{R})$ by Crandall [5]. Let the initial data f be in $C_c^\infty(\mathbb{R})$. Then by the classical theory (of Lax, Oleinik, ...), a shock will develop and for some $t_0 > 0$, $u(t_0) = u(t_0, \cdot)$ will have (at least) one downward jump as a function of $x \in \mathbb{R}$. An useful lead in finding an invariant set for (1.2) may be obtained by using a result of Burch [3]. Let $F \in C^2(\mathbb{R}^N)$, $F(0) = 0$, and suppose

$$\sum_{i,j=1}^{N} \frac{\partial^2 F(x)}{\partial x_i \partial x_i}\xi_i\xi_j \geq 0 \qquad (\text{ and } \not\equiv 0)$$

holds for all $x, \xi \in \mathbb{R}^N$. Consider the Hamilton - Jacobi equation

$$u_t + F(\nabla_x u) = 0, \qquad u(0) = g \tag{1.3}$$

in $X = BUC(\mathbb{R}^N)$, the bounded uniformly continuous real functions on $\mathbb{R}^N$. This is of the form (1.1) with A defined by $Au = -F(\nabla_x u))$ m-dissipative on X (by [3]). For $k > 0$ define

$$\begin{aligned} B_k = \{u \in X : \text{ for all } x, y \in \mathbb{R}^N, \\ |u(x)| \leq k, \quad |u(x) - u(y)| \leq k|x-y|, \\ u(x+y) + u(x-y) - 2u(x) \leq k|y|^2\}. \end{aligned}$$

Thus $u \in B_k$ implies that u satisfies a semiconcavity condition, incorporating a one sided bound on the second order difference quotient. Burch proved that

$$T(t)(B_k) \subset B_k \text{ for all } t, k > 0.$$

Here T is the semigroup determined by A via the Crandall-Liggett theorem. For $N = 1$ and u the solution of (1.3), $v = u_x$ satisfies (1.2) with $\varphi = F$ and $v(0, x) = g'(x) = f(x)$. If $f \in C_c^\infty(\mathbb{R})$, then $\int^x f = g \in B_k$ for some $k > 0$. Hence, by Burch's result, $u(t) \in B_k$ for all $t > 0$. Thus, for all $t > 0$, $v(t, \cdot)$, the solution of the conservation law (1.2), remains bounded (in x) with no discontinuities other than downward jumps.

The preceding discussion leads naturally to the Favard class. Again, let A be m-dissipative on X, so that A determines a contraction semigroup T on $D = \overline{\mathfrak{D}(A)}$ (as before). The Yosida approximation A_λ of A is defined to be $A_\lambda = \lambda^{-1}(I - (I - \lambda A)^{-1})$ for $\lambda > 0$. (See, e.g. [1], [2], [13].) The *Favard Class* (or *generalized domain*), $\widehat{\mathfrak{D}}(A)$, of A, was introduced by U. Westphal [17] and later independently by M. Crandall [6]. There are three equivalent ways to define it, namely

$$\begin{aligned} \widehat{\mathfrak{D}}(A) &= \{f \in \overline{\mathfrak{D}(A)} : \overline{\lim_{\lambda \downarrow 0}} \|A_\lambda f\| < \infty\} \\ &= \left\{f \in \overline{\mathfrak{D}(A)} : \|T(t)f - f\| \leq M_f t \text{ for some } M_f > 0 \text{ and } 0 < t < 1\right\} \\ &= \{f \in \overline{\mathfrak{D}(A)} : \text{ for some sequence } f_n \in \mathfrak{D}(A) \text{ and } g_n \in Af, \ g_n \to f \\ &\qquad \text{and } Af_n \text{ is bounded (as } n \to \infty)\}. \end{aligned}$$

The theory of the Favard class is summarized by the following three facts:

i) $\mathfrak{D}(A) \subset \widehat{\mathfrak{D}}(A) \subset \overline{\mathfrak{D}(A)}$,

ii) $\widehat{\mathfrak{D}}(A) = \mathfrak{D}(A)$ if X is reflexive,

ii) $T(t)(\widehat{\mathfrak{D}}(A)) \subset \widehat{\mathfrak{D}}(A)$ for each $t > 0$.

Thus $\widehat{\mathfrak{D}}(A)$ is the invariant set which can be used to explain relevant spatial regularity of a problem. The problem with this method is that $\widehat{\mathfrak{D}}(A)$ is very difficult to compute explicitly. Of course, $\widehat{\mathfrak{D}}(A)$ gives less regularity than $\mathfrak{D}$ (A) does, but it gives useful information whenever one can calculate it.

We illustrate these concepts in the simple case $A = d/dx$, $(T(t)f)(x) = f(x + t))$, and $X = BUC(\mathbb{R})$. Then

$$\begin{aligned} \mathfrak{D}(A) &= \{u \in X : u \in C^1(\mathbb{R}),\ u' \in X\}, \\ \widehat{\mathfrak{D}}(A) &= \{u \in X : \text{ for some } M > 0, \\ |u(x) - u(y)| &\leq M|x - y| \text{ for all } x, y \in \mathbb{R}\} \\ &= \text{Lip } (\mathbb{R}) \cap BUC(\mathbb{R}). \end{aligned}$$

In the case of the single conservation law (1.2), $\widehat{\mathfrak{D}}(A)$ is not known. Similarly $\widehat{\mathfrak{D}}(A)$ is now known for the Hamilton-Jacobi equation (1.3), although it is conjectured that $\widehat{\mathfrak{D}}(A) = \bigcup_{k>0} B_k$ is this case. From Burch's results, $\widehat{\mathfrak{D}}(A) \supset \bigcup_{k>0} B_k$ follows.

Our purpose in this paper is to characterize $\widehat{\mathfrak{D}}(A)$ precisely in a nontrivial case. This will be the first such result in the literature. Of concern is a nonlinear parabolic equation with degeneracy at the spatial boundary. A variety of boundary conditions is allowed.

The main result is Theorem 1. It is stated in Section 2 and proved in Section 3. Section 4 contains some remarks and extensions.

§1. A Nonlinear Parabolic Problem

The equation of concern is

$$\partial u/\partial t = \varphi(x, \partial u/\partial x)\partial^2 u/\partial x^2$$

for $t \geq 0$ and $0 \leq x \leq 1$; and initial condition and various boundary conditions will be imposed at $x = j$ for $j = 0, 1$. Let X be the real space $C[0,1]$. The Dirichlet and nonlinear boundary conditions at $j = 0, 1$ are

$$(BC_j^D) \qquad u(j) = 0,$$

$$(BC_j) \qquad (-1)^j u'(j) \in \beta_j(u(j)).$$

Here β_j is a strictly increasing maximal monotone graph in $\mathbb{R}^2$ containing the origin. Thus $0 \in \beta_j(0)$, and $y_i \in \beta_j(x_i)$, $i = 1, 2$, and $x_1 < x_2$ implies $y_1 < y_2$.

Let Y_{BC} be $C^1[0,1]$ [resp. X; $C^1(0,1] \cap X$; $C^1[0,1) \cap X$] when the boundary conditions are (BC_0), (BC_1) [resp. $(BC_0^D), (BC_1^D)$; $(BC_0^D), (BC_1)$; (BC_0), (BC_1^D)]. Y_{BC} will be incorporated into the definition of the domain of A.

Let $\varphi \in C([0,1] \times \mathbb{R})$ satisfy $\varphi(x,\xi) > 0$ for all $(x,\xi) \in (0,1) \times \mathbb{R}$, $\varphi(x,\xi) \geq \varphi_0(x)$ for all $(x,\xi) \in (0,1) \times \mathbb{R}$. Here $\varphi_0 \in C[0,1]$, $\varphi_0(x) \geq 0$ for $x \in [0,1]$ and $\frac{1}{\varphi_0} \in L^1(0,1)$. Define

$$Au(x) = \varphi(x, u'(x))u''(x)$$

for u in $\mathfrak{D}(A) = \{v \in C^2(0,1) \cap Y_{BC} : u$ satisfies the boundary conditions (one specified at $x = 0$, one at $x = 1$) and $Au \in X\}$.

Theorem 1: *A is m-dissipative on X.*

This result is due to J. A. Goldstein and C.-Y. Lin [16] in the special case when $\varphi_0 > 0$ on $(0,1)$ and $\frac{1}{\varphi_0} \in L^2(0,1)$. G.R. Goldstein [10] extended [16] to the present case (and in various other directions as well).

Before stating our main result we introduce one additional hypothesis.

[H] *There exist $\varepsilon_0 > 0, C_0 > 0$ such that*

$$\varphi(x, \xi + a) \leq C_0 \varphi(x, \xi)$$

for all $x \in [0,1]$ and all $\xi \in \mathbb{R}$ and all a with $|a| < \varepsilon_0$.

This hypothesis is not very restrictive. A special case in which [H] holds is

$$\varphi(x,\xi) = \varphi_0(x)\psi(x,\xi)$$

where φ_0 vanishes only at $0,1$ and ψ is bounded away from 0. Another example is

$$\varphi(x,\xi) = [x(1-x)]^\alpha\{\left|x-\frac{1}{2}\right|^{2\beta} + (\xi^2+1)^{-1}\}^{\frac{1}{2}}$$

for $0 < \alpha, \beta < 1$ and $(x,\xi) \in [0,1] \times \mathbb{R}$. Here

$$\varphi_0(x) = [x(1-x)]^\alpha|x-\frac{1}{2}|^\beta.$$

Noting that

$$\frac{1}{5}\varphi(x,\xi) \le \varphi(x,\xi+a) \le 5\varphi(x,\xi)$$

for $|a| < 1$ and all $(x,\xi) \in [0,1] \times \mathbb{R}$, we see that [H] holds with $C_0 = 5$ and $\varepsilon_0 = 1$. In this case φ_0 necessarily vanishes at an interior point.

The following theorem is the main result. (But see also Section 4.)

Theorem 2. *Define A, $\mathfrak{D}(A)$ as above and suppose that [H] holds. Then the Favard class of A is*

$$\widehat{\mathfrak{D}}(A) = \{u \in C^1(0,1) \cap Y_{BC} : u' \in AC[0,1],\ \varphi(x,u')u'' \in L^\infty(0,1)$$
$$\text{and } u \text{ satisfies the boundary conditions associated with } A\}.$$

§3. Proof of Theorem 2

Let Z be the set defined in the statement of the theorem. We must show

i) $\widehat{\mathfrak{D}}(A) \subset Z$,

ii) $\widehat{\mathfrak{D}}(A) \supset Z$.

Proof of i). Let $u \in \widehat{\mathfrak{D}}(A)$. Choose $v_n \in \mathfrak{D}(A)$ such that $\|v_n - u\| \to 0$ and $\|Av_n\| \le M_1$, for some $M_1 > 0$ and all $n = 1,2,3,\ldots$. Here $\|\cdot\|$ is the norm in X, i.e. the supremum norm.

Let $M_0 = \|\varphi_0^{-1}\|_1$ (the L^1 norm). Then

$$\|v_n''\| \le \|\frac{\varphi(x, v_n')v_n''}{\varphi_0}\|_1 \le M_0 M_1$$

for all $n \ge 1$. Using the estimate

$$\|v_n'\| \le 4\|v_n\| + \|v_n''\|_1$$

(See Dorroh and Rieder [8]), it follows that $\{v_n\}$ is a bounded sequence in the Sobolev space $W^{2,1}(0,1)$. Moreover, an oscillation argument gives a relative compactness result for the sequence $\{v_n\}$.

Define two moduli of continuity, ω_L, ω_C by

$$\omega_L(f;\delta) = \sup\{\int_E |f(x)|dx : E \text{ is a subinterval of } [0,1), \quad |E| = \int_E dx < \delta\};$$
$$\omega_C(f,\delta) = \sup\{|f(x) - f(y)| : x, y \in [0,1), |x-y| \le \delta\}.$$

Let $\{f_n\}$ be a sequence of integrable functions on $[0,1]$. Notice that the statement that for each $\varepsilon > 0$ there is a $\delta > 0$ such that $\omega_L(f_n, \delta) < \varepsilon$ (resp. $\omega_C(f_n) < \varepsilon$) for all n is equivalent to the statement that $\{f_n\}$ is uniformly integrable (resp. $\{f_n\}$ is equicontinuous) on $[0,1]$. Dorroh and Rieder [8] proved that for $0 < \delta \le \frac{1}{2}$ and $f \in Y_{BC} \cap C^2(0,1)$ with $f'' \in L^1(0,1)$,

$$\|f'\| \le \frac{2}{\delta}\|f\| + \omega_L(f'', \delta). \tag{3.1}$$

It follows that, given $\varepsilon > 0$, we have

$$\omega_L(v_n'', \delta) \le M_1 \omega_L(\varphi_0^{-1}, \delta) < \varepsilon \tag{3.2}$$

for some $\delta = \delta(\varepsilon) > 0$ sufficently small since φ_0^{-1} is integrable an $[0,1]$. Moreover,

$$\omega_C(v_n', \delta) \le \omega_L(v_n'', \delta) \le \|v_n''\|_1 \le M_0 M_1$$

for all $\delta \in (0, \frac{1}{2}]$ and

$$\omega_C(v_n', \delta) < \varepsilon \tag{3.3}$$

for $\delta = \delta(\varepsilon) > 0$ sufficiently small by (3.2). Hence $\{v_n'\}$ is a pointwise bounded equicontinuous sequence in $C[0,1]$, so by the Arzela-Ascoli theorem, there is a subsequence, which we again denote by $\{v_n'\}$, which converges uniformly to a continuous function w. Using the boundedness of $\{Av_n\}$ it follows that (at least for some subsequence)

$$v_n \to u \text{ in } X,$$

$$v_n' \to u' \text{ in } X,$$

$$v_n'' \to u'' \text{ a.e.},$$

$$\omega_L(u'', \delta_1) \leq M_0 M_1 \text{ for } 0 < \delta_1 \leq \frac{1}{2},$$

and

$$\omega_L(u'', \delta) < \varepsilon$$

for some δ sufficiently small. It also follows that

$$\|\varphi(x, u')u''\|_\infty \leq M_1 < \infty.$$

It remains to show that u satisfies the boundary conditions. For the (BC_j^D) case, this follows from the uniform convergence of v_n to u. In the case (BC_j), we have

$$(-1)^j v_n'(j) \in \beta_j(v_n(j)),$$

and $v_n(j) \to u(j)$, $\quad v_n'(j) \to u'(j)$. Hence

$$(-1)^j u'(j) \in \beta(u(j))$$

since the graph β_j is closed. This completes the proof of i).

We note that hypothesis [H] was not needed in the proof of i). However it is required for ii).

Proof of ii). Let $u \in Z$. By standard density results of real analysis, one may choose f_n in $C[0,1]$ such that

a) $f_n \to u''$ a.e. and in $L^1(0,1)$,

b) $|f_n(x)| \leq 2|u''(x)| + 1$ a.e. for all $n \geq 1$,

c) $\sup \|\varphi(x, u_n')u_n''\|_\infty \leq M < \infty$.

To see this, define

$$u_n(x) = a_n + b_n x + \int_0^x \int_0^y f_n(s)dsdy; \tag{3.4}$$

then

$$u_n' = b_n + \int_0^x f_n(s)ds, \text{ for all } x \in (0,1),$$

$$u_n''(x) = f_n(x) \quad \text{a.e.}$$

There are uniquely determined $a, b \in \mathbb{R}$ such that

$$u(x) = a + bx + \int_0^x \int_0^y u''(s)dsdy.$$

(Namely, $a = u(0)$, $\quad b = u'(0)$.) In the definition of f_n above, choose a_n, b_n in (3.4) so that

$$a_n \to a, \; b_n \to b$$

as $n \to \infty$. Part c) (i.e.

$$\sup_n \|\varphi(x, u_n')u_n''\|_\infty < \infty)$$

follows from hypothesis [H] together with the assumption that $u \in Z$.

To show that $u \in \widehat{\mathfrak{D}}(A)$ it remains to check that the boundary conditions hold so that $f_n \in \mathfrak{D}(A)$ for all n.

Case 1: $\quad (BC_0), \; (BC_1)$.

Under these nonlinear boundary condtions, $b \in \beta_0(a)$ must hold as must

$$-(b + c) \in \beta_1(a + b + d),$$

where

$$c = \int_0^1 u''(s)ds, \quad d = \int_0^x \int_0^y u''(s)dsdy.$$

We want

$$b_n \in \beta_0(a_n), \quad -(b_n + c_n) \in \beta_1(a_n + b_n + d_n) \tag{3.5}$$

to hold for all n, where

$$c_n = \int_0^1 f_n(s)ds, \quad d_n = \int_0^x \int_0^y f_n(s)dsdy. \tag{3.6}$$

If both boundary conditions (BC_j) holds, then $-c_n \in \gamma_n(a_n)$ where

$$\gamma_n(s) := \beta_1((I+\beta_0)(s)+d_n)+\beta_0(s).$$

Let also

$$\gamma(s) := \beta_1((I+\beta_0)(s)+d)+\beta_0(s).$$

By the strict monotonicity of β_0 and β_1 the maximal monotone graphs γ_n, γ have the property that Range $(\gamma) =$ Range$(\gamma_n) = J$, where J is the open interval

$$J = (\inf_{\mathbb{R}} \beta_1 + \inf_{\mathbb{R}} \beta_0, \quad \sup_{\mathbb{R}} \beta_1 + \sup_{\mathbb{R}} \beta_0).$$

Note that J is independent of n. Since $c \in J$ and $c_n \to c$, it follows that $c_n \in J$ for sufficiently large n. For such n, a_n exists and is uniquely determined by the strict monotonicity of β_0 and β_1. Next for these n there is a $b_n \in \beta_0(a_n)$ such that (see (3.5))

$$-c_n \in \beta_1(a_n+b_n+d_n)+b_n.$$

This gives the construction of the desired pair (a_n, b_n) for n large enough. As noted above, a_n is uniquely determined. Also, b_n is uniquely determined provided β_1 is single-valued. But even if this is not the case, from $b_n \in \beta_0(a_n)$ and $a_n \to a, \quad b \in \beta_0(a)$, we may choose $b_n \in \beta_0(a_n)$ for large enough n such that (3.5)and (3.6) hold, and $b_n \to b$ as $n \to \infty$.

Case 2: $(BC_0^D), (BC_1^D)$.

Since $u_n(0) = a_n, \; u_n(1) = a_n + b_n + d_n$ must hold, simply take $a_n = 0 = a, \quad b_n = -d_n \to -d = b$.

Case 3: $(BC_0^D), (BC_1)$.

Case 4: $(BC_0), (BC_1^D)$.

These are handled in a similar manner. Actually, Case 1 is the hardest to prove.

This completes the proof. ∎

§4. Concluding Remarks and Further Results

For J an interval, let $C_0(J)$ be the closure of the space of continuous functions which are compactly supported in J. Then

$$\begin{aligned} C_0([0,1]) &= C[0,1], \\ C_0(0,1] &= \{u \in C[0,1] : u(0) = 0\}, \\ C_0[0,1) &= \{u \in C[0,1] : u(1) = 0\}, \\ C_0(0,1) &= \{u \in C[0,1] : u(0) = u(1) = 0\}. \end{aligned}$$

The closure D of $\mathfrak{D}(A)$ is $C[0,1]$ [resp. $C_0(0,1]; C_0[0,1); C_0(0,1)$] when the boundary conditions are $(BC_0), (BC_1)$ [resp. $(BC_0^D), (BC_1); (BC_0), (BC_1^D); (BC_0^D), (BC_1^D)$].

It is of course of great interest to extend the idea of this paper to more general contraction semigroups. A specific candidate for extension is the n-dimensional extension of the one dimensional operator considered in this paper.

Now let $Y = L^\infty(0,1)$ and define

$$(\widehat{A}u)(x) = \varphi(x, u'(x))u''(x) \quad (x \in (0,1))$$

for $u \in \mathfrak{D}(\widehat{A}) := \widehat{\mathfrak{D}}\,(A) = \{u \in C^1(0,1) \cap Y_{BC} : u' \in AC[0,1], \widehat{A}u \in Y\}$. In other words $\widehat{A}$ is the natural extension of A from $C[0,1]$ to $L^\infty(0,1)$, and the natural maximal domain given to $\widehat{A}$ is the Favard class $\widehat{\mathfrak{D}}\,(A)$ of A.

Theorem 3. *The operator $\widehat{A}$ satisfies the hypotheses of the Crandall-Liggett theorem.*

Thus $\widehat{A}$ is dissipative and satisfies $\mathscr{R}\,(I - \lambda\widehat{A}) \supset \overline{\mathfrak{D}(\widehat{A})}$ for all $\lambda > 0$, whence $\widehat{A}$ determines a contraction semigroup

$$\widehat{T} = \{\widehat{T}(t) : t \geq 0\} \text{ on } Y_0 = \overline{\mathfrak{D}\,(\widehat{A})} \subset C[0,1],$$

since $\overline{\mathfrak{D}\ (\widehat{A})} = \overline{\mathfrak{D}\ (A)}$; this is the set D computed in the first paragraph of this section. The semigroup $\widehat{T}$ gives unique mild solutions to

$$\partial u/\partial t = \varphi(x, \partial u/\partial x)\partial^2 u/\partial x^2,$$
$$u(x,0) = f(x)$$

together with the boundary conditions at $x = 0, 1$ for $f \in \overline{\mathfrak{D}\ (\widehat{A})} = D$. For $\lambda > 0$

$$\mathscr{R}\ (I - \lambda\widehat{A}) \supset \mathscr{R}\ (I - \lambda A) = C[0,1] \supset \overline{\mathfrak{D}\ (\widehat{A})},$$

and so the range condition holds (by Theorem 1). The main point of Theorem 3 is that $\widehat{A}$ is dissipative on Y. The proof depends on a careful examination of the duality map of Y. This involves using finitely (but not countably) additive set functions defined on the Borel sets of $[0,1]$ and with values in $[0,1]$. The relevant ideas in the case of $\ell^\infty (= L^\infty(\mathbb{N}))$ are explained in [14]; see [15] for the case of $L^\infty(0,1)$. A full proof of Theorem 3 will be given in [11].

Theorem 3 is not only of intrinsic interest; it paves the way for Theorem 4. Thus far we have emphasized *spatial* regularity. The following result gives additional regularity in time.

Theorem 4. *Let $\widehat{A}$, $\widehat{\mathfrak{D}}\ (A)$ be as in Theorems 2,3. Then for all $f \in \widehat{\mathfrak{D}}\ (A)$, the unique mild solution $u(t) = \widehat{T}(t)f$ of $u' = \widehat{A}u, u(0) = f$ satisfies*

$$weak^* \frac{d}{dt} u(t) = \widehat{A}(u(t))$$

for $t \geq 0$.

The weak* derivative assertion means that for all $h \in L^1(0,1)$,

$$t \to (u(t), h) = \int_0^1 u(t,x)h(x)dx$$

is locally absolutely continuous on $[0, \infty)$ to $\mathbb{R}$ and satisfies

$$\frac{d}{dt}(u(t), h) = (\widehat{A}(u(t)), h) \text{ a.e.}$$

The proof of Theorem 4 will also be given in [11]. Note that result makes sense since $L^\infty(0,1)$ is a dual space whereas $C[0,1]$ is not. It also requires the extension $\widehat{A}$ of Theorem 3.

Acknowledgements

The first two named authors gratefully acknowledge the partial support of NSF grants. This work was largely carried out during two visits by the third named author to LSU. These visits were supported by NSF and by LEQSF grants.

REFERENCES

1. V. Barbu, *Nonlinear Semigroups and Differential Equations in Banace Spaces,* Noordhoff, 1976.

2. Ph Benilan, M.G. Crandall and A. Pazy, *Nonlinear Evolution Governed by Accretive Operators,* in preparation.

3. B.C. Burch, A semigroup approach to the Hamilton-Jacobi equation, *J. Diff. Eqns.* 23 (1977), 107-124.

4. P. Butzer and H. Berens, *Semigroups of Operators and Approximation,* Springer, 1967.

5. M.G. Crandall, The semigroup approach to first order quasilinear equations in several space variables, *Israel J. Math.* 12 (1972), 108-132.

6. M.G. Crandall, A generalized domain for semigroup generators, *Proc. Amer. Math. Soc.* 37 (1973), 434-440.

7. M.G. Crandall and T.M. Liggett, Genration of semigroups of nonlinear transformations on general Banach spaces, *Amer. J. Math* 93 (1971), 265-298.

8. J. R. Dorroh and G. R. Rieder, A singular quasilinear parabolic problem in one space dimention, *J. Diff. Eqns.* 91 (1991), 1-23.

9. N. Dunford and J.T. Schwartz, *Linear Operators, Part I,* Interscience, 1958.

10. G.R. Goldstein, Nonlinear diffusion with nonlinear boundary conditions, *Math.*

Meth. Appl. Sci. 20 (1993), 1-20.

11. G.R.Goldstein, J.A. Goldstein and S. Oharu, in preparation.

12. J.A. Goldstein, *Semigroups of Linear Operators and Applications,* Oxford U. Press, 1985.

13. J. A. Goldstein, *Semigroups of Nonlinear Operators and Applications*, in preparation.

14. I. Hada, K. Hashimoto and S. Oharu, On the duality map of ℓ^∞, *Tokyo J. Math.* 2 (1979), 71-97.

15. K. Hashimoto and S. Oharu, On the duality mapping of $L^\infty(0,1)$, to appear.

16. J.A. Goldstein and C.-Y. Lin, Singular nonlinear parabolic boundary value problems in one space dimension, *J. Diff. Eqns.* 68 (1987), 429-443.

17. U. Westphal, Sur la saturation pour des semigroupes nonlineares, *C.R. Acad. Sc. Paris* 274 (1972), 1351-1353.

Gisèle Ruiz Goldstein and Jerome A. Goldstein
Department of Mathematics
Louisiana State University
Baton Rouge, LA, 70803, USA

Shinnosuke Oharu
Department of Mathematics
University of Hiroshima
Higashi-Hiroshima, 724, Japan

S HEIKKILÄ

On first order discontinuous scalar differential equations

0. Introduction

We shall first describe how existence, uniqueness, extremality and comparison results derived for a Carathéodory type differential equations (cf. [2,4,6,8,9,10]) can be generalized to the initial value problem (IVP)

$$x' = q(x)g(t,x), \qquad x(0) = c, \tag{1}$$

where q is positive-valued, measurable and essentially bounded, $\frac{1}{q}$ is locally essentially bounded, $g(\cdot,x)$ is measurable, $g(t,\cdot)$ is right-continuous and upper semicontinuous, and g is bounded by a Lebesgue integrable function of t. Modification and generalization of the approach used in [1] when $q(z) \equiv 1$, combined with methods and results developed in [5,6] makes possible to treat the IVP (1) under the above hypotheses.

The so obtained results and a generalized monotone iterative method developed in [6] are then applied to show that the IVP

$$x' = q(x)f(t,x,x), \qquad x(0) = c, \tag{2}$$

has extremal solutions which are increasing in c and in qf, when the functions involved in the differential equation admit different types of discontinuities in each of their variables. Measurability is assumed in the independent variable t. With respect to x we allow L^∞-type of discontinuity in q, left-discontinuity (as above for $g(t,\cdot)$) in the second variable of f, and monotonizable discontinuity in the last variable of f. Complete proofs of the results described in this paper are to be found in [3, 5, 6, 7].

1. On solvability of the IVP (1)

In this chapter we shall consider existence of unique or extremal solutions to the IVP (1) on an interval $J = [0,T]$, $T > 0$, and their dependence on data.

1.1 Hypotheses and preliminaries. In the following Z denotes a null-set in J. Let $AC(J)$ denote the space of all absolutely continuous functions $x: J \to \mathbb{R}$, and $L_+(J)$ the space of Lebesgue integrable functions $x: J \to \mathbb{R}_+$.

A function $x \in AC(J)$ is said to be a *solution* of (1) if

$$x'(t) = q(x(t))g(t, x(t)) \text{ for almost all (a.a.) } t \in J, \text{ and } x(0) = c.$$

Assume that $q: \mathbb{R} \to (0, \infty)$ has property

(Q) q is measurable and essentially bounded, and $\frac{1}{q}$ is locally essentially bounded,

and that $g: J \times \mathbb{R} \to \mathbb{R}$ satisfies conditions

(A) $g(\cdot, z)$ is measurable for each $z \in \mathbb{R}$ and $g(t, \cdot)$ right-continuous and upper semicontinuous for all $t \in J \setminus Z$;

(B) there is $m \in L_+(J)$ such that $|g(t, z)| \leq m(t)$ for all $z \in \mathbb{R}$ and $t \in J \setminus Z$.

Applying a "partial separation of variables" the IVP (1) can be converted to an integral equation (cf. [3]).

Lemma 1.1. *If the hypotheses (Q), (A) and (B) hold and $c \in \mathbb{R}$, then x is a solution of the IVP (1) if and only if x is a solution of the integral equation*

$$\int_c^{x(t)} \frac{dz}{q(z)} = \int_0^t g(s, x(s))ds, \quad t \in J, \tag{1.1}$$

on the set

$$C_c^w(J) = \{x \in AC(J) \mid x(0) = c, \ |x(t) - x(\bar{t})| \leq |w(t) - w(\bar{t})|, \ \bar{t}, t \in J\}, \tag{1.2}$$

where

$$w(t) = |c| + \int_0^t \|q\|_\infty m(s)ds, \quad t \in J. \tag{1.3}$$

The following result is proved in [3] by modifying the method used in [1] when $q(z) \equiv 1$ to fit in this more general situation.

Lemma 1.2. *Assume that conditions (Q), (A) and (B) hold, and let $c \in \mathbb{R}$ be given. Then for each $n = 1, 2, \ldots$ there is a partition $P_n = \{t_i^n\}_{i=0}^{l_n}$ of J so that $P_n \subset P_{n+1}$, $t_{i+1}^n - t_i^n \leq T2^{-n+1}$, $i = 0, \ldots, l_n - 1$ and $l_n \leq 4^n$, and a function $x_n \in C_c^w(J)$ such that*

$$\int_c^{x_n(t)} \frac{dv}{q(v)} = \int_0^t g_n(s)ds, \ t \in J, \tag{1.4}$$

where

$$g_n(s) = \sup\{g(s,x) \mid |x - x_n(t_i^n)| \le w(s) - w(t_i^n)\},\ s \in [t_i^n, t_{i+1}^n] \setminus Z,\ 0 \le i \le l_n - 1, \tag{1.5}$$

and that

$$x_n(t) \le x_{n-1}(t), \quad t \in J,\ n = 2, 3, \dots. \tag{1.6}$$

1.2. Existence of extremal solutions. Assume now that $AC(J)$ is ordered pointwise. A solution $y \in AC(J)$ of (1) is called *maximal* if $x \le y$ for each solution x of (1), and *minimal* if the reverse inequality holds. If both these solutions exist we call them *extremal solutions.*

Theorem 1.1. *If conditions (Q), (A) and (B) hold, then the IVP (1) has for each $c \in \mathbb{R}$ extremal solutions.*

Proof. In view of lemma 1.2 the sequence $(x_n)_{n=1}^{\infty}$, defined by (1.4), (1.5), is contained in the set $C_c^w(J)$, which is uniformly bounded and equicontinuous. Thus $(x_n)_{n=1}^{\infty}$ has by Ascoli-Arzela theorem a subsequence which converges uniformly on J. Because $(x_n)_{n=1}^{\infty}$ is by (1.6) decreasing, this entire sequence converges uniformly on J. Denote

$$\bar{x}(t) = \lim_{n\to\infty} x_n(t), \quad t \in J. \tag{a}$$

Because each x_n belongs to $C_c^w(J)$, then also $\bar{x}$ belongs to $C_c^w(J)$. By using the given hypotheses and the properties listed in lemma 1.2 it can be shown (cf. [3]) that

$$\lim_{n\to\infty} g_n(t) = g(t, \bar{x}(t)), \tag{b}$$

where g_n is defined by (1.5). From (1.4) it then follows, as $n \to \infty$, applying (b) and the dominated convergence theorem to the right hand side, and (a) and absolute continuity of $x \mapsto \int_c^x \frac{dv}{q(v)}$ to the left hand side of (1.4), that $\bar{x}$ is a solution of the integral equation (1.1). This implies by lemma 1.1 that $\bar{x}$ is a solution of the IVP (1). Moreover, it can be shown (cf. [3]) that $\bar{x}$ is the maximal solution of (1).

Denote by S the set of all the solutions of (1). Since $S \subset C_c^w(J)$ by lemma 1.1, then S is equicontinuous and uniformly bounded, whence one can construct a decreasing sequence $(y_n)_{n=1}^{\infty}$ in S which converges on a dense subset of J to the function

$$\underline{x}(t) = \inf\{x(t) \mid x \in S\}, \qquad t \in J. \tag{a}$$

Because each y_n satisfies by lemma 1.1 the integral equation

$$\int_c^{y_n(t)} \frac{dz}{q(z)} = \int_0^t g(s, y_n(s))\, ds. \quad t \in J,$$

this implies that the limit function x of $(y_n)_{n=1}^{\infty}$ is a solution of the integral equation (1.1). Each y_n belongs to $C_c^w(J)$, whence also $x \in C_c^w(J)$, so that $\underline{x}$ is a solution of (1) by lemma 1.1. Since x equals to $\underline{x}$ on a dense subset of J, the definition (a) of $\underline{x}$ implies that x is the minimal solution of (1) in $C_c^w(J)$, and that $x = \underline{x}$. But $C_c^w(J)$ contains by lemma 1.1 all the solutions of (1), whence $\underline{x}$ is the least of all the solutions of (1). □

1.3. Comparison and uniqueness results. A function $y \in AC(J)$ is said to be a *lower solution* of (1) if $y'(t) \leq q(y(t))g(t, y(t))$ for a.a. $t \in J$, and $y(0) \leq c$, and an *upper solution* if the reversed inequalities are satisfied.

The following result is proved in [3].

Lemma 1.3. *Assume that conditions (Q), (A) and (B) are valid.*
a) If y is a lower solution and $\bar{x}$ the maximal solution of (1), then $y \leq \bar{x}$.
b) If y is an upper solution and $\underline{x}$ the minimal solution of (1), then $\underline{x} \leq y$.

Applying the results of lemma 1.3 it is easy to prove (cf. [3]) that under the hypotheses (Q), (A) and (B) the extremal solutions of the IVP (1) are increasing with respect to c and qg. As for the uniqueness of the solution of (1) we have (cf. [3])

Proposition 1.1. *The IVP (1) has for each $c \in \mathbb{R}$ a unique solution if conditions (Q) and (A) and (B) are valid, and if*

(C) *$g(t, z) - g(t, y) \leq h(t, z - y)$ for all $t \in J \setminus Z$ and $y, z \in \mathbb{R}$, $y \leq z$, where $h\colon J \times \mathbb{R}_+ \to \mathbb{R}_+$ is a Carathéodory function, $h(t, \cdot)$ is increasing for all $t \in J \setminus Z$, and the IVP $u' = \|q\|_\infty h(t, u)$, $u(0) = u_o$ has upper solutions when $u_o \geq 0$, and zero-function is its only solution when $u_o = 0$.*

Remarks 1.1. If the IVP (1) has lower and upper solutions α, β, if (A), (B) hold in $\Omega = \{(t, z) \in J \times \mathbb{R} \mid \alpha(t) \leq z \leq \beta(t)\}$, and if (Q) is valid, then the results of theorem 1.1 and lemma 1.3 hold for solutions of (1) in the order interval $[\alpha, \beta] = \{x \in AC(J) \mid \alpha \leq x \leq \beta\}$ (cf. [3]).

Condition (C) holds if there is $p \in L_+(J)$ such that $g(t, y + z) - g(t, y) \le p(t)z$ for a.a. $t \in J$ and for all y, $z \in \mathbb{R}$, and in particular, if $g(t, \cdot)$ is decreasing for a.a. $t \in J$.

2. General discontinuous ODE

We shall consider in this section the existence of extremal solutions to

$$x' = q(x)f(t, x, x), \qquad x(0) = c, \tag{2}$$

on an interval $J = [0, T]$, $T > 0$, and their dependence on data.

2.1. Hypotheses and preliminaries. In the following Z denotes a null set in J. Assume that $f \colon J \times \mathbb{R} \times \mathbb{R} \to \mathbb{R}$ satisfies conditions

(f0) $f(\cdot, x, y(\cdot))$ is measurable for all $x \in \mathbb{R}$ and $y \in AC(J)$;

(f1) $\limsup_{z \to x-} f(t, z, y) \le f(t, x, y) = \lim_{z \to x+} f(t, z, y)$ for all $t \in J \setminus Z$ and x, $y \in \mathbb{R}$;

(f2) there is $\varphi \colon J \times \mathbb{R} \to \mathbb{R}$ such that $\varphi(\cdot, x)$ is Lebesgue integrable, $\varphi(t, \cdot)$ is increasing and continuous and $f(t, x, \cdot) + \varphi(t, \cdot)$ is increasing for all $t \in J \setminus Z$ and $x \in \mathbb{R}$,

(f3) $|f(t, x, y)| \le p(t)\, h(|x|, |y|)$ for all $t \in J \setminus Z$ and x, $y \in \mathbb{R}$, where $p \in L_+(J)$, $h \colon \mathbb{R}_+^2 \to (0, \infty)$ is increasing in both of its arguments and $\int_o^\infty \frac{dz}{h(z,z)} = \infty$.

These hypotheses and condition (Q) imply that the IVP (2) has for each $c \in \mathbb{R}$ the extremal solutions. In the proof of this result we need

Lemma 2.1. *Given an order interval $[\alpha, \beta]$ in $AC(J)$ and an increasing mapping $G \colon [\alpha, \beta] \to [\alpha, \beta]$, assume there is $\gamma \in L_+(J)$ such that $|(Gx)'(t)| \le \gamma(t)$ for all $x \in [\alpha, \beta]$ and for a.a. $t \in J$. Then G has the least fixed point x_* and the greatest fixed point x^*, and*

$$x_* = \min\{x \mid Gx \le x\}, \quad x^* = \max\{x \mid x \le Gx\}. \tag{2.1}$$

Proof. The assertions can be proved by using a generalized monotone iteration method developed in [6]. For instance, x_* is obtained as the supremum of a countable well-ordered chain in $[\alpha, \beta]$, which equals to the iteration sequence $(G^n\alpha)_{n=o}^\infty$ if G is left-continuous. □

2.2. Existence of extremal solutions. Our main result is

Theorem 2.1. *Given $f: J \times \mathbb{R}^2 \to \mathbb{R}$, $q: \mathbb{R} \to (0, \infty)$ and a null set Z of J, assume that conditions (f0)–(f3) and (Q) hold. Then for each fixed $c \in \mathbb{R}$ the IVP (2) has the extremal solutions, and all the solutions of (2) belong to the order interval $[\alpha, \beta]$, given by*

$$\alpha(t) = c + |c| - v(t),\ \beta(t) = c - |c| + v(t),\ t \in J, \tag{2.2}$$

where $v \in AC(J)$ is the solution of the IVP

$$v' = \|q\|_\infty p(t)h(v,v), \qquad v(0) = |c|. \tag{2.3}$$

Proof. Let $c \in \mathbb{R}$ be given. Condition (f3) implies by lemma 1.5.3 of [6] an existence and uniqueness of v. From (f3) it follows that α and β are lower and upper solutions of (2). Let $y \in [\alpha, \beta]$ be given. Conditions (f0)–(f3) imply that the function $g(t,z) = f(t,z,y(t)) + (\varphi(t,y(t)) - \varphi(t,z))$ satisfies conditions (A) and (B) in $\Omega = \{(t,z) \in J \times \mathbb{R} \mid \alpha(t) \le z \le \beta(t)\}$. Hence, by theorem 1.1 and remarks 1.1 the IVP

$$x' = q(x)[f(t,x,y(t)) + \varphi(t,y(t)) - \varphi(t,x)], \quad x(0) = c, \tag{2.4}$$

has the extremal solutions in $[\alpha, \beta]$.

We now define a map $G: [\alpha, \beta] \to [\alpha, \beta]$ by

$$Gy = x, \quad y \in [\alpha, \beta], \tag{2.5}$$

where x is the maximal solution of (2.4) in $[\alpha, \beta]$. Applying lemma 1.2 and remarks 1.1 it is easy to show (cf. [7]) that G satisfies the hypotheses of lemma 2.1. Thus G has the greatest fixed point of x^*. From (2.4) and (2.5) it follows that x^* is also a solution of the IVP (2) in $[\alpha, \beta]$.

If x is any solution of (2) in $[\alpha, \beta]$, then it satisfies also the IVP (2.4) with $y = x$. But Gx is the maximal solution of (2.4) with $y = x$, whence $x \le Gx$. This and (2.1) imply that $x \le x^*$. Thus x^* is the maximal solution of (2) in $[\alpha, \beta]$. Similar reasoning shows that the IVP (2) has the minimal solution x_* in $[\alpha, \beta]$.

If x is a solution of (2), then it satisfies also the integral equation

$$x(t) = c + \int_0^t q(x(s))f(s,x(s),x(s))ds, \quad t \in J. \tag{2.6}$$

Applying this, (f3) and lemma 1.5.3 of [6] it can be shown (cf. [7]) that $x \in [\alpha, \beta]$. Thus x_* and x^* are the extremal solutions of (2). □

As for the dependence on data we have the following result, which is proved in [7].

Proposition 2.1. *If conditions (Q) and (f0)–(f3) hold, then the extremal solutions of the IVP (2) are increasing with respect to c and to qf.*

Example 2.1. Define

$$\varphi(z) = \operatorname{sgn}(z)|z|\ln(|z|+1), \; z \in \mathbb{R},$$

let D be the Dirichlet function

$$D(z) = \begin{cases} 1 & \text{if } z \text{ is irrational,} \\ 0 & \text{if } z \text{ is rational,} \end{cases}$$

and denote by $[z]$ the greatest integer $\leq z$. It is easy to show that the hypotheses of theorem 2.1 hold for the IVP

$$x' = D(t)(1 + D(x))[([x] - x)\varphi(x + [t + x]) - D(t)x], \quad x(0) = c, \qquad (2.7)$$

when $J = [0, 1]$. Thus (2.7) has for each $c \in \mathbb{R}$ the extremal solutions, which are increasing with respect to c.

Remarks 2.1. If the IVP (2) has lower and upper solutions α, β, and (f3) is replaced by $|f(t, y, z)| \leq N(t)$ for all $t \in J \setminus Z$ and x, $y \in [\alpha(t), \beta(t)]$, then then the results of theorem 2.1 and proposition 2.1 hold for solutions of (2) in the order interval $[\alpha, \beta] = \{x \in AC(J) \mid \alpha \leq x \leq \beta\}$ (cf. [7]).

If f in above theorem 2.1 and proposition 2.1 is nonnegative-valued, then local essential boundedness of $\frac{1}{q}$ in condition (Q) can be weakened to its local Lebesgue integrability. On the other hand, if $q(z) = \begin{cases} 1, & z = 0, \\ 0, & z \neq 0, \end{cases}$ it can be shown (see [6], example 2.1.2) that the IVP

$$x' = q(x), \qquad x(0) = 0$$

has no solution on any interval $J = [0, T]$, $T > 0$.

References

1. Biles, D.C., *Continuous dependence of nonmonotonic discontinuous differential equations*, Trans. Amer. Math. Soc. **339, 2** (1993), 507-524.
2. Carathéodory, C, *Vorlesungen über Reelle Funktionen*, Chelsea Publishing Company, New York, 1948.

3. Carl, S, Heikkilä, S. and Kumpulainen, M., *On solvability of nonmonotonic discontinuous scalar differential equations*, Math., Univ. of Oulu, Preprint (1994).
4. Coddington, E. A. and Levinson, N, *Theory of Ordinary Differential Equations*, McGraw-Hill, New York-Toronto-London, 1955.
5. Heikkilä, S., *On discontinuously perturbed Carathéodory type differential equations*, Nonlinear Analysis (to appear).
6. Heikkilä, S. and Lakshmikantham, V, *Monotone Iterative Techniques for Discontinuous Nonlinear Differential Equations*, Marcel Dekker Inc., New York-Basel, 1994.
7. Heikkilä, S. and Lakshmikantham, V, *A unified theory for first order discontinuous scalar differential equations*, Nonlinear Analysis (to appear).
8. Lakshmikantham, V and Leela, S, *Differential and Integral Inequalities I*, Academic Press, New York-London, 1969.
9. McShane, E. J, *Integration*, Princeton Univ. Press, Princeton, New Jersey, 1974.
10. Walter, W, *Differential and Integral Inequalities*, Springer-Verlag, Berlin - Heidelberg - New York, 1970.

Department of Mathematical Sciences
University of Oulu, 90570 Oulu, Finland

M HIEBER

Heat kernel estimates and analytic semigroups on L^1 spaces

1 Introduction

Let $\Omega \subset \mathbb{R}^n$ be an open set and let $T_p = (T_p(t))_{\geq 0}$ be a family of consistent semigroups on $L^p(\Omega)$, $1 \leq p < \infty$. Suppose that T_{p_0} is an analytic semigroup on $L^{p_0}(\Omega)$ of angle φ for some $p_0 \in (1, \infty)$. We are interested in finding conditions under which the semigroups T_p are analytic too. In order to convey the basic phenomena, assume, for the time being, that T_{p_1} and T_{p_2} are contraction semigroups for given $p_1, p_2 \in [1, \infty)$ and that T_{p_0} is an analytic semigroup of contractions for some $p_0 \in (p_1, p_2)$. Then, by standard arguments, T_p is analytic for all $p \in (p_1, p_2)$. Notice, however, that T_{p_1} is not analytic, in general. The aim of this note is twofold: first we present a result saying that the "endpoint" semigroup T_{p_1} is analytic provided T_{p_0} satisfies an upper Gaussian estimate of order m. The case $p_1 = 1$ is of course of particular interest and has received recently some attention. In this context we refer to the papers [Ou], [A-E] and [Da].

Our approach applies in particular to the semigroups generated by elliptic differential operators of order m on $\mathbb{R}^n$ or by second order elliptic differential operators A subject to rather general boundary conditions. In fact, a famous result of Agmon, Douglis and Nirenberg [A-D-N] combined with Agmon's trick [Ag] implies that, in the latter case, the L^p realization of such a boundary value problem generates an analytic semigroup on $L^p(\Omega)$, $1 < p < \infty$, provided the top-order coefficients of A belong to $BUC(\Omega)$. Observe that their method does not extend to the space $L^1(\Omega)$. Assuming slightly more regularity on the coefficients of A, namely Hölder continuity, our result implies that the solutions of this kind of problems are governed by analytic semigroups also on $L^1(\Omega)$.

Gaussian estimates for semigroups are, generally speaking, rather difficult to obtain. We therefore present as a second aim of this paper a characterization of analytic semigroups admitting a Gaussian estimate in terms of pointwise upper bounds on the kernel of a certain power of the resolvent. For detailed proofs of the results presented below we refer to [Hi2]. Finally, we note that further applications of Gaussian estimates to evolution equations may be found in [Are], [Da], [Hi1] and [H-K-M].

2 Main results

Let $\Omega \subset \mathbb{R}^n$ be an open set, $p_0 \in [1, \infty)$ and let T be a C_0-semigroup on $L^{p_0}(\Omega)$ with generator A. In the following we always identify $L^{p_0}(\Omega)$ with a subspace of $L^{p_0}(\mathbb{R}^n)$ by extending functions by zero. Let $n \in \mathbb{N}, m \in \mathbb{N}\backslash\{1\}$ and define a constant $c_{mn} > 0$ such that $\frac{1}{c_{mn}} \int_{\mathbb{R}^n} exp(\frac{-|x|^{\frac{m}{m-1}}}{4})dx = 1$. Moreover, define the family $(G_{p_0}(t))_{t \geq 0}$ of operators on $L^{p_0}(\mathbb{R}^n)$ by $G_{p_0}(t)f := k_t * f$, where

$$k_t(x) := \frac{1}{c_{mn}} \frac{1}{t^{n/m}} exp(\frac{-|x|^{\frac{m}{m-1}}}{4t^{\frac{1}{m-1}}}) \qquad (t > 0, x \in \mathbb{R}^n).$$

Generalizing a notation of Arendt [Are] we introduce the following definition. We say that the semigroup T satisfies an *upper Gaussian estimate of order* m if there exist constants $a \geq 0, M, b > 0$ such that

$$|T(t)f| \leq Me^{at}G_{p_0}(bt)|f| \qquad (t \geq 0)$$

for all $f \in L^{p_0}(\Omega)$. Notice that G_{p_0} coincides with the Gaussian semigroup on $L^{p_0}(\mathbb{R}^n)$ provided $m = 2$. Furthermore, we assume that E and F are Banach spaces and that there exists a topological vector space G such that $E \hookrightarrow G$ and $F \hookrightarrow G$. Then two operators $S_E \in \mathcal{L}(E)$ and $S_F \in \mathcal{L}(F)$ are called *consistent* if $S_E x = S_F x$ for all $x \in E \cap F$. We call two semigroups T_E and T_F on E and F consistent if $T_E(t)$ and $T_F(t)$ are consistent for all $t \geq 0$.

Assume now that T is a C_0-semigroup on $L^{p_0}(\Omega)$ which satisfies an upper Gaussian estimate of order $m \neq 1$. Then it is not difficult to verify that there exist consistent semigroups T_p on $L^p(\Omega)$, $(1 \leq p < \infty)$, such that $T = T_{p_0}$ and

$$|T_p(t)f| \leq Me^{at}G_p(bt)|f| \qquad (f \in L^p(\Omega), t \geq 0). \tag{2.1}$$

Considering $e^{-at}T(t)$ instead of $T(t)$, we may always assume that (2.1) is satisfied with $a = 0$. Our first result deals in particular with the L^1-analyticity of the consistent semigroup T_1 on $L^1(\Omega)$. More precisely, the following holds.

Theorem 2.1. *Suppose that T is a a bounded analytic C_0-semigroup on $L^{p_0}(\Omega)$ of angle φ satisfying a Gaussian estimate of order m. Then T_p is an analytic C_0-semigroup of angle φ on $L^p(\Omega)$ for all $p \in [1, \infty)$.*

For a proof we refer to [Hi2;Thm.2.3]. We remark that the above Theorem 2.1 generalizes in particular a recent result of Ouhabaz [Ou] saying that T_p is an analytic semigroup of angle $\pi/2$ on $L^p(\Omega)$ for $p \in [1, \infty)$ whenever A_2 is self-adjoint and T admits an upper Gaussian estimate of order 2.

In the following we give two examples to which our theorem applies.

Example 2.2. *Elliptic boundary value problems on $L^1(\Omega)$*

Let Ω be a bounded domain in $\mathbb{R}^n$ such that $\partial\Omega \in C^{2+\rho}$ for some $\rho \in (0, 1)$. Consider a differential operator A of the form

$$A(x, \partial) := - \sum_{1 \leq i,j \leq N} a_{ij}(x)\partial_i\partial_j + \sum_{1 \leq i \leq N} a_i(x)\partial_i + a_0(x)$$

where $a_{ij}, a_i, a_0 \in BUC^{\rho}(\Omega)$ and

$$\sum_{1 \leq i,j \leq N} a_{ij}(x)\xi_i\xi_j \geq c|\xi|^2$$

for all $x \in \mathbb{R}^N$, $\xi = (\xi_1, \ldots, \xi_N) \in \mathbb{R}^N$ and some constant $c > 0$. Let $B(x, \partial) := b(x) \cdot \nabla + b_0(x)$ be boundary operators such that $b = (b_1, \ldots, b_n), b_i, b_0 \in C^{\rho}(\Omega)$ and $b(x) \cdot \nu(x) \geq c_0 > 0$, where $\nu(x)$ is the unit outward normal vector to $\partial\Omega$ at the point $x \in \partial\Omega$. Given $p \in (1, \infty)$, the operator

$$D(\mathcal{A}_p) := \{u \in W_p^2(\Omega); Bu = 0\} \qquad \mathcal{A}_p u := Au$$

is called the L^p-realization of the boundary value problem (A, B). Set

$$\varphi_A := \max_{x \in \overline{\Omega}, \xi \in S^{n-1}} arctg \frac{|Im a_\pi(x, \xi)|}{Re a_\pi(x, \xi)},$$

where a_π denotes the symbol of the principal part of A. Let $\varphi \in (\varphi_A, \pi/2)$. Then a famous result of Agmon, Douglis and Nirenberg [A-D-N] combined with Agmon's trick [Ag] yields that $-\mathcal{A}_p$ generates an analytic semigroup T_p on $L^p(\Omega), 1 < p < \infty$ of angle $\pi/2 - \varphi$. For details we refer to [Am]. Furthermore, it is shown in [Iv] and [So] that the semigroup T_p generated by $-\mathcal{A}_p$ satisfies an upper Gaussian estimate of order 2. Denote by T_1 the consistent semigroup on $L^1(\Omega)$. Then Theorem 2.1 implies the following result.

Proposition 2.3. *Let $1 < p < \infty$ and let T_p be the analytic C_0-semigroup on $L^p(\Omega)$ of angle $\pi/2 - \varphi$ defined as above. Then T_1 is an analytic semigroup on $L^1(\Omega)$ of angle $\pi/2 - \varphi$.*

Example 2.4. *Elliptic operators on $L^p(\mathbb{R}^n)$ with Hölder continuous coefficients*

Let $A = \sum_{|\alpha| \leq m} a_\alpha(x) D^\alpha$ and assume that $a_\alpha \in BUC^{\rho}(\mathbb{R}^n, \mathbb{C})$ for some $\rho \in (0, 1)$ and all α with $|\alpha| \leq m$. Suppose that there exists a constant $\delta > 0$ such that

$$\sup_{|\xi|=1} Re \sum_{|\alpha|=m} a_\alpha(x)(i\xi)^\alpha < -\delta \qquad \text{for all} \quad x \in \mathbb{R}^n.$$

Given $p \in (1, \infty)$, we define the L^p-realization $\mathcal{A}_p$ of A by

$$(2.2) \qquad \begin{aligned} D(\mathcal{A}_p) &:= W_p^m(\mathbb{R}^n) \\ \mathcal{A}_p f &:= Af \qquad \text{for all} \quad f \in D(\mathcal{A}_p). \end{aligned}$$

Then it follows from [A-H-S;Cor.9.5] that $\mathcal{A}_p$ generates an analytic C_0-semigroup T_p on $L^p(\mathbb{R}^n)$ $(1 < p < \infty)$ of some angle $\varphi \in (0, \pi/2]$. Furthermore, it was shown by Friedman [Fr;Thm.9.4.2] that T_p satisfies an upper Gaussian estimate of order m. Denote by T_1 the consistent semigroup on $L^1(\mathbb{R}^n)$. Then by Theorem 2.1 the following holds.

Proposition 2.5. *The semigroup T_1 is an analytic C_0-semigroup on $L^1(\mathbb{R}^n)$ of angle φ.*

Considering powers of the resolvent rather than the resolvent itself in Theorem 2.1 we are able to characterize analytic semigroups admitting a Gaussian bound in terms of a pointwise upper bound on the kernel of a certain power of the resolvent. More specifically, we introduce the following notation. We call an operator $S \in \mathcal{L}(L^p(\Omega), L^q(\Omega))$, $(1 \le p, q \le \infty)$, an *integral operator*, if there exists a measurable function $K : \Omega \times \Omega \to \mathbb{C}$ such that for all $f \in L^p(\Omega)$, $K(x, \cdot)f(\cdot) \in L^1(\Omega)$ x-a.e. and

$$(Sf)(x) = \int_\Omega K(x,y)f(y)dy \qquad x - a.e.$$

In that case S is represented by the kernel K and we write $S \sim K$. If in addition $|K|$ defines also an integral operator in $\mathcal{L}(L^p(\Omega), L^q(\Omega))$, then S is called a *regular integral operator*. It follows by standard arguments (cf. [Sch]) that $T_p(t)$ is an integral operator provided T_p satisfies an upper Gaussian estimate of order m.

Theorem 2.6. *Let T be a bounded analytic C_0-semigroup on $L^{p_0}(\Omega)$ of angle φ with generator A. Then the following assertions are equivalent.*

a) *T satisfies an upper Gaussian estimate of order m with $a = 0$.*

b) *There exist an even integer $l > \frac{n}{m} + 1$ and constants $C, c > 0$ such that $(\lambda - A)^{-l}$ is a regular integral operator whose kernel $K_R^l(\lambda, \cdot, \cdot)$ satisfies*

$$|K_R^l(\lambda, x, y)| \le C|\lambda|^{\frac{n}{m} - l} e^{-c|\lambda|^{\frac{1}{m}}|x-y|}$$

for all $x, y \in \Omega$ and all $\lambda \in \{z \in \mathbb{C}\backslash\{0\}; |arg z| < \theta\}$, where $\theta \in (\pi/2, \varphi + \pi/2)$.

For a proof of Theroem 2.6 we refer to [Hi2;Thm.2.5]. We remark that the above condition b) can be verified for certain classes of operators such as uniformly elliptic differential operators A in divergence form with L^∞-coefficients acting in $L^2(\mathbb{R}^n)$, where $n \ge 3$. Hence, via Theorem 2.6, one obtains an alternative proof of a classical result due to Aronson [Aro] saying that the semigroup T on $L^2(\mathbb{R}^n)$ generated by A satisfies an upper Gaussian estimate of order 2.

References

[Ag] S. Agmon, *On the eigenfunctions and on the eigenvalues of general elliptic boundary value problems.* Comm. Pure Appl. Math. 15 (1962), 119-147.

[A-D-N] S. Agmon, A. Douglis, L. Nirenberg, *Estimates near the boundary for solutions of elliptic partial differential equations satisfying general boundary conditions.* Comm. Pure Appl. Math. 12, (1959), 623-727.

[Am] H. Amann, *Linear and Quasilinear Parabolic Problems.* Book in preparation (1994).

[A-E] H. Amann, J. Escher, *Strongly continuous dual semigroups.* preprint, (1994).

[A-H-S] H. Amann, M. Hieber, G. Simonett, *Bounded H_∞-calculus for elliptic operators.* Differential Integral Equations 7, (1994), 613-653.

[Are] W. Arendt, *Gaussian estimates and interpolation of the spectrum in L^p.* Differential Integral Equations 7 (1994), 1153-1168 .

[Aro] D.G. Aronson, *Non-negative solutions of linear parabolic equations.* Ann. Sci. Norm. Sup. Pisa (3) 22, (1968), 607-694.

[Da] E.B. Davies, *L^p spectral independence and L^1 analyticity.* Preprint (1994).

[Fr] A. Friedman, *Partial Differential Equations of Parabolic Type.* Prentice Hall, New Jersey, 1964.

[Hi1] M. Hieber, *Heat kernel estimates and bounded H^∞-calculus on L^p spaces.* In: Partial Differential Equations; Models in Physics and Biology, S. Nicaise, G. Lumer, B.-W. Schulze (eds.), Akademie Verlag, Berlin, (1994), to appear.

[Hi2] M. Hieber, *Gaussian estimates and holomorphy of semigroups on L^p spaces.* (1994), submitted.

[H-K-M] M. Hieber, P. Koch-Medina, S. Merino, *Linear and semilinear parabolic equations on $BUC(\mathbb{R}^n)$.* (1994), submitted.

[Iv] S.D. Ivasisien, *Green's matrices of boundary value problems for Petrovskii parabolic systems of general form. I.* Math. USSR Sbornik 42, (1982), 93-144.

[Ou] E. Ouhabaz, *Propriétés d'ordre et de contractivité des semi-groupes avec applications aux opérateurs elliptiques.* Ph.D. Thesis, Besançon, (1992).

[Sch] H.H. Schaefer, *Banach Lattices and Positive Operators.* Springer Verlag, Berlin, 1974.

[So] V.A. Solonnikov, *On boundary value problems for linear parabolic systems of differential equations of general form* . Trudy Mat. Inst. Steklov 83 (1965); English transl., Proc. Steklov Inst. Math. 83 (1965).

Universität Zürich
Mathematisches Institut
Winterthurerstr.190
CH-8057 Zürich
Switzerland

O HIJAB

Range characterization, hyper-Markov semigroups and Hermite polynomials

Let $L^2(\mathbf{R})$ denote the complex Hilbert space of functions on $\mathbf{R}$ that are square-integrable against Lebesgue measure dx. Let

$$p_t(x) = \frac{1}{\sqrt{2\pi t}} e^{-x^2/2t}, x \in \mathbf{R},$$

be the Gaussian of variance $t > 0$ and let $L^2(\mathbf{R}, p_t)$ denote the complex Hilbert space of functions on $\mathbf{R}$ that are square-integrable against the probability measure $p_t(x)dx$.

A well-known fact, at least one hundred years old, is that the Hermite polynomials provide an orthogonal basis for $L^2(\mathbf{R}, p_t)$. Here we present a new proof of this result — the Hermite Theorem — and in the process present a range characterization and an identity for the heat semigroup on $\mathbf{R}$ that are also apparently new.

These results were obtained in seeking an analytic proof of Leonard Gross's extension [G1] of the Hermite theorem to Lie groups G of compact type. Gross's original proof was probabilistic; subsequently the author [Hi1] obtained a mostly analytic proof of Gross's theorem. However a portion of the proof in [Hi1] relied on results in [G1] and hence was not completely analytic. After this B. K. Driver [Dr], building on results of B. C. Hall [Ha], succeeded in obtaining a "complex-variable" proof of Gross's theorem. Recently [Hi2] the author has incorporated Driver's ideas into the setting of [Hi1] to obtain a completely analytic "real-variable" proof of Gross's theorem.

In this paper we will not discuss the Lie Group case and instead generalize aspects of the proof in [Hi2] to an abstract semigroup setting. In §1 we give a range characterization for a contraction semigroup on a Hilbert space H in terms of a simple identity. In §2 we introduce a new class of contraction semigroups on $H = L^2$, the hyper-Markov semigroups, and derive an identity for them which implies the identity in §1. In §3 we use the identity in §2 to derive the Hermite theorem.

Supported by NSF Grant #DMS-9121317. To appear in the proceedings of the "International Conference on Evolution Equations", held at the University of Strathclyde, Glasgow, July 1994

§1. Range Characterization of Semigroups in Hilbert Space

Let $\{P_t : t \geq 0\}$ be a continuous contraction semigroup of self-adjoint operators on a complex Hilbert space H and let A denote its infinitesimal generator. Then A is a nonpositive self-adjoint operator, $A \leq 0$, and $P_t = e^{tA}$ for each $t > 0$.

We seek to characterize the range of P_t *for each* $t > 0$ *fixed.* As a warm-up, let us first derive the well-known fact that each operator P_t is necessarily injective. Indeed if $P_t x = 0$ then $0 = \langle P_t x, x\rangle = \langle P_{t/2}P_{t/2}x, x\rangle = \left\|P_{t/2}x\right\|^2$ hence $P_{t/2}x = 0$. Iterating yields $P_{t2^{-n}}x = 0$ for $n \geq 1$; sending $n \to \infty$ we obtain $x = 0$.

Denote the intersection of the domains of A^n, $n \geq 1$, by C^∞. Then for each $t > 0$ the range of P_t is contained in C^∞. We say $\{x_n\} \subset C^\infty$ converges *in* C^∞ to $x \in C^\infty$ if $A^k x_n \to A^k x$ as $n \to \infty$ for all $k \geq 0$. For example if $x \in C^\infty$ then $(P_t x - x)/t \to Ax$ in C^∞ as $t \downarrow 0$.

Proposition. *Let* $D = \sqrt{-2A}$ *and fix* $t > 0$. *If* $x \in H$ *then* $y = P_t x \in C^\infty$ *and*

$$\|x\|^2 = \sum_{n=0}^{\infty} \frac{t^n}{n!}\|D^n y\|^2. \tag{1}$$

Conversely, if $y \in C^\infty$ *and the series in (1) is finite, then there is a unique* $x \in H$ *satisfying* $y = P_t x$.

Proof. Let $\{E(\lambda) : \lambda \geq 0\}$ denote a spectral resolution of $-A$ and suppose $y = P_t x$. Then $A = -D^2/2$ and hence

$$\begin{aligned}
\sum_{n=0}^{\infty} \frac{t^n}{n!}\|D^n y\|^2 &= \sum_{n=0}^{\infty} \frac{t^n}{n!}\langle D^n y, D^n y\rangle \\
&= \sum_{n=0}^{\infty} \frac{t^n}{n!}\langle (-2A)^n y, y\rangle \\
&= \sum_{n=0}^{\infty} \frac{t^n}{n!}\langle (-2A)^n P_t x, P_t x\rangle \\
&= \sum_{n=0}^{\infty} \frac{t^n}{n!}\langle (-2A)^n P_{2t} x, x\rangle \\
&= \sum_{n=0}^{\infty} \frac{t^n}{n!}\int_0^\infty (2\lambda)^n e^{-2t\lambda} d\left(\|E(\lambda)x\|^2\right) \\
&= \int_0^\infty d\left(\|E(\lambda)x\|^2\right) = \|x\|^2.
\end{aligned}$$

Conversely, suppose the series is finite and for each $N \geq 1$ set $x_N = B_N y$ where $B_N = \psi_N(-A)$ and $\psi_N(\lambda) = \sqrt{\sum_{n=0}^N (2t\lambda)^n/n!}$. Since $y \in C^\infty$, the sequence

$\{x_N\}$ is well-defined. Then $\|x_N\|^2 = \langle B_N y, B_N y\rangle = \langle B_N^2 y, y\rangle$ which equals the N-th partial sum of the series in (1). Thus $\{x_N\}$ is bounded in H hence $\{x_N\}$ subconverges weakly to a limit x. Since P_t is linear, $\{P_t x_N\}$ subconverges weakly to $P_t x$. On the other hand $\{e^{-t\lambda}\psi_N(\lambda)\}$ increases to 1 as $N \to \infty$ for all $\lambda \geq 0$ hence $P_t x_N = e^{tA}\psi_N(-A)y \to y$ weakly as $N \to \infty$. Thus $y = P_t x$. □

Example (Heat Semigroup). Take $H = L^2(\mathbf{R})$ and $Af = f''/2$ for $f \in C_0^\infty(\mathbf{R})$. Then the closure of A is the infinitesimal generator of a continuous contraction semigroup of self-adjoint operators on H and C^∞ consists of the L^2 functions f whose Fourier transform $\hat{f}$ is rapidly decreasing. In fact the semigroup is $P_t f = p_t * f$, $f \in L^2(\mathbf{R})$, where $*$ denotes convolution. Hence we obtain the result that a function $g \in L^2(\mathbf{R})$ is a convolution of the form $g = p_t * f$ for $f \in L^2(\mathbf{R})$ iff $g \in C^\infty(\mathbf{R})$ and

$$\sum_{n=0}^{\infty} \frac{t^n}{n!} |g^{(n)}|^2$$

is in $L^1(\mathbf{R})$. Moreover we obtain the identity

$$\|f\|^2 = \sum_{n=0}^{\infty} \frac{t^n}{n!} \left\|(p_t * f)^{(n)}\right\|^2. \tag{2}$$

Of course using the Fourier transform $g \mapsto \hat{g}$, the identity (2) falls out as an immediate consequence of the MacLaurin expansion of $t \mapsto |\hat{p}_t|^2$.

§2. Hyper-Markov Semigroups on L^2

Let $(X, \mathcal{F}, \mu)$ be a measure space. In this section by "a semigroup" we shall mean a continuous contraction semigroup of self-adjoint operators on $H = L^2(X, \mathcal{F}, \mu)$. Below $f \geq g$ means $f \geq g$ a.e.-μ and $L^p = L^p(X, \mathcal{F}, \mu)$.

A semigroup is *Markov* if for all $t > 0$

- $f \in L^2$ and $f \geq 0$ imply $P_t f \geq 0$ and $\int_X P_t f d\mu = \int_X f d\mu$.

Thus a semigroup is Markov if it preserves positivity and total mass.

Let $\{P_t : t \geq 0\}$ be a semigroup. We wish to associate to each function f functions $\Gamma_n(f)$, $n \geq 0$, depending quadratically on f, in such a way that the identity

$$P_t(|f|^2) = \sum_{n=0}^{\infty} \frac{t^n}{n!} \Gamma_n(P_t f), t \geq 0, \tag{3}$$

holds.

For each $n \geq 0$ let ($i = \sqrt{-1}$)

$$\Gamma_n(f, g) \equiv \frac{1}{4}\left(\Gamma_n(f+g) - \Gamma_n(f-g) + i\Gamma_n(f+ig) - i\Gamma_n(f-ig)\right),$$
$$\Gamma_n(f, g) = \overline{\Gamma_n(g, f)},$$

denote the bilinear form associated to the quadratic form $\Gamma_n(f)$. Then formal differentiation of (3) with respect to t without regard to rigor shows that (3) holds iff

$$\Gamma_0(f) = |f|^2$$
$$\Gamma_{n+1}(f) = A(\Gamma_n(f)) - \Gamma_n(f, Af) - \Gamma_n(Af, f), n \geq 0.$$

Let $\{P_t : t \geq 0\}$ be a semigroup. An *algebra core* is a linear subspace $\mathcal{D} \subset C^\infty \cap L^1$ such that

- $\mathcal{D}$ is dense in L^2;
- $\mathcal{D}$ is closed under pointwise multiplication of functions;
- $f_n, g_n, f, g \in \mathcal{D}$ and $f_n \to f$, $g_n \to g$, in C^∞ implies $f_n g_n \to fg$ in C^∞;
- $A(\mathcal{D}) \subset \mathcal{D}$ and $P_t(\mathcal{D}) \subset \mathcal{D}$ for $t > 0$.

Then $\Gamma_n(f) \in \mathcal{D}$ is well-defined for $f \in \mathcal{D}$ and $n \geq 0$.

Example. Let $X = \{-1, +1\}$, $\mu(\pm 1) = \frac{1}{2}$, and $Af(\pm 1) = \pm(f(-1) - f(+1))$ for $f \in L^2$. Then A is the infinitesimal generator of a Markov semigroup, $\mathcal{D} = C^\infty = L^2$ is an algebra core, and

$$\Gamma_n(f)(\pm 1) = 4^{n-1}|f(+1) - f(-1)|^2, n \geq 1.$$

Example. The heat semigroup is Markov, the Schwartz space $\mathcal{D} = \mathcal{S}(\mathbf{R})$ is an algebra core, and

$$\Gamma_n(f) = |f^{(n)}|^2, f \in \mathcal{D}, n \geq 0.$$

Example (Hermite Semigroup). Let $X = \mathbf{R}$, $d\mu = p_{1/2}dx$, and $Af = f''/2 - x^2 f'$ for f in the space of polynomials $\mathcal{P}(\mathbf{R})$. Then the closure of A is the infinitesimal generator of a Markov semigroup, $\mathcal{D} = \mathcal{P}(\mathbf{R})$ is an algebra core, and

$$\Gamma_0(f) = |f|^2$$
$$\Gamma_1(f) = |f'|^2$$
$$\Gamma_2(f) = |f''|^2 + 2|f'|^2$$
$$\Gamma_3(f) = |f'''|^2 + 6|f''|^2 + 4|f'|^2$$
$$\text{etc.}$$

If A is bounded on L^2 then one can easily majorize the series in (3) to conclude that for $f \in \mathcal{D}$ (3) does in fact hold rigorously. If A is unbounded then there is no reason for (3) to hold: We need additional information. This is provided by the following idea.

Definition. The semigroup $\{P_t : t \geq 0\}$ is *hyper-Markov* if

- the semigroup is Markov,
- there is an algebra core $\mathcal{D}$,
- $\Gamma_n(f) \geq 0$ for all $n \geq 0$, $f \in \mathcal{D}$.

The above examples are all hyper-Markov. So is the heat semigroup on a compact Lie group [G1], [G2], [H1], [H2].

Let $\{P_t : t \geq 0\}$ be a Markov semigroup with an algebra core $\mathcal{D}$. Then one always has $\Gamma_0(f) \geq 0$, $\Gamma_1(f) \geq 0$ for $f \in \mathcal{D}$. The operator Γ_2 has deeper significance: Its sign controls the behavior of the semigroup on L^p. For example, in certain situations, P. A. Meyer [M] and D. Bakry [Ba] have shown that the nonnegativity of Γ_2 implies the boundedness of the associated Riesz transform on L^p, $1 < p < \infty$. Moreover D. Bakry and M. Emery [BE] have shown that a strong positivity condition on Γ_2 implies the hypercontractivity of the semigroup i.e. a logarithmic Sobolev inequality [G3] holds for A.

Here is an example of a Markov semigroup with an algebra core that is not hyper-Markov.

Example. Consider $X = \mathbf{R}$, $d\mu = e^{-2V}dx$, and $Af = f''/2 - V'(x)f'$ for $f \in C_0^\infty(\mathbf{R})$, where $V \in C^\infty(\mathbf{R})$ is an even monic polynomial of degree at least 2. Then the closure of A is the infinitesimal generator of a Markov semigroup. Let $\mathcal{T}(\mathbf{R})$ denote the space of infinitely differentiable functions f such that $f^{(n)}$ has polynomial growth for all $n \geq 0$. Then $\mathcal{T}(\mathbf{R})$ is an algebra core. Here

$$\begin{aligned}
\Gamma_0(f) &= |f|^2 \\
\Gamma_1(f) &= |f'|^2 \\
\Gamma_2(f) &= |f''|^2 + 2V''(x)|f'|^2 \\
&\text{etc.}
\end{aligned}$$

We conclude that if V is not convex, then the semigroup is not hyper-Markov.

Lemma. *Suppose $\{P_t : t \geq 0\}$ is a Markov semigroup with an algebra core $\mathcal{D}$. Then*

$$\int_X \Gamma_n(f)d\mu = \langle (-2A)^n f, f\rangle = \|D^n f\|^2, f \in \mathcal{D}, n \geq 0. \tag{4}$$

Proof. It follows [Da] from the Markov property that f, Af, $A^2 f$ in $L^2 \cap L^1$ imply $\int_X Af d\mu = 0$. In particular this holds for $f \in \mathcal{D}$. Then (4) follows by induction on $n \geq 0$ using $\int_X A(\Gamma_n(f))d\mu = 0$. □

Theorem. *If $f \in \mathcal{D}$ and the semigroup is hyper-Markov, then (3) holds.*

Proof. First by induction on $n \geq 0$ one shows that

$$\frac{\Gamma_n(P_t f) - \Gamma_n(f)}{t} \to \Gamma_n(f, Af) + \Gamma_n(Af, f), n \geq 0,$$

in C^∞ as $t \to 0$. Hence $(d/dt)\Gamma_n(P_t f) = \Gamma_n(AP_t f, P_t f) + \Gamma_n(P_t f, AP_t f)$, $n \geq 0$. Now let $u_N(t) \in \mathcal{D}$ denote the N-th partial sum of the series in (3). Then $u_N(0) = |f|^2$ and differentiation yields

$$\left(\frac{\partial}{\partial t} - A\right) u_N(t) = -\frac{t^N}{N!}\Gamma_{N+1}(P_t f) \leq 0.$$

Since the semigroup is Markov, it follows that $u_N(t) \leq P_t(|f|^2)$. Sending $N \to \infty$, we obtain that the right side in (3) is less than or equal to the left side in (3). But now integrate both sides in (3) over X. By the Lemma and the Proposition we have equality. The result follows. □

In fact the method of proof yields a hierarchy of identities of which (3) is the first.

Theorem. *If $f \in \mathcal{D}$ and the semigroup is hyper-Markov, then*

$$P_t(\Gamma_N(f)) = \sum_{n=0}^{\infty} \frac{t^n}{n!}\Gamma_{n+N}(P_t f), t \geq 0,$$

for all $N \geq 0$.

Proof. The proof is almost identical to the above. □

Since we now know (3) holds for the heat semigroup, we obtain

$$P_t(|f|^2)(x) = \sum_{n=0}^{\infty} \frac{t^n}{n!}|(P_t f)^{(n)}(x)|^2 \tag{5}$$

a.e. for $f \in \mathcal{S}(\mathbf{R})$. In fact a little more work shows that (5) holds for all $f \in \mathcal{P}(\mathbf{R})$ and for all $x \in \mathbf{R}$.

§3. The Hermite Theorem

For each $n \geq 0$ the n-th Hermite polynomial H_n is given by

$$p_t^{(n)}(x) = (-1)^n H_n(x) p_t(x), x \in \mathbf{R}.$$

Actually these are the Hermite polynomials "of variance t". The usual ones are obtained by setting $t = 1$.

The Fock space $\exp(t\mathbf{C})$ over $\mathbf{C}$ is similar to the usual Hilbert space ℓ^2 but has a different norm. Let $\exp(t\mathbf{C})$ denote the set of sequences $\alpha = (\alpha_1, \alpha_2, \dots)$ of complex numbers satisfying

$$\|\alpha\|^2 = \sum_{n=0}^{\infty} \frac{n!}{t^n}|\alpha_n|^2 < \infty. \tag{6}$$

Then $\exp(t\mathbf{C})$ is a complex Hilbert space in a natural way whose corresponding norm is displayed in (6). If α has finitely many nonzero terms we say α is a finite sequence.

In quantum mechanics the state of a system can be described in two ways: As a particle i.e. as an element of $\exp(t\mathbf{C})$ and as a wave i.e. as an element of $L^2(\mathbf{R}, p_t)$. The wave-particle duality of quantum mechanics states that these descriptions are equivalent. The Hermite theorem is a precise mathematical interpretation of this fact. See [BSZ].

Given a finite sequence $\alpha \in \exp(t\mathbf{C})$, set

$$H_\alpha(x) = \sum_{n=0}^{\infty} \alpha_n H_n(x).$$

Then H_α is a polynomial.

Hermite Theorem. *The Hermite map $\alpha \mapsto H_\alpha$ extends to a linear isometry of $\exp(t\mathbf{C})$ onto $L^2(\mathbf{R}, p_t)$.*

There are many ways of verifying this; the method we describe here will use the identity (5).

To begin instead of working with the Hermite map, we work with its adjoint $K : L^2(\mathbf{R}, p_t) \to \exp(t\mathbf{C})$. Given $f \in \mathcal{P}(\mathbf{R})$ define $\alpha = Kf$ by setting

$$\alpha_n = \frac{t^n}{n!}\langle f^{(n)}, p_t\rangle_{L^2(\mathbf{R})}, n \geq 0.$$

Then α is a finite sequence.

Lemma. *For any $f \in \mathcal{P}(\mathbf{R})$ and finite sequence $\beta \in \exp(t\mathbf{C})$,*

$$\langle Kf, \beta\rangle_{\exp(t\mathbf{C})} = \int_{-\infty}^{\infty} f(x)\overline{H_\beta(x)}p_t(x)dx.$$

Proof. Integration by parts. □

If we establish the isometry of K, then by the Lemma we obtain the isometry of H hence the Hermite theorem. But by (6) K is an isometry from $\mathcal{P}(\mathbf{R}) \subset L^2(\mathbf{R}, p_t dx)$ into $\exp(t\mathbf{C})$ iff

$$\sum_{n=0}^{\infty} \frac{t^n}{n!}|\langle f^{(n)}, p_t\rangle_{L^2(\mathbf{R})}|^2 = \int_{-\infty}^{\infty} |f(x)|^2 p_t(x)dx \tag{7}$$

and (7) is obtained from (5) by inserting $x = 0$ since $P_t(f^{(n)}) = (P_t f)^{(n)}$. This establishes the isometry of K on polynomials; hence K extends to an isometry of $L^2(\mathbf{R}, p_t)$ into $\exp(t\mathbf{C})$.

The final step is to show K is onto $\exp(t\mathbf{C})$. For this it is enough to show K is onto a dense subset of $\exp(t\mathbf{C})$. But this is immediate since for f a polynomial of degree $N \geq 0$ the sequence $\alpha = Kf \in \exp(t\mathbf{C})$ satisfies $\alpha_n = 0$ for $n \geq N + 1$ and $\alpha_N \neq 0$. Thus the range of K includes all finite sequences in $\exp(t\mathbf{C})$. Since we now know K is an isometry, the Lemma implies H extends to an isometry. □

REFERENCES

[Ba] D. Bakry, *Transformations de Riesz pour les Semi-groupes Symétriques*, Lecture Notes in Mathematics 1123, Springer-Verlag, 1985.

[BE] D. Bakry and M. Emery, *Diffusions Hypercontractives*, Lecture Notes in Mathematics 1123, Springer-Verlag, 1985.

[BSZ] J. C. Baez, I. E. Segal, & Z. Zhou, *Introduction to Algebraic and Constructive Quantum Field Theory*, Princeton University Press, Ewing, NJ, 1992.

[Da] E. B. Davies, *One-Parameter Semigroups*, Academic Press, 1980.

[Dr] B. K. Driver, *On the Kakutani-Itô-Wiener-Gross and the Segal-Bargmann-Hall isomorphisms*, To Appear, Journal of Functional Analysis.

[G1] L. Gross, *Uniqueness of Ground States for Schrödinger Operators Over Loop Groups*, J. Functional Analysis **112** (1993), 373-441.

[G2] ———, *The Homogeneous Chaos over Compact Lie Groups*, "Stochastic Processes, A Festschrift in Honour of Gopinath Kallianpur" (S. Cambanis *et al.*, Eds.), Springer-Verlag, New York, 1993, pp. 117-123.

[G3] ———, *Logarithmic Sobolev Inequalities*, Amer. J. Math. **97** (1976), 1061-1083.

[Ha] B. C. Hall, *The Segal-Bargmann "Coherent State" Transform for Compact Lie Groups*, To Appear, J.. Functional Analysis.

[Hi1] O. Hijab, *Hermite Functions on Compact Lie Groups, I*, To Appear, J. Functional Analysis.

[Hi2] ———, *Hermite Functions on Compact Lie Groups, II*, To Appear, J. Functional Analysis.

[M] P. A. Meyer, *Sur La Théorie de Littlewood-Paley-Stein*, Lecture Notes in Mathematics 1123, Springer-Verlag, 1985.

DEPARTMENT OF MATHEMATICS, TEMPLE UNIVERSITY, BROAD & MONTGOMERY, PHILADELPHIA, PA 19122

E-mail address: hijab@math.temple.edu

S JIANG

Exponential stability of spherically symmetric solutions to the equations of a viscous polytropic ideal gas

1 Introduction

The motion of a viscous polytropic ideal gas (in $\mathbb{R}^n$, $n = 2, 3$) is described by the following equations in Eulerian coordinates (cf. [2, 12])

$$
\begin{aligned}
\frac{\partial \rho}{\partial t} + \text{div}\,(\rho \mathbf{v}) &= 0, \\
\rho\left[\frac{\partial \mathbf{v}}{\partial t} + (\mathbf{v}\cdot\nabla)\mathbf{v}\right] &= \mu\Delta\mathbf{v} + (\lambda+\mu)\nabla(\text{div}\,\mathbf{v}) - R\nabla(\rho\theta), \\
c_V\rho\left[\frac{\partial \theta}{\partial t} + (\mathbf{v}\cdot\nabla)\theta\right] &= \kappa\Delta\theta - R\rho\theta(\text{div}\,\mathbf{v}) + \lambda(\text{div}\,\mathbf{v})^2 + 2\mu D\cdot D. \qquad (1.1)
\end{aligned}
$$

Here ρ, θ, and $\mathbf{v} = (v_1, \cdots, v_n)^T$ $(n = 2, 3)$ are the density, the absolute temperature and the velocity respectively, R, c_V and κ are positive constants; λ and μ are the constant viscosity coefficients, $\mu > 0$, $\lambda + 2\mu/n \geq 0$; $D = D(\mathbf{v})$ is the deformation tensor

$$
D_{ij} := \frac{1}{2}\left(\frac{\partial v_i}{\partial x_j} + \frac{\partial v_j}{\partial x_i}\right) \qquad \text{and} \qquad D\cdot D := \sum_{i,j=1}^{n} D_{ij}^2.
$$

Let $\Omega := \{x \in \mathbb{R}^n \mid a < |x| < b\}$ $(b, a > 0)$ denote an annular domain in $\mathbb{R}^n$ $(n = 2, 3)$. We shall consider the initial boundary value problem of (1.1) in the region $\{t > 0, x \in \Omega\}$ with the following initial and boundary conditions

$$
\rho(x,0) = \rho^0(x), \quad \mathbf{v}(x,0) = \mathbf{v}^0(x), \quad \theta(x,0) = \theta^0(x), \qquad x \in \bar{\Omega}, \qquad (1.2)
$$

$$
\mathbf{v}|_{\partial\Omega} = 0, \qquad \left.\frac{\partial \theta}{\partial \nu}\right|_{\partial\Omega} = 0, \qquad t \geq 0, \qquad (1.3)
$$

where ν denotes the exterior normal vector.

The global existence and asymptotic behavior of smooth solutions to initial boundary value problems and the Cauchy problem of (1.1) have been investigated by many authors. In one dimension, it is well known that global smooth solutions exist for smooth (large) initial data, and converge to a (constant) steady state in the case of bounded domains as $t \to \infty$. In more than one dimension the global existence and the asymptotic behavior of smooth soltuions have been investigated for general domains only in the case of sufficiently small initial data (see [8, 13], [3]–[4] for initial boundary value problems, [7] for the Cauchy problem; also see [9, 11] and the references cited therein).

For large initial data the global existence of solutions to (1.1) has been studied in the case of a bounded annular domain. Nikolaev [10] in 1983 considered the initial boundary value problem of (1.1) with vanishing velocity and constant temperature on the boundary and proved that for (smooth) spherically symmetric initial data a (smooth) spherically symmetric solution exists globally in time if the initial density and temperature are strictly positive. Recently, Yashima and Benabidallah [14]–[15] dealt with the case of non-negative initial density and temperature. They showed the global existence of spherically symmetric solutions to (1.1). The asymptotic behavior of the (spherically symmetric) solutions, however, is not discussed in [10], [14]–[15].

The aim of the present work is to study the asymptotic behavior of the spherically symmetric solutions to (1.1)–(1.3). We will show that the spherically symmetric solutions of (1.1)–(1.3) decay to a constant state exponentially as time goes to infinity.

2 Exponential decay

We first derive the spherically symmetric form of (1.1). Spherically symmetric solutions to (1.1) have the form

$$v_i(x,t) = \frac{x_i}{r} v(r,t), \;\; i = 1, \cdots, n, \qquad \rho(x,t) = \rho(r,t), \qquad \theta(x,t) = \theta(r,t), \qquad (2.1)$$

where $x = (x_1, \cdots, x_n)^T \in \mathbb{R}^n$ $(n = 2,3)$, $r := |x|$. Assuming that $\rho^0(x) = \rho_0(r)$, $\mathbf{v}^0(x) = x v_0(r)/r$ and $\theta^0(x) = \theta_0(r)$, we thus reduce the system (1.1)–(1.3) to the following equations for $\rho(r,t)$, $v(r,t)$ and $\theta(r,t)$ of the form

$$\rho_t + (\rho v)_r + \frac{(n-1)}{r} \rho v = 0,$$

$$
\begin{aligned}
\rho\left(v_t + v v_r\right) &= (\lambda + 2\mu)\left(v_{rr} + \frac{(n-1)}{r} v_r - \frac{(n-1)}{r^2} v\right) - R(\rho v)_r, \\
c_V \rho\left(\theta_t + v\theta_r\right) &= \kappa\theta_{rr} + \kappa\frac{(n-1)}{r}\theta_r - R\rho\theta\left(v_r + \frac{(n-1)}{r} v\right) \\
&+ \lambda\left(v_r + \frac{(n-1)}{r} v\right)^2 + 2\mu v_r^2 + 2\mu\frac{(n-1)}{r^2} v^2
\end{aligned}
\tag{2.2}
$$

with the initial and boundary conditions

$$
\begin{aligned}
\rho(r,0) &= \rho_0(r), \quad v(r,0) = v_0(r), \quad \theta(r,0) = \theta_0(r), \qquad r \in [a,b], \\
v(a,t) &= v(b,t) = 0, \quad \theta_r(a,t) = \theta_r(b,t) = 0, \qquad t \geq 0.
\end{aligned}
\tag{2.3}
$$

To show the time-asymptotic behavior it is convenient to transform the system (2.2) to that in Lagrangian coordinates. Let

$$
L := \int_a^b s^{n-1}\rho_0(s)ds > 0. \tag{2.4}
$$

We denote the Lagrangian mass coordinates by (x,t) and the specific volume by $u := 1/\rho$. Then (2.2)–(2.3) in the new variables (x,t) read:

$$
u_t = \left(r^{n-1}v\right)_x, \tag{2.5}
$$

$$
v_t = r^{n-1}\left[(\lambda+2\mu)\frac{(r^{n-1}v)_x}{u} - R\frac{\theta}{u}\right]_x, \qquad x \in (0,L),\ t > 0, \tag{2.6}
$$

$$
\begin{aligned}
c_V\theta_t &= \kappa\left[\frac{r^{2n-2}\theta_x}{u}\right]_x + \frac{1}{u}\left[(\lambda+2\mu)(r^{n-1}v)_x - R\theta\right]\left(r^{n-1}v\right)_x \\
&\quad -2\mu(n-1)\left(r^{n-2}v^2\right)_x
\end{aligned}
\tag{2.7}
$$

with the initial and boundary conditions

$$
u(x,0) = u_0(x), \quad v(x,0) = v_0(x), \quad \theta(x,0) = \theta_0(x), \qquad x \in [0,L], \tag{2.8}
$$

$$
v(0,t) = v(L,t) = 0, \quad \theta_x(0,t) = \theta_x(L,t) = 0, \quad t \geq 0. \tag{2.9}
$$

Here $u_0 = 1/\rho_0$, $r \equiv r(x,t)$ is defined by

$$
r(x,t) := r_0(x) + \int_0^t v(x,\tau)d\tau, \quad r_0(x) := \left\{a^n + n\int_0^x u_0(y)dy\right\}^{1/n}, \quad n = 2,3; \tag{2.10}
$$

and (without danger of confusion) we have still used $\{u(x,t), v(x,t), \theta(x,t)\}$ to denote $\{u(r(x,t),t), v(r(x,t),t), \theta(r(x,t),t)\}$.

As mentioned in the introduction Nikolaev [10], Yashima and Benabidallah [13, Proposition 1] established the existence of global solutions to (2.5)–(2.9). It is proved in [9, 13-14] that if

$$u_0,\ u_0',\ v_0,\ v_0',\ v_0'',\ \theta_0,\ \theta_0',\ \theta_0'' \in C^\alpha[0,L] \text{ for some } \alpha \in (0,1),$$
$$u_0(x),\ \theta_0(x) > 0 \text{ on } [0,L], \tag{2.11}$$

and the initial data are compatible with the boundary conditions (2.9), then there exists a unique solution $\{u(x,t), v(x,t), \theta(x,t)\}$ with positive u and θ to (2.5)–(2.9) on $[0,L] \times [0,\infty)$ such that for every $T > 0$

$$u,\ u_x,\ u_t,\ u_{xt},\ v,\ v_x,\ v_t,\ v_{xx},\ \theta,\ \theta_x,\ \theta_t,\ \theta_{xx}\ \in C^{\alpha,\alpha/2}(Q_T),$$
$$u_{tt},\ v_{xt},\ \theta_{xt} \in L^2(Q_T). \tag{2.12}$$

Here $C^\alpha[0,L]$ stands for the Banach space of functions on $[0,L]$ which are uniformly Hölder continuous with exponent α and $C^{\alpha,\alpha/2}(Q_T)$ for the Banach space of functions on $Q_T := [0,L] \times [0,T]$ which are uniformly Hölder continuous with exponent α in x and $\alpha/2$ in t.

Denote

$$u^* := \int_0^L u_0(x)dx, \qquad \theta^* := \frac{1}{c_V L}\int_0^L \left\{c_V\theta_0 + \frac{v_0^2}{2}\right\}(x)dx;$$
$$r^*(x) := (a^n + nu^*x)^{1/n}, \quad x \in [0,L]. \tag{2.13}$$

We assume that λ and μ satisfy

$$n\lambda + 2\mu > 0. \tag{2.14}$$

Then our main result reads:

Theorem 2.1 *Assume that (2.11) and (2.14) are satisfied. Let $\{u(x,t), v(x,t), \theta(x,t)\}$ be a solution of (2.5)–(2.9) in the function class indicated in (2.12). Then $\{u(x,t) - u^*, v(x,t), \theta(x,t) - \theta^*\}$ and $r(x,t) - r^*(x)$ converge to zero in $H^1(0,L)$ and $H^2(0,L)$ respectively as $t \to \infty$. Moreover, there are positive constants γ, T_0, C, independent of t, such that*

$$\|u(t) - u^*\|_{H^1} + \|v(t)\|_{H^1} + \|\theta(t) - \theta^*\|_{H^1} + \|r(t) - r^*\|_{H^2} \le Ce^{-\gamma t} \quad \text{for any } t \ge T_0.$$

Remark 2.1 *An analogous theorem holds when (1.3) is replaced by the following boundary conditions:*

$$\mathbf{v}|_{\partial\Omega} = 0, \qquad \theta|_{\partial\Omega} = 1, \qquad t \geq 0.$$

Remark 2.2 *Theorem 2.1 remains valid for any $n > 3$ or/and for the case when (2.11) is replaced by*

$$u_0,\ u_0',\ v_0,\ v_0',\ \theta_0,\ \theta_0' \in L^2(0,L), \qquad u_0(x),\ \theta_0(x) > 0 \ on\ [0,L].$$

The decay constant γ may depend on the initial data, λ, μ, R, c_V, κ, n, a and b.

The proof of Theorem 2.1 is essentially based on a careful examination of a priori estimates which are shown to be independent of t. The difficulties arise from the dependence on the time and spatial variables of the coefficients in the the equations (2.5)–(2.7), but can be overcome in our approach by modifying an idea of Kazhikhov [6, 1] for the one-dimensional case and establishing an additional estimate embodying the dissipative effects of viscosity and thermal diffusion. The proof is rather long and technical; see [5] for the details.

References

[1] Antontsev, S.N., A.V. Kazhikhov, A.V. and Monakhov, V.N., Boundary Value Problems in Mechanics of Nonhomogeneous Fluids, North-Holland, Amsterdam, New York, 1990.

[2] Batchelor, G.K., An Introduction to Fluid Dynamics, Cambridge Univ. Press, London, 1967.

[3] Deckelnick, K., *Decay estimates for the compressible Navier-Stokes equations in unbounded domains,* Math. Z. **209** (1992), 115-130.

[4] Deckelnick, K., *L^2 Decay for the compressible Navier-Stokes equations in unbounded domains,* Comm. PDE **18** (1993), 1445-1476.

[5] Jiang, S., *Exponential stability of spherically symmetric solutions to the equations of a viscous polytropic ideal gas,* Preprint (1994).

[6] Kazhikhov, A.V., *To a theory of boundary value problems for equations of one-dimensional nonstationary motion of viscous heat-conduction gases,* Boundary Value Problems for Hydrodynamical Equations (in Russian), **No. 50**, Inst. Hydrodynamics, Siberian Branch Akad., USSR., 1981, pp. 37-62.

[7] Matsumura, A. and Nishida, T., *The initial value problem for the equations of motion of compressible and heat-conductive fluids,* Proc. Japan Acad. Ser. A **55** (1979), 337-342.

[8] Matsumura, A. and Nishida, T., *Initial boundary value problems for the equations of motion of compressible and heat-conductive fluids,* Commnu. Math. Phys. **89** (1983), 445-464.

[9] Matsumura, A. and Padula, M., *Stability of stationary flow of compressible fluids subject to large external potential forces,* SAACM **2** (1992), 183-202.

[10] Nikolaev, V.B., *On the solvability of mixed problem for one-dimensional axisymmetrical viscous gas flow,* Dinamicheskie zadachi Mekhaniki sploshnoj sredy, **63** Sibirsk. Otd. Acad. Nauk SSSR, Inst. Gidrodinamiki (1983). (Russian)

[11] Padula, M., *Stability properties of regular flows of heat-conducting compressible fluids,* J. Math. Kyoto Univ. **32** (1992), 401-442.

[12] Serrin, J., *Mathematical Principles of classical fluid mechanics,* Handbuch der Physik VIII/1, Springer-Verlag; Berlin, 1972, pp. 125-262.

[13] Valli, A. and Zajaczkowski, W.M., *Navier-Stokes Equations for compressible fluids: global existence and qualitative propertives of the solutions in general case,* Commnu. Math. Phys. **103** (1986), 259-296.

[14] Yashima, H.F. and Benabidallah, R., *Equation á symétrie sphérique d'un gaz visqueux et caloriférе avec la surface libre,* Preprint Dip. Mat. Pisa 2.88 (1992).

[15] Yashima, H.F. and Benabidallah, R., *Unicité de la solution de l'équation monodimensionnelle ou á symétrie sphérique d'un gaz visqueux et calorifére,* Preprint Dip. Mat. Pisa 2.100 (1992).

Song JIANG

Institut für Angewandte Mathematik

der Universität Bonn

Wegelerstrasse 10

53115 Bonn

Germany

M JUNG

Functional calculi in semigroup theory

Multiplicative perturbations were considered in semigroup theory by many authors who used various techniques in their approach (see the references for a list). Results were also extended to the field of integrated semigroups. Some techniques involve the use of functional calculi, and this article will present an excerpt of these. Most proofs may be found elsewhere and are omitted in that case. Theorem 1 may be found in [9] by A. Holderrieth and Theorems 2, 5, and 7 and its related corollaries may be found in [10] or [11] by the author.

Let X be a Banach space. A (strongly continuous) semigroup will always be understood to act on this space. Bounded, linear operators are understood to be defined everywhere. We denote the spectrum of an operator B with $\sigma(B) := \{\lambda \in \mathbf{C} : (\lambda I - B) \text{ is not a bijection}\}$. $S(\alpha) := \{z \in \mathbf{C} : z = re^{i\phi}, r > 0, \phi \in (-\alpha, \alpha)\}$ is called a sector of the complex plane with angle α.

The first calculus presented is actually not a functional calculus in the true sense, it is just an extension of the Laplace transform to operator valued functions, of which we consider only consider one case. However, it nicely shows the spirit of arguments that are also used later. Let A generate the semigroup $T(\cdot)$ and B be continuous and commuting with $T(\cdot)$. By making, formally, the substitution $\lambda \mapsto \lambda B$ we gain by the well known resolvent equation for semigroups:

$$(\lambda I - BA)^{-1} = B^{-1}(\lambda B - A)^{-1} = B^{-1} \int_0^\infty e^{-t\lambda B^{-1}} T(t)dt.$$

One then tries to find sufficient conditions for this integral to converge. Since a similar Laplace formula holds for integrated semigroups one may prove theorems of the following sort.

Theorem 1 *Let A generate an integrated semigroup $T(\cdot)$ and let B be a bounded, linear operator that commutes with $T(\cdot)$, such that $\|e^{itB^{-1}}\| = O(t^n)$ for some $n \in \mathbf{N}$. Then BA generates an integrated semigroup.*

The second calculus to be reviewed is used with holomorphic semigroups. The functional calculus available for such semigroups was used for instance by R. DeLaubenfels and F. Neubrander (private communication) in the commuting case. There one uses

$$(\lambda I - BA)^{-1} = \int_\Gamma (\lambda I - \omega B)^{-1}(\omega I - A)^{-1} d\omega$$

where Γ is a suitable curve (see [1]). But the standard calculus available for all bounded operators may also be used. With it one can achieve results even for the non-commuting case. In contrast with the previous results, the semigroup itself is "constructed" as opposed to its resolvent. In the commuting case this is easy, while in the non-commuting case a Trotter-Kato approximation may be used to obtain the result.

Theorem 2 *Let A be the generator of a semigroup $T(\cdot)$ bounded holomorphic in the sector $S(\alpha)$. Let B be a bounded, linear operator with $\sigma(B) \subset S(\alpha)$ and Γ a curve around $\sigma(B)$ inside $S(\alpha)$. If $\mathbf{C}_+$ is not a subset of $\sigma(AB)$ and there exists an $M \geq 1$, such that all powers of*

$$F(t) = \int_\Gamma T(\lambda t)(\lambda I - B)^{-1} d\lambda \tag{1}$$

are bounded in norm by M for all $t \geq 0$, then AB generates a bounded semigroup.

Note, that in case that $T(\cdot)$ and B commute, $F(t)$ already presents the semigroup. The following theorem requires more assumptions on the semigroup, but gives a result not requiring the technical condition of the above theorem on $F(t)$. It shows, that the condition is in fact not too hard to check in certain examples. It also generalizes Theorem 7 (s. b.) for the special case $\alpha = \pi/2$. However, we emphasize, that the boundedness condition on $F(t)$ above cannot be omitted altogether as there are counterexamples to the resulting theorem (see [9]).

Proposition 3 *Let A generate a holomorphic semigroup $T(\cdot)$ of angle $\pi/2$ that is bounded in the right half plane. If B is a bounded, linear operator with $\sigma(B) \subset S(\alpha)$, then BA and AB will also generate bounded analytic semigroups, which are both of (at least) angle $\pi/2 - \alpha$.*

Proof: Suppose $\|T(\lambda)\| \leq M$ for $M \geq 1$. We consider $A_\delta := A - \delta I$ as generator of $T_\delta(\cdot)$ for $\delta > 0$. We shall prove that the assumption of Theorem 2 is fulfilled. For each $r > 0$ and with $|\lambda - r| < r$ the power series

$$T_\delta(\lambda t) = \sum_{n=0}^{\infty} \frac{(\lambda - r)^n}{n!} t^n A_\delta^n T_\delta(rt)$$

converges uniformly, since $\frac{d^n T_\delta(\lambda t)}{d\lambda^n}(r) = t^n A_\delta^n T_\delta(rt)$. Choose $r > 0$ and Γ to be a curve in such a manner, that Γ lies inside the circle with radius r and around the spectrum $\sigma(B)$. Applying this to Formula 1 yields

$$F_\delta(t) = \sum_{n=0}^{\infty} \frac{(B - r)^n}{n!} t^n A_\delta^n T_\delta(rt)$$

for all $t \geq 0$ and sufficiently large $r > 0$. We now use the estimate

$$\frac{\|A_\delta^n T_\delta(t)\|}{n!} \leq (\frac{Ce^{-\delta t}}{t})^n,$$

which holds for these holomorphic semigroups (see [13]), to get

$$\begin{aligned}\|F_\delta(t)\| &\leq \sum_{n=0}^{\infty} \|B-r\|^n (\frac{Ce^{-\delta rt}}{r})^n \\ &\leq \sum_{n=0}^{\infty} C^n e^{-n\delta rt} \|1-r^{-1}B\|^n \\ &\leq \frac{1}{1-Ce^{-\delta rt}\|1-r^{-1}B\|}.\end{aligned}$$

Here C is independent of r and t, and r is chosen large enough so the series converges uniformly. The last term tends to 1 as $r \to \infty$. We conclude that $\|F_\delta(t)\| \leq 1$, since $F_\delta(t)$ is independent of r. But $F_\delta(t)$ obviously converges to $F(t)$, uniformly on compact intervals, as $\delta \to 0$. Therefore $\|F(t)\| \leq 1$ and the premises of Theorem 2 holds. But this is not only true for B itself, but also for all $e^{i\phi}B$ with $\phi \in (-\pi/2 + \alpha, \pi/2 - \alpha)$. Thus the semigroup generated by BA is holomorphic as claimed. To see that AB also generates a semigroup use [3], Theorem 1. That this semigroup is also holomorphic in the desired sector is easy to see by applying the theorem to $e^{i\phi}AB$ and observing the bounds obtained.

Corollary 4 *Consider the Laplacian Δ in $L_2(\mathbf{R}^n)$. Then for any $h \in L_\infty(\mathbf{R}^n)$ with essential range in $S(\alpha)$ that is bounded away from zero, the operators $M_h\Delta$ and ΔM_h generate contraction semigroups holomorphic in $S(\pi/2-\alpha)$. (M_h denotes the multiplication operator associated with h.)*

We now turn to a special functional calculus for the perturbing operator B. It is the most prominent one and used widely for self-adjoint operators. One can see, that it is mostly useful in the case where B and the semigroup commute. If $\{E_\lambda\}_{\lambda\in[a,b]}$ is the spectral measure associated with B, then we use

$$F(t) = \int_a^b T(\lambda t) dE_\lambda. \tag{2}$$

Theorem 5 *Let A be the generator of a semigroup $T(\cdot)$ in the Hilbert space H. Let B be a bounded, linear, positive semi-definite Operator (in H), such that B commutes with $T(\cdot)$. BA then generates a semigroup.*

Corollary 6 *(a) If A generates a contraction semigroup, then BA generates a contraction semigroup.*

(b) If A generates a group and B is just self-adjoint (not necessarily positive semidefinite), then BA generates a group.

Another theorem that is closely related to the spectral schemes used above, but does not use functional calculi is the following. It improves a result found in [9] giving less cumbersome bounds about the sector, in which the perturbed semigroup is holomorphic.

Theorem 7 *Let A be normal in the Hilbert space H and the generator of a bounded semigroup, holomorphic in $S(\alpha)$. Let B be a bounded, linear operator with $\|B\| < \sin\beta$ with $\alpha > \beta > 0$, then $(I+B)A$ and $A(I+B)$ generate semigroups, holomorphic in $S(\alpha-\beta)$.*

This theorem can be generalized to include generators A in arbitrary Banach spaces, for which $\|A(\lambda I - A)^{-1}\| = \sup\{|\mu| : \mu \in \sigma(A(\lambda I - A)^{-1})\}$. This is the case e. g. for multiplication operators that generate semigroups in L_p spaces.

References

[1] DeLaubenfels, Bounded, Commuting Multiplicative Perturbations of Strongly Continuous Group Generators, *Houst. J. Math.* 17 (1991), 299–310

[2] Desch, W., I. Lasieka and W. Schappacher, Feedback Boundary Control Problems for Linear Semigroups, *Israel J. Math.* 51 (1989), 177–207

[3] Desch, W. and W. Schappacher, Some Generation Results for Perturbed Semigroups, in *Lecture Notes in Pure and Applied Mathematics* Vol. 116, Marcel Dekker, New York (1989), 125–152

[4] Dorroh, J. R., Contraction Semigroups in a Banach Space, *Pac. J. Math.* 19 (1966), 35–38

[5] Dorroh, J. R. and A. Holderrieth, Multiplicative Perturbation of Semigroup Generators, in *LSU Sem. Notes in Func. Ana. and PDES,* Dept. Math./Louis. State Univ., Baton Rouge (1990-1991),59–68

[6] Gustafson, K. A note on Left Multiplication of Semigroup Generators, *Pac. J. Math.* 24 (1968) 463–465

[7] Gustafson, K. and G. Lumer, Multiplicative Perturbations of Semigroup Generators, *Pac. J. Math.* 41 (1972) 731–742

[8] Gustafson, K. and K. Sato, Some Perturbation Theorems for Nonnegative Contraction Semigroups, *J. Math. Soc. Japan* 21 (1969) 200–204

[9] Holderrieth, A., Multiplicative Perturbations, Doctorate Thesis, Eberhard-Karls-Universiät Tübingen, Tübingen, 1992

[10] JUNG, M., Multiplikative Störungen von C_0-Halbgruppen, Doctorate Thesis, Technische Universität Berlin, Berlin, 1994

[11] JUNG, M., Some Perturbation Results for Semigroups, Preprint, Technische Universität Berlin, Berlin, 1994

[12] LUMER, G., New Singular Multiplicative Generation Results via Homotopy-like Perturbations, *Arch. Math.* 53 (1989) 52–60

[13] PAZY, A., Semigroups of Linear Operators and Applications to Partial Differential Equations, Springer, Berlin, 1983

R M KAUFFMAN

Functional analysis and spherical functions

The theory of generalized eigenfunction expansions for a single operator is well-established. For a recent paper giving a simple formulation of the abstract theory, as well as some applications to Schrödinger operators, see Poerschke-Stolz-Weidmann [7]; for another exposition concentrating particularly on Schrödinger operators see Simon [8]. For many applications, however, one studies simultaneous generalized eigenfunction expansions for a family of commuting operators. A general theory of these expansions is given in [6]. In this note, we discuss a celebrated eigenfunction expansion in geometry, the spherical function expansion of Harish-Chandra, to analyze which portion of the theory of that expansion is geometry, and which follows from general functional-analytic principles, in particular the general theory of eigenfunction expansions given in [6]. As an illustration of these ideas we analyze the Bessel function expansion in R^2. For the proofs of these results, the reader is referred to [6], which also contains a good deal of additional material.

A modern exposition of the spherical function expansion is given in the book by Helgason [3], which, along with his classic book [2], gives a self-contained exposition of all the necessary background material. A generalization of this expansion is given in the well-known paper of Helgason [4]. The relation of that expansion theory to the expansion given in [6] is a subject for future research.

Theorem 25, for eigenfunctions of a single operator, first appeared in Edmunds and Kauffman [1].

Definition 1 *A locally convex topological vector space is said to be a* **nuclear space** *if, for any convex balanced neighborhood V of 0, there exists another convex balanced neighborhood $U \subseteq V$ of 0 such that the canonical mapping $T : X_U \to X_V$ is nuclear. A* **nuclear operator** *from a locally convex topological vector space X into a Banach space Y is an operator of the form*

$$Tx = s - \lim_{n\to\infty} \sum_{j=1}^{n} c_j f_j(x) y_j$$

where $\{f_j\}$ *is an equicontinuous sequence of continuous linear functionals on* X, $\{y_j\}$ *is a bounded sequence of elements of* Y, *and* $\{c_j\}$ *is a sequence of non-negative real numbers such that* $\sum_{j=1}^{\infty} c_j < \infty$. *The spaces* X_U *and* $\hat{X}_U$ *are defined as follows: let* U *be a convex balanced neighborhood of* 0 *in* X. *Let* κ_U *be the Minkowski functional on* U. *Let* $N_U = \{x \in X : \lambda x \in U \ \forall \lambda > 0\}$. *Then* N_U *is a closed subspace of* X, *and the quotient space* $\frac{X}{N_U}$ *is a normed linear space* X_U *under the norm induced by* κ_U . $\hat{X}_U$ *is the completion of* X_U.

Theorem 2 *A locally convex topological vector space* X *is nuclear if and only if for any convex balanced neighborhood* V *of 0, the natural mapping* I_V *from* X *into* $\hat{X}_V$ *is nuclear.*

Lemma 3 *Let* ω *be a trace class operator in a Hilbert space* h, *such that the null space of* ω *is trivial. Let* $X = \cap_{n=1}^{\infty} range\,(\omega^n)$. *Give* X *the seminorms* $\rho_n(x) = \|(\omega^n)^{-1}x\|$. *Then* X *is nuclear.*

Definition 4 *Let* $\mathcal{A}$ *be a von Neumann algebra; that is, an algebra of operators on a Hilbert space* h *which is closed under adjoint, and which is complete in the strong operator topology. A normal operator* A *is said to be* **affiliated** *with* $\mathcal{A}$ *if the spectral projections for* A *all lie in* $\mathcal{A}$. *(This is an equivalent form of the usual definition; see [5].)*

Remark: An operator is normal if and only if it is affiliated with a commutative von Neumann algebra; this is a theorem which is apparently due to Murray and is discussed in [5].

Definition 5 *A* **differential expression** *is defined to be a continuous linear operator on* $C_0^{\infty}(M)$ *which does not increase supports.*

Definition 6 *Let* D *be a family of differential expressions on a Riemannian manifold* Q. D *is called a* **regular family** *if:*

1. $D = \{\tau_i\}_{i=1}^{k}$ *is a family of formally commuting differential expressions on* Q, *such that the closure* $(\tau_i)_0$ *of the restriction of* τ_i *to* $C_0^{\infty}(Q)$ *is normal;*

2. *for any* $\tau_i \in D$, *the formal adjoint expression* $\tau_i^{+} \in D$; *(recall that* τ^{+} *is defined by the relation* $[\tau f, g] = [f, \tau^{+} g]$ *for all* $f, g \in C_0^{\infty}(Q)$*);*

3. *the smallest von Neumann algebra $\mathcal{A}$ with which each operator $(\tau_i)_0$ is affiliated is commutative;*

4. *there is an elliptic differential expression ℓ, such that the closure ℓ_0 of the restriction of ℓ to $C_0^\infty(Q)$ is affiliated with $\mathcal{A}$ (and is therefore normal).*

Definition 7 *A* **strict inductive limit** *V of a sequence $\{V_n\}$ of locally convex topological vector spaces, where $V_n \subset V_{n+1}$ and the containment is algebraic and topological, is the topological vector space formed by giving $\cup_{n=1}^{\infty} V_n$ the topology such that a set is open if and only if its intersection with each V_n is open.*

Definition 8 *Let $\mathcal{F}$ be a family of normal operators on a Hilbert space h. Suppose that W is a nuclear space which is a strict inductive limit of separable Frechet spaces, and that W is contained in the domain of every element of $\mathcal{F}$. Further suppose that $W \subset \mathsf{h} \subset W'$, where the containment is algebraic and topological, so that the embedding of W into W' uses the inner product of h. Finally suppose that W is dense in h. Then we say that W is a* **space of attainable states** *for $\mathcal{F}$. (In this note, the main example given will be $C_0^\infty(M)$, where M is a complete Riemannian manifold, but other examples, analogous to the rapidly decreasing functions on R^n, may also be shown to exist in geometry; see [6].)*

Definition 9 *Suppose W is a Montel space, in the sense that closed and bounded subsets are compact (the above W is such a space). The space $C(W, W')$ is defined as follows: give W' the topology of uniform convergence on bounded subsets of W; and give the continuous linear transformations from W into W' the topology defined by the set of all seminorms $\kappa_{A,B}$, where A, B are bounded convex balanced subsets of W, and*

$$\kappa_{A,B}(T) = \sup_{x \in A, y \in B} |T(x)(y)| .$$

Definition 10 *Let D be a regular family of differential expressions on a Riemannian manifold Q. A space of* **regular attainable states** *for D is a topological vector space W with the following properties:*

1. *W is a space of attainable states for* $\{(\tau_i)_0\}_{i=1}^k$;

2. $W \subset C^\infty(Q)$;

3. $C_0^\infty(Q)$ *is topologically contained in* W.

Definition 11 *(This is an informal definition; for a formal treatment see Helgason [2].)* **Riemannian globally symmetric spaces** *are connected Riemannian manifolds* M *where for any point* $p \in M$ *there is a global isometry of* M *(for which* p *is a necessarily isolated fixed point) which when restricted to a neighborhood of* p *reverses the direction of all the geodesics through* p*; in other words, the geodesic with tangent vector* X *at* p *goes to the geodesic with tangent vector* $-X$ *at* p*. Such spaces have many interesting properties; for example, they are complete, and for any two points* x *and* y *of* M*, there is an isometry of* M *taking* x *to* y*. (This property is one of the reasons why these spaces have physical interest.) However, the curvature of* M *need not be constant, because the sectional curvatures for different tangent planes through* p *may be unequal. If* M *is globally symmetric, then* M *is diffeomorphic to* G/K*, where* G *is the identity component of the isometry group of* M*, and* K *is the subgroup fixing* p*.* K *is necessarily compact. If* G *is semisimple (which means that it is naturally a semi-Riemannian manifold with metric arising from the Killing form) and* K *is a maximal compact subgroup, then* M *is said to be of noncompact type. These spaces have nonpositive sectional curvature, and have the interesting property that a geometric property, the maximum dimension of any flat totally geodesic submanifold of* M*, called the rank of* M*, is the number of generators of the ring of all differential expressions* τ *on* M *which commute formally with* G*, in the sense that for any* $g \in G$*, and any* C^∞ *function* f*,* $\tau f \circ g = \tau(f \circ g)$*.*

Definition 12 *Let* M *be a Riemannian globally symmetric space. The family of* C^∞ *differential expressions* τ *such that for any* $g \in G$*, and* $f \in C^\infty(M)$*,* $\tau(f \circ g) = \tau(f) \circ g$*, will be called the family of* **invariant differential expressions***, and denoted by* $D(G/K)$*.*

Remark: In the language of this note, we use the phrase "differential expression" instead of the more usual phrase "differential operator" to highlight the fact that in general there are many Hilbert or Banach space operators associated with a given differential expression. However, for an invariant differential expression τ on a Riemannian globally symmetric space M, there is in general only one reasonable operator τ_0

in $L_2(M)$; this is the closure of the restriction of τ to $C_0^\infty(M)$, which is shown in [6] to be a normal operator in $L_2(M)$. This is part of the assertion of the following theorem.

Theorem 13 *$C_0^\infty(M)$ is a space of regular attainable states for $D(G/K)$.*

Definition 14 *The* **extended Gelfand transform** *$\widehat{GT}$ is a ring homomorphism between the operators affiliated with a commutative von Neumann algebra $\mathcal{A}$ and the normal functions on the maximal ideal space of $\mathcal{A}$. A real valued function on an extremely disconnected compact Hausdorff space X (this means that X has the property that the closure of every open set is open; the maximal ideal space of $\mathcal{A}$ has this property) is said to be* **normal** *if it is either continuous at a point x or converges to ∞ or $-\infty$ at x, for each point x of X, and if furthermore the set of discontinuities is a meager set. (Thus bounded normal functions are continuous). It not so obvious how to multiply two normal functions, or to multiply two unbounded operators affiliated with $\mathcal{A}$, but this can nevertheless be done (See Kadison and Ringrose [5]). When the transform $\widehat{GT}$ is restricted to $\mathcal{A}$, it is the usual Gelfand transform, denoted by GT, taking $\mathcal{A}$ isometrically onto $C(X)$, where X is the maximal ideal space of $\mathcal{A}$.*

Definition 15 *Let $\mathcal{A}$ be a commutative von Neumann algebra with identity, of operators on h. By the Gelfand representation theorem, $\mathcal{A}$ is isometric to $C(X)$, where X is the maximal ideal space of $\mathcal{A}$. For any $e \in \mathsf{h}$, we define the measure μ_e on X as the linear functional J on $C(X)$ such that $J(GT(A)) = [Ae, e]$, where $[\ ,\ ]$ denotes the inner product of h.*

Theorem 16 *Let W be a space of regular attainable states for $D(G/K)$. Let e be a cyclic vector for the restriction of $\mathcal{A}(D(G/K))$ to the subspace of $L_2(M)$ which is invariant under the action of K. Then there exists a meager subset Λ of the maximal ideal space X of $\mathcal{A} = \mathcal{A}(D(G/K))$, and a continuous mapping $\lambda \to P_\lambda$ from $X\backslash\Lambda$ into $C(W, W')$ such that the following holds:*

1. *for $A \in \mathcal{A}$,*

$$A = \int_{X(\mathcal{A})} \overline{GT(A)}(\lambda) P_\lambda d\mu_e(\lambda)$$

where the integral converges in the sense that for any continuous seminorm σ *of* $C(W, W')$, $GT(A)\sigma(P_\lambda)$ *is in* $L_1(\mu_e)$; *(since* $GT(A)$ *is bounded and continuous, this is an assertion only about* P_λ;

2. *for all* λ *in the complement of* Λ, *the net* $\{P(\Delta)\}$ *converges to* P_λ *in* $C(W, W')$ *as* Δ *takes on all values of the directed set of open sets about* λ *in* $X(\mathcal{A})$;

3. *Any* $F \in \underline{range}(P_\lambda)$ *is an element of* $C^\infty(M)$ *and satisfies the equation* $\tau f = \widehat{GT(N)(\lambda)}F$ *for all* $\lambda \in X\backslash\Lambda$ *and all* $N = \tau_0$, *with* $\tau \in D(G/K)$; *where* $\widehat{GT}(N)(\lambda)$ *is the extended Gelfand transform of* N *evaluated at* λ, *which is finite and well-defined at all points of* $X\backslash\Lambda$;

4. *For any bounded subset* B *of* $\mathcal{A}$, *and any neighborhood* Γ *of* 0 *in* $C(W, W')$ *there exists a finite partition* $\{\Delta_i\}_{i=1}^k$ *of* $X\backslash\Lambda$, *such that for any elements* $\lambda_i \in \Delta_i$, *and any* $A \in B$,

$$A - \sum_{i=1}^{k} \overline{GT(A)}(\lambda)P_{\lambda i}\mu_e(\Delta_i) \in \Gamma.$$

Definition 17 *Let* Q *be a Riemannian manifold,* ρ *be the natural measure on* Q, $D = \{\tau_i\}_{i=1}^k$ *be a regular family of differential expressions on* Q, *and* W *a set of regular attainable states for* D. *We further suppose that* κ *is a subgroup of the group of unitary operators in* $L_2(Q, \rho)$ *equipped with the strong operator topology such that* κ *is compact, and such that if* $U \in \kappa$ *then* U *takes* W *continuously into* W, *and the mapping* $U \to U\phi$ *is a continuous function from* κ *into* W, *for any* $\phi \in W$, *and that this is also true when* $W = C_0^\infty(Q)$. *(Note that these hypotheses imply that the strong and weak operator topologies on* κ *are the same, since a 1-1 continuous mapping on a compact space is a homeomorphism. Hence, in particular, the mapping taking* U *to* U^{-1} *is continuous on* κ.*) In addition, we suppose:*

1. *if* ℓ *is as in Definition 6, then any* f *in the domain of* $\ell_0 = H$ *is in* $L_\infty(Q)$;

2. $\kappa \subset \mathcal{A}'$, *the set of all bounded operators commuting with* $\mathcal{A}$;

3. *if F and G are elements of $C^\infty(Q) \cap W'$ such that for all $U \in \kappa$, if U^t is the transpose of U considered as a continuous linear transformation from W into W, $U^t(F) = F$, $U^t(G) = G$, and $\tau_i F = \lambda_i F$ and $\tau_i G = \lambda_i G$ for all $\tau_i \in D$, where each λ_i is a complex number, then F and G are linearly dependent.*

Example 18 *If $N > n/4$, where n is the dimension of the Riemannian globally symmetric space M, and $\mathcal{L}$ is the Laplacian on M, and $H_{\mathcal{L}}$ is the closure of the restriction of $\mathcal{L}$ to $C_0^\infty(M)$, then any $f \in D\left((H_{\mathcal{L}})^N\right)$ is in $L_\infty(M)$. Hence if $\{\tau_i\}_{i=1}^k$ generate the ring $D(G/K)$, and $W = C_0^\infty(M)$, and $\mathcal{A}$ is the smallest von Neumann algebra with which each $(\tau_i)_0$ is affiliated, and $D = \{(\tau_i)_0\}$, and κ is $\{U_g : g \in K\}$, then the hypotheses of the previous definition are satisfied with $\ell = \mathcal{L}^N$.*

Definition 19 *An* **invariant eigenfunction expansion** *on a Borel subset $\Delta \subset R^k$ associated with the regular family of differential expressions $\{\tau_i\}_{i=1}^k$ on the Riemannian manifold Q with a space W of regular attainable states is defined to be a complete positive Borel measure μ on Δ together with an almost everywhere (with respect to μ) defined mapping $\beta \to G_\beta$ from the subset Δ of R^k into $\{G \in W' \cap C^\infty(Q) : U^t(G) = G \ \forall\, U \in \kappa\}$, such that $\mu(C)$ is finite for any compact subset C of Δ, and such that the following hold:*

1. *the mapping taking β to G_β is scalarwise measurable from Δ into W', in the sense that for any element $\phi \in W$, $c_\phi(\beta) = G_\beta(\phi) = \int_M \phi(x)\, \bar{G}_\beta(x)\, d\mu(x)$ is a Borel measurable function of β;*

2. *the mapping V from ϕ to c_ϕ is an isometry from W_I, with the norm of $L_2(Q)$, onto a dense subspace of $L_2(\Delta, \mu)$; V therefore extends to a unitary operator, also denoted by V, from $L_{2,I}$ onto $L_2(\Delta, \mu)$;*

3. *for almost every $\beta = <\beta_1, ..., \beta_k> \in \Delta$ with respect to μ, there exists an element F of $C^\infty(Q) \cap W'$ such that $\tau_i F = \beta_i F$.*

Definition 20 *Let $M = G/K$ be a Riemannian globally symmetric space. A spherical function F is an element of $C^\infty(M)$ which satisfies $\tau F = \lambda(\tau) F$ for all $\tau \in D(G/K)$, where the mapping $\tau \to \lambda(\tau)$ is a ring homomorphism from $D(G/K)$ to the complex numbers, and which is invariant under K in the sense that $F \circ g = F$ for all $g \in K$.*

Theorem 21 *Let M be a Riemannian globally symmetric space, and $\lambda \to P_\lambda$ be as in Theorem 16. Let the sequence $\phi_n \in C_0^\infty(M)$ converge to $\theta \in L_2(M)$. Then, for all $\mu \in X$ in the complement of a meager set, where X is the maximal ideal space of $\mathcal{A}(D(G/K))$, $P_\mu \phi_n$ converges in W' to a limit which is independent of the approximating sequence ϕ_n, depending only on θ.*

Definition 22 *Let $\theta \in L_2(M)$. For all λ in the complement of a meager set, define $F_{\lambda,\theta}$ to be the above limit.*

Theorem 23 *Let X be as above. Let S be a meager subset of X, with closure $\overline{S}$. Let $\theta \in \mathsf{h}$. Then $\mu_e\left(\overline{S}\right) = 0$.*

Remark: We now use the above formalism to construct an eigenfunction expansion in terms of spherical functions for any Riemannian globally symmetric space. The following theorem also states that any other method of constructing such an expansion yields essentially the same result. Finally, it gives the conclusion that any such expansion converges in a sense which corresponds to absolute and uniform convergence in the case of a series.

Theorem 24 *Let M be an n-dimensional Riemannian globally symmetric space $M = G/K$, where as usual G is the identity component of the isometry group of M and K is the subgroup of G fixing a given point. Let $\kappa = \{U_g : g \in K\}$, where $U_g(f) = f \circ g$. Let $L_{2,I}$ be the subset of $L_2(M)$ of functions which are invariant under K. Let $\{\tau_i\}_{i=1}^k$ be a set of generators of the ring $D(G/K)$. Let $\mathcal{A}_I$ be the smallest von Neumann algebra in $L_2(M)$ containing $\mathcal{A}(D(G/K))$ and P_I, the orthogonal projection onto I. Let ν_I be the clopen (closed and open) subset of the maximal ideal space of $\mathcal{A}_I$ such that the Gelfand transform of P_I is the characteristic function of ν_I; define χ on ν_I by $\chi(\lambda) = < \widehat{GT}((\tau_1)_0)(\lambda), ..., \widehat{GT}((\tau_k)_0)(\lambda) >$. Let Δ be the range of χ. Then:*

1. *For any positive integer m, $L_{2,I}$ has a cyclic vector $e \in D((H_{\mathcal{L}})^m)$, where $H_{\mathcal{L}}$ is the self adjoint operator corresponding to the Laplacian;*
2. *Δ is a countable union of compact sets;*
3. *Δ is the spectrum of the restriction of $H_{\mathcal{L}}$ to $L_{2,I}$;*

4. *if* $F_{\chi(\lambda)}$ *is defined to be* $F_{\lambda,e}$ *(this is defined using Definition 22), and* μ *is defined to be the positive measure on* Δ *such that* $\mu(S) = \mu_e(\chi^{-1}(S))$, *with* μ_e *defined as above on the maximal ideal space of* $\mathcal{A}$, *then* $F_{\chi(\lambda)}$ *is a spherical function for almost every* $\beta \in \Delta$ *with respect to* μ*;*

5. *the mapping* $\beta \to F_\beta$ *is an invariant eigenfunction expansion on* Δ*;*

6. *Let* $\beta \to G_\beta$ *be an invariant eigenfunction expansion for* $\tau_1, ..., \tau_k, K$ *on the Borel subset* Γ *of* R^k, *with an associated measure* γ *on* Γ, *such that under the associated unitary transformation from* $L_{2,I}$ *to* $L_2(\Gamma, \gamma)$, $(\tau_i)_0$ *is unitarily equivalent to multiplication by* β_i. *Then* μ *and* γ *are mutually absolutely continuous on* $\Gamma \cap \Delta$, *which is a set of full measure with respect to each;*

7. *the* G_β *are spherical functions for almost every* β *with respect to* γ *and* $G_\beta = \alpha(\beta) F_\beta$ *for a Borel measurable function* α *such that* $|\alpha(\beta)|^2 d\gamma = d\mu$*;*

8. *for any* $g \in L_1(M)$, $\gamma(\{\beta : gG_B \notin L_1(M)\}) = 0$*; (this follows from the known fact that for almost every* β *the spherical function* G_β *is actually bounded, but it is also a consequence of the general theory of [6].*

9. *for any* $f \in D\left((H_{\mathcal{L}})^N\right)$, *where* $N > n/4$, $\int_\Delta |c_f(\beta)| G_\beta(x) d\gamma \in L_\infty(Q)$.

Remark: The last conclusion is true for any invariant eigenfunction expansion, as the following theorem states (see [6] for the proof in this context, which imitates the proof of the corresponding assertion for a single operator in [1]).

Theorem 25 *Suppose* $\beta \to F_\beta$ *is an invariant eigenfunction expansion on* Δ. *Let* $f \in D(H) \cap L_1(Q, \rho)$. *The function* $h : h(x) = \int_\Delta |c_f(\beta) F_\beta(x)| d\mu$ *is in* $L_\infty(Q, \rho)$.

- To see what the theory says in a well known situation, we use it to study the generalized eigenfunction expansion for the polar coordinate Laplacian in R^2.

Let $\mathsf{h} = L_2(R^2)$, and let $D = H_{\mathcal{L}}$, where $H_{\mathcal{L}}$ is the closure in $L_2(R^2)$ of the restriction of the Laplacian to $C_0^\infty(R^2)$. R^2 is a symmetric space. Let κ be the unitary group of rotations. Let W be the rapidly decreasing functions on R^2. It clear that W is a space of regular attainable states for D. Hence, Theorem 24 guarantees an invariant eigenfunction expansion in terms of spherical functions. What are these? The functional calculus for operators affiliated with a von Neumann algebra guarantees that except for a meager set the range of $\widehat{GT}(H_{\mathcal{L}})$ is contained in the spectrum of $H_{\mathcal{L}}$. By direct calculation, one can see that the generalized eigenfunctions of $H_{\mathcal{L}}$ which are invariant under κ, and which correspond to $\lambda < 0$, are all multiples of $J_0\left(\sqrt{-\lambda}r\right)$, where J_0 is the Bessel function of order 0 of the first kind. In fact, since the generalized eigenfunctions must be C^∞, they are well-behaved at the origin; one can either observe that the second solution to Bessel's equation is unbounded, by calculating it, or more simply one can calculate the Wronskian of any two linearly independent solutions and show that it blows up at the origin, so that Bessel functions of the first kind are the only solutions which can show up in our theory. The subspace $L_{2,I}$ is just the functions of r alone. We obtain from Theorem 25 the conclusion that for any $f \in D(H_1^\alpha)$, where $\alpha > \frac{1}{2}$, its Bessel expansion in terms of the above eigenfunctions converges absolutely and uniformly, no matter what spectral measure is used. From Theorem 16 and a simple limiting process, we see that the expansion converges to f. Let e be a cyclic vector for $L_{2,I}$, which exists by Theorem 24. With respect to the measure, using Theorem 24, we see that for any $g \in L_1(R^2)$ the measure of the set of all λ such that $J_0\left(\sqrt{-\lambda}r\right) f$ is not in $L_1(R^2)$ is a set of measure 0; hence, in particular, there must be some $\lambda < 0$ such that $J_0\left(\sqrt{-\lambda}r\right) f \in L_1(R^2)$. From this, we easily see that $J_0(r) f \in L_1(R^2)$ for all $f \in L_1(R^2)$. This implies the well-known fact that $J_0 \in L_\infty(R^2)$.

Remark: The Fourier-Bessel expansion in R^2, which expands non-radial functions, is also an invariant eigenfunction expansion, but to treat this expansion goes beyond the scope of this note. It is treated as part of the theory of invariant eigenfunction expansions in [6].

References

[1] D. E. Edmunds and R. M. Kauffman, "Continuous Spectrum Eigenfunction Expansions and the Cauchy Problem in L_1", *Proc. Royal Soc. London* (A) **441** (1993) 407-422.

[2] S. Helgason, *Differential Geometry, Lie Groups and Symmetric Spaces,* (Academic Press, New York, 1978).

[3] S. Helgason, *Groups and Geometric Analysis,* (Academic Press, New York, 1984).

[4] S. Helgason, "A duality for symmetric spaces with applications to group representations", *Advances in Mathematics* **5** (1970) 1-154.

[5] Richard V. Kadison and John R. Ringrose, *Fundamentals of the Theory of Operator Algebras,* (Academic Press, New York, 1983).

[6] R. M. Kauffman, "Eigenprojection Representations and Riemannian Symmetric Spaces" (to appear).

[7] T. Poerschke, G. Stolz and J. Weidmann, "Expansions in Generalized Eigenfunctions of Selfadjoint Operators", *Math. Zeitschrift* **202** (1989) 397-408.

[8] Barry Simon, "Schrödinger Semigroups", *Bull. Amer. Math. Soc.* **7**, 3 (1982) 447-526.

V KHATSKEVICH, S REICH AND D SHOIKHET

Ergodic type theorems for nonlinear semigroups with holomorphic generators

1. INTRODUCTION. Let X be a reflexive Banach space and let $T : X \to X$ be a linear operator all the powers of which are uniformly bounded. By the Mean Ergodic Theorem [Y] the Cesaro means of the powers of T converge strongly, pointwise on X , to a linear projection of X onto the fixed point set of T. An analogous result, where the Cesaro means are replaced by the corresponding integral, holds in the continuous semigroup case [HP]. In recent years it has turned out that there are nonlinear mean ergodic theorems for nonexpansive mappings and semigroups in Hilbert and in "nice" Banach spaces (see, for example, [P],[R2],[GR], and the references mentioned there). However, in the nonlinear case only weak convergence of the Cesaro means can be established in general. In both the linear and the nonlinear cases, strong convergence can be obtained for the resolvents of the generators of the semigroups. In the linear case this is, in fact, a consequence of the mean ergodic theorem (cf.[S]), while in the nonlinear case a completely different argument is required [R1].

Our intention in this paper is to study the corresponding questions for discrete and continuous semigroups of holomorphic mappings in reflexive Banach spaces. Our first main result (Theorem 1 below) is a continuous analogue of the Mazet-Vigué discrete ergodic theorem [MV]. Our second main result (Theorem 2) is a convergence theorem for the resolvents of generators. It has precisely the same form both in the discrete and in the continuous cases.

2. NONLINEAR SEMIGROUPS. Let D be a subset of a Banach space X.

A family $S=\{F_t\}$, where either $t\in R^+(=[0,\infty))$ or $t\in N$ $(=\{0,1,2,\dots\})$, of self mappings F_t of D is called a (one parameter) semigroup if

$$F_{s+t} = F_s \circ F_t , \qquad s,t\in R^+ \ (s,t\in N), \tag{1}$$

and

$$F_0 = I|_D , \tag{2}$$

where I is the identity mapping on X.

A semigroup $S=\{F_t\}$, $t\in R^+$, is said to be continuous if the vector-valued function $F_t x$: $R^+ \to D$ is continuous in t for each $x\in D$.

If $t\in N$ we say that the semigroup S is discrete. In other words, a discrete semigroup $S=\{F_t\}$, $t\in N$, is the family of iterates of a self-mapping $F=F_1$: $D \to D$.

Definition 1. *Let $S=\{F_t\}$, $t\in R^+$, be a continuous semigroup on D. If the strong limit*

$$f(x) = \lim_{t\to 0+} \frac{x-F_t(x)}{t} \tag{3}$$

exists for each $x\in D$, then it will be called the generator of the semigroup S.

It follows by the semigroup properties (1) and (2) that in this case F_t is a solution of the right hand Cauchy problem

$$\frac{\partial^+ F_t(x)}{\partial t} + f(F_t(x)) = 0, \qquad F_0(x)=x \; . \tag{4}$$

Let D and $\tilde{D}$ be domains in X. We shall denote by $Hol(D,\tilde{D})$ the set of holomorphic mappings from D into $\tilde{D}$ and by $Lip(\bar{D},\tilde{D})$ the set of Lipshitzian mappings from $\bar{D}$, the closure of D, into $\tilde{D}$. The set $Hol(D,\tilde{D}) \cap Lip(\bar{D},\tilde{D})$ will be denoted by $HL(\bar{D},\tilde{D})$.

Definition 2. *A mapping $f\in Hol(D,X)$ is said to be a semi-plus complete vector field if the Cauchy problem*

$$\frac{\partial F(x)}{\partial t} + f(F_t(x)) = 0, \qquad F_0(x)=x \; , \tag{5}$$

has a solution $\{F_t(x)\} \subset D$, $t\geq 0$, $x\in D$.

The semigroup properties (1) and (2) imply the following fact.

Proposition 1. *Let* $f \in Hol(D,X)$ *be the generator of a continuous semigroup, and assume that the convergence in (3) is uniform on each compact subset of D. Then* f *is a semi-plus complete vector field.*

It is clear that a semi-plus complete vector field is the generator of a continuous semigroup. Moreover, if it is bounded we have additional information on the convergence in formula (3).

Proposition 2. *Let* D *and* $\tilde{D}$ *be bounded domains in* X, *and let* $f \in Hol(D,\tilde{D})$ *be a semi-plus complete vector field. Then the net*

$$f_t(x) = \frac{x-F_t(x)}{t}$$

in (3) converges to f *in the topology of local uniform convergence on* D.

The topology of local uniform convergence is discussed, for example, in [IS].

Proof. Let U be an arbitrary closed subset of D. Since f is bounded on D, it follows by the Cauchy inequalities that $f \in Lip(U,D)$. Hence on some disk $\Omega \subset \mathbb{C}$ centered at $0 \in \mathbb{C}$, there is a unique solution $\Phi(t,x)$ of the Cauchy problem

$$\frac{\partial\ \Phi(t,x)}{\partial t} + f(\Phi(t,x)) = 0, \qquad \Phi(0,x)=x\ , \tag{6}$$

$(t,x) \in \Omega \times U$, which is holomorphic and bounded on $\Omega \times U$. Moreover, $\Phi(t,x) = F_t(x)$ on $(\Omega \cap R^+) \times U$. Thus we have

$$\Phi(t,x)=x+tf(x)+\omega(t,x)$$

for $(t,x) \in \Omega \times U$, where $\omega(t,x)$ is holomorphic in $t \in \Omega$ and bounded for each $x \in U$. By the Schwarz lemma we obtain

$$\|\omega(t,x)\| \le |t|^2 \sup_{(t,x)\in\Omega\times U} \|\omega(t,x)\|\ \varepsilon^{-2}\ ,$$

where ε is the radius of Ω. Then for $t \in \Omega \cap R^+$ we have the inequality

$$\|f_t(x)-f(x)\| \le t \sup_{(t,x)\in\Omega\times U} \|\omega(t,x)\| \, \varepsilon^{-2} ,$$

which proves the proposition.

Now we consider the stationary point set $\mathfrak{F}_D$ of a semigroup $S=\{F_t\}$ with a holomorphic generator. This set is defined as the common fixed point set of $\{F_t\}$ for all t, i.e. $\mathfrak{F}_D=\bigcap \mathrm{Fix} F_t$, $t\in R^+$ $(t\in N)$.

If the generator f is semi-plus complete, then it follows from the uniqueness of the Cauchy problem solutions that the stationary point set of S coincides with the null point set of f, i.e.

$$\mathfrak{F}_D = \mathrm{Null}_D f \tag{7}$$

(see, for example, [A]). Note that actually this also holds for the more general case, when f is a generator in the sense of Definition 1.

The following example shows that formula (7) is no longer true for the closure of D even in the case when f is continuous on $\bar{D}$.

Example. Let D be the unit disk in the complex plane $\mathbb{C}$, i.e. $D=\{x\in\mathbb{C}: |x|<1\}$. Consider $f(x)=x-1+\sqrt{1-x}$. It is clear that $f\in \mathrm{Hol}(D,\mathbb{C})$ and that it is continuous on $\bar{D}$. In addition, $\mathrm{Null}_D f=\{0,1\}$.

However, the Cauchy problem (5) has the solution F_t: $\bar{D} \to \bar{D}$, $t\ge 0$, defined by the formula:

$$F_t x = 1-[1-e^{-t/2} + e^{-t/2} \sqrt{1-x}]^2,$$

and for all $t>0$ we have

$$F_t(1) = 1-[1-e^{-t/2}]^2 < 1.$$

Thus $\mathfrak{F}_{\bar{D}} \ne \mathrm{Null}_{\bar{D}} f$.

3. ERGODIC TYPE THEOREMS. In order to formulate our main results we need some additional notions.

Definition 3. *Let D be a domain in X. We shall say that the sequence $\{g_n\}_{n=1}^{\infty}$ (respectively, the curve $\{g_t\}$, $t \ge 0$), $g_n \in Hol(D,X)$ ($g_t \in Hol(D,X)$), Φ_a-converges (Φ_a^*-converges) to $g \in Hol(D,X)$ with respect to a point $a \in D$, if the sequence (curve) of linear operators $\{g'_n(a)\}_{n=1}^{\infty}$ ($\{g'_t(a)\}$, $t \ge 0$) strongly converges to $g'(a)$ and there is a subsequence $\{g_{n_k}\}_{k=1}^{\infty}$ ($\{g_{t_n}\}$), which converges (weakly converges) to g pointwise in D.*

We write in this case that g is a Φ_a-(weak Φ_a-) limit of $\{g_n\}_{n=1}^{\infty}$, (of g_t) or

$$g = \Phi_a\text{-}\lim_n g_n \quad (\Phi_a^*\text{-}\lim_n g_n)$$

$$(g = \Phi_a\text{-}\lim_{t\to\infty} g_t \quad (\Phi_a^*\text{-}\lim_{t\to\infty} g_t)).$$

Definition 4. *A mapping $\Pi \in Hol(D,D)$ with $\mathfrak{F}_D = Fix_D \Pi \neq \emptyset$ is said to be a quasi-retraction of D onto $\mathfrak{F}_D$ if the sequence of iterates $\{\Pi^n\}_{n=1}^{\infty}$ strongly converges in D to some retraction Ψ of D onto $\mathfrak{F}_D$, i.e. $\Psi = \lim_{n\to\infty} \Pi^n$, $\Psi^2 = \Psi \in Hol(D,D)$.*

Theorem 1. *Let D be a bounded convex domain in a reflexive Banach space X. Let $S=\{F_t\}$, $t \in R^+$, be a continuous one-parameter semigroup of holomorphic self-mappings of D, whose generator exists and is a semi-plus complete vector field. Suppose that the stationary point set $\mathfrak{F}_D$ of S in D is not empty. Then*

1) $\mathfrak{F}_D$ is a complex analytic connected submanifold of D ;

2) For each $a \in \mathfrak{F}_D$, there exists

$$\Pi = \Phi_a^*\text{-}\lim_{t\to\infty} \frac{1}{t} \int_0^t F_\tau d\tau \tag{8}$$

which is a quasi-retraction of D onto $\mathfrak{F}_D$. Furthermore, $\Pi'(a)=P$ is a linear projection of X onto the tangent subspace to $\mathfrak{F}_D$ at the point a.

Remark 2. This theorem is a continuous analogue of the Mazet-Vigué discrete ergodic theorem [MV] for a semigroup $S = \{F_t\}$, $t\in N$, defined by the iterates $\{F^n\}$ of a self mapping $F_1=F$ which belongs to Hol(D,D). Instead of formula (8) they consider

$$\Phi^*_a\text{-}\lim_n \frac{1}{n}\sum_{k=0}^{n-1} F^k . \qquad (8')$$

However, in spite of the analogy we cannot consider the formula (8') as a special case of (8), because it is not clear whether there is a continuous semigroup $\{F_t\}$, $t\in R^+$, such that $F_n=F^n$. It seems that this question goes back to G.Koenigs (see, for example, [H]).

Nevertheless, using relation (7) we are able to establish another ergodic type theorem in terms of the null-point set of generators which has precisely the same form both in the discrete and in the continuous cases. In addition, we use this assertion to prove Theorem 1. We need another definition.

Definition 5. *We say that* $f \in Hol(D,X)$ *belongs to the class GH(D) if* f *is bounded and either* f *is a semi-plus complete vector field, or* $F = I-f$ *is a self-mapping of D. In other words,* f *generates a continuous or a discrete semigroup on D (or both).*

Theorem 2. *Let D be a bounded convex domain in a reflexive Banach space X, and let* $f \in GH(D)$. *Suppose that* $\mathrm{Null}_D f \neq \emptyset$. *Then*

1) $\mathrm{Null}_D f$ *is a complex analytic connected submanifold of D which is tangent to ker* $f'(a)$;

2) *For each* $r\geq 0$ *there exists a single-valued resolvent* $J_r=(I + rf)^{-1}\in Hol(D,D)$, *and* $\mathrm{Fix}_D J_r = \mathrm{Null}_D f$, *for all* $r>0$;

3) *If* $a \in \mathrm{Null}_D f$, *then there exists a mapping*

$$J = \Phi^*_a\text{-}\lim_{r\to\infty} J_r$$

which is a quasi-retraction of D onto $\mathrm{Null}_D f$.

Proof of Theorem 2. Step 1. First we consider assertions 1) and 2) for the discrete case.

Let $f \in Hol(D,X)$ and assume that $F=I-f$ is a self mapping of D.

Assertion 1) was proved by Mazet and Vigué [MV]. As a matter of fact, it also follows immediately from assertion 3) and a property of holomorphic retracts [C].

Turning to assertion 2), we fix $y \in D$, $r>0$, and consider the equation

$$(I + rf)x = y \ . \tag{9}$$

Setting $t=r/(r+1)$ we can rewrite (9) in the equivalent form

$$x = (1-t)y + tFx \ . \tag{10}$$

Since $t \in (0,1)$ and D is convex, the mapping $x \longmapsto (1-t)y + tFx$ maps D "strictly inside" (see, for example, [EH]) D. By the Earle-Hamilton fixed point theorem [EH], the equation (10) has a unique solution $x = G_t y$, which holomorphically depends on $y \in D$. Returning to $r = t/(1-t)$ and setting $J_r = G_{r/(1+r)}$ we have that $J_r \in Hol(D,D)$ and that $x = J_r y$ is the unique solution of (9). It is easy to see that

$$Fix_D J_r = Null_D f \tag{11}$$

for $r>0$.

Step 2. Here we establish that assertion 2) holds for the continuous case too.

Let $f \in Hol(D,X)$ be a semi-plus complete vector field, and let $\{F_t\}$, $t \in R^+$, be the semigroup generated by f. Setting $f_t=(I-F_t)/t$ we have by Proposition 2 that $\{f_t\}$ converges to f as t tends to 0^+ in the topology of local uniform convergence over D. Since $F_t \in Hol(D,D)$, by step 1 we have that for each $r>0$ there exists the resolvent $J_{r,t} = (I + rf_t)^{-1} = (I + r_1 tf_t)^{-1}$, where $r_1 = r/t$. In addition, it follows by (11) and (7) that for each $r>0$,

$$Null_D f \subseteq Fix_D F_t = Fix_D J_{r,t} \ . \tag{12}$$

Now choose any sequence $t_n \to 0^+$ and denote $T_n = I + rf_{t_n}$. For this sequence of mappings we have the following properties:

(i) If $a \in \mathrm{Null}_D f$, then $T_n(a)=a$;

(ii) $\{T_n\} \subset \mathrm{Hol}(D,X)$ converges to $T=I+rf$ in the topology of local uniform convergence on D ;

(iii) For each n there exists $T^{-1} = J_{r,t_n} \in \mathrm{Hol}(D,D)$.

We want to show that $\{T_n^{-1}\}$ converges to $T^{-1} \in \mathrm{Hol}(D,D)$, where $T = (I + rf)$. Indeed, let $A_n = (T_n)'(a)$ and $A = T'(a)$. It follows by (i),(ii) and the Cauchy inequalities that $\{A_n\}$ converges to A in the operator topology. In addition, because $T_n(a)=a$, by the chain rule we obtain that $A_n^{-1}=(T_n^{-1})'(a)$. Once again, using the Cauchy inequalities and (iii), we have that $\| A_n^{-1}\|$ is uniformly bounded. Therefore the linear operator A is invertible, and hence T is locally invertible in some neighborhood U of the point a .Thus there is a neighborhood $V=T(U)\subset D$ of the point a, such that $T^{-1}\in \mathrm{Hol}(V,U)$. Take an arbitrary $y\in V$. Then there is $x \in U$ such that $y = Tx = \lim_{n\to\infty} T_n x$. Setting $T_n x = y_n$, we have $\| y_n - y \| \to 0$, as $n \to \infty$, and $x = T_n^{-1} y_n = T^{-1} y$. Hence,

$$\| T^{-1}y - T_n^{-1}y \| = \| T_n^{-1}y_n - T_n^{-1}y \| \le K \| y_n - y \| \to 0$$

where $K = \sup_{z\in\bar{V}}\| (T_n^{-1})'(z) \| < \infty$, because D is bounded. By the Vitali theorem we obtain that $T_n^{-1} = J_{r,t_n}$ converges to $T^{-1} = (I + rf)^{-1}$ on the whole of D , and $T^{-1}\in \mathrm{Hol}(D,D)$. But $T^{-1}(a) = a \in D$, and T^{-1} is biholomorphic on D. Hence $T^{-1}(D)\subset D$. By the way we note that formula (11) holds for the continuous case, too. Again using the Mazet-Vigué theorem we obtain assertion 1) also for the continuous case.

Step 3. Now we prove the last assertion for both cases. We have now, if $f \in GH(D)$, that for each $r>0$ the mapping $J_r = (I + rf)^{-1}$ exists, belongs to $\mathrm{Hol}(D,D)$, and $\mathrm{Fix}_D J_r = \mathrm{Null}_D f$. Since X is reflexive there is a sequence $r_n \to \infty$ such that the sequence $\{J_{r_n}\}$ weakly converges to $J \in \mathrm{Hol}(D,D)$. But for $a\in \mathrm{Null}_D f$ we have $J(a)=a$, and hence $J \in \mathrm{Hol}(D,D)$. Let now $B = f'(a)$. It is easy to see that

$$e^{tB} = (F_t)'(a), \tag{13}$$

and thus B generates a uniformly continuous semigroup e^{tB} because of the Cauchy inequalities and the boundedness of D. In addition, by the chain rule $J'_r(a) = (I + rB)^{-1}$. It is known (see, for example, Shaw [S]) that for all $x \in X$ the strong limit

$$P = \lim_{t\to\infty} \frac{1}{t}\int_0^t e^{\tau B}\, d\tau = \lim_{r\to\infty} (I + rB)^{-1} \tag{14}$$

exists and is a projection onto $\ker B$. It is clear that $J'(a)=P$ and thus $J=\Phi_a^*\text{-}\lim J_r$. It follows from Vesentini's theorem (see [V1], [V2]) that J is a quasi-retraction onto $\mathrm{Fix}_D J$. But $\mathrm{Null}_D f \subseteq \mathrm{Fix}_D J$ are complex analytic connected manifolds with the same tangent space $PX = \ker B$ at a. Therefore they are identical. This concludes the proof of Theorem 2.

Note that the last considerations, including (14), prove also Theorem 1.

REFERENCES

[A] *Abate,M.,* The infinitesimal generators of semigroups of holomorphic maps, *Annali Mat. Pura ed Applicata (IV),* 161 (1992), 167–180.

[C] *Cartan, H.,* Sur les retractions d'une variété, *C.R. Acad. Sc. Paris,* 303 (1986), 715–716.

[EH] *Earle, C.J. and Hamilton, R.S.,* A fixed-point theorem for holomorphic mappings, *Proc. Symposia Pure Math.,* vol. 16, *Amer. Math. Soc. Providence, R.I.* (1970), 61–65.

[GR] *Goebel, K. and Reich, S.,* Uniform Convexity, Hyperbolic Geometry and Nonexpansive Mappings, *Pure and Applied Math., Marcel Dekker Inc., New York,* (1984).

[H] *Harris, T.E.,* The Theory of Branching Processes, *Springer-Verlag, Berlin,* (1963).

[HP] *Hille, E., and Phillips, R.S.,* Functional Analysis and Semigroups, *American Mathematical Society, Providence, RI,* (1957).

[IS] *Isidro, J. M., Stacho, L. L.,* Holomorphic Automorphism Groups in Banach Spaces: An Elementary Introduction, *Mathematics Studies* 105, North-Holland, Amsterdam, (1985).

[MV] *Mazet, P. and Vigué, J.-P.,* Points fixes d'une application holomorphe d'un domaine borné dans lui-même, *Acta Math.,* 166 (1991), 1-26.

[P] *Pazy, A.,* Remarks on nonlinear ergodic theory in Hilbert space, *Nonlinear Anal.*, 3 (1979), 863-871.

[R1] *Reich, S.,* Strong convergence theorems for resolvents of accretive operators in Banach spaces, *J. Math. Anal. Appl.,* 75 (1980), 287-292.

[R2] *Reich, S.,* A note on the mean ergodic theorem for nonlinear semigroups, *J. Math. Anal. Appl.*, 91 (1983), 547-551.

[S] *Shaw, S.-Y.,* Ergodic projections of continuous and discrete semigroups, *Proc. Amer. Math. Soc.*, 78 (1980), 69-74.

[V1] *Vesentini, E.,* Iterates of holomorphic mappings, *Uspekhi Mat. Nauk,* 40 (1985), 13-16.

[Ve2] *Vesentini, E.,* Su un teorema di Wolff e Denjoy. *Rend. Sem. Mat. Fis. Milano* 53 (1983), 17-25.

[Y] *Yosida, K.,* Mean ergodic theorem in Banach space, *Proc. Imp. Acad. Tokyo* 14 (1938) 292-294.

V. Khatskevich
Department of Mathematics
University of Haifa
Mount Carmel, 31905 Haifa
Israel

S. Reich
Department of Mathematics
The Technion - Israel Institute of Technology
32000 Haifa
Israel

D. Shoikhet
Department of Applied Mathematics
International College of Technology
P.O.Box 78, 20101 Karmiel
Israel

T MANDAI

Asymptotic solutions for exceptional cases of characteristic Cauchy problems to Fuchsian partial differential equations

We construct asymptotic solutions of characteristic Cauchy problems to Fuchsian partial differential equations when some characteristic indices become exceptional.

1. Introduction

M.S.Baouendi and C.Goulaouic ([1]) considered a Fuchsian partial differential operator with weight $m-k$

$$P = t^k\partial_t^m + \sum_{l=1}^{k} b_{m-l}(x)t^{k-l}\partial_t^{m-l} + \sum_{j+|\alpha|\le m, j<m} t^{\max\{j-m+k+1,0\}}c_{j,\alpha}(t,x)\partial_t^j\partial_x^\alpha, \quad (1.1)$$

where m is a positive integer, k is a non-negative integer such that $0 \le k \le m$, $b_{m-l}(x)$ are holomorphic in a neighborhood of $x=0 \in \boldsymbol{C}^n$ and $c_{j,\alpha}(t,x)$ are holomorphic in a neighborhood of $(t,x)=(0,0) \in \boldsymbol{C}\times\boldsymbol{C}^n$. In the category of holomorphic functions, they showed the unique solvability of the characteristic Cauchy problem

$$\text{(CP)} \quad \begin{cases} Pu &= f(x,t), \\ \partial_t^j u|_{t=0} &= g_j(x) \quad (j=0,1,\ldots,m-k-1). \end{cases}$$

under the condition

$$\text{(A)} \quad \mathcal{C}^{(P)}(0;\lambda) \ne 0 \quad \text{for} \quad \lambda \in (m-k)+\boldsymbol{N} := \{\, m-k, m-k+1, \ldots\},$$

where $\mathcal{C}^{(P)}(x;\lambda) := (\lambda)_m + \sum_{l=1}^{k} b_{m-l}(x)(\lambda)_{m-l}$ with $(\lambda)_j := \prod_{l=0}^{j-1}(\lambda-l)$. The polynomial $\mathcal{C}^{(P)}(x;\lambda)$ of λ is called the *indicial polynomial* of P, and a root of $\mathcal{C}^{(P)}(x;\lambda)=0$ is called a *characteristic index* of P at x. A characteristic index λ is called to be *exceptional*, if $\lambda \in (m-k)+\boldsymbol{N}$. If the condition (A) is not satisfied, that is, if some characteristic indices are exceptional, then it is called the *exceptional case*, and the Cauchy problem does not necessarily have holomorphic solutions for every holomorphic Cauchy data.

The research was supported in part by Grant-in-Aid for Scientific Research (No.05640168, No.06640222), Ministry of Education, Science and Culture (Japan).

H. Tahara ([3],[4] etc.) also considered the characteristic Cauchy problems for Fuchsian *hyperbolic* equations in the category of C^∞ functions on real domains under the same condition (A).

In this talk, we consider the case when this condition (A) is not satisfied (*Exceptional Case*), and construct asymptotic solutions. We can easily get "exact" solutions from these asymptotic solutions, using already known results.

For simplicity, we concentrate on Fuchsian operators, though the construction of these asymptotic solutions can be applied to a class of operators wider than that of Fuchsian operators.

NOTATIONS :

(i) The set of all nonnegative integers is denoted by $\boldsymbol{N}$. Put $l+\boldsymbol{N} := \{j \in \boldsymbol{N} : j \geq l\}$ for $l \in \boldsymbol{N}$.

(ii) The real part of a complex number z is denoted by $\operatorname{Re} z$.

(iii) Put $\vartheta := t\partial_t$.

(iv) For a bounded domain Ω in $\boldsymbol{C}^n$, we denote by $\mathcal{O}(\Omega)$ the set of all holomorphic functions on Ω.

(v) $\mathcal{R}(\boldsymbol{C}^*)$ denotes the universal covering of $\boldsymbol{C}^* := \boldsymbol{C} \setminus \{0\}$.

(vi) For a commutative ring R, the ring of polynomials of λ with the coefficients belonging to R is denoted by $R[\lambda]$. The order of $a(\lambda) \in R[\lambda]$ is denoted by $\operatorname{ord}_\lambda a$. Also, the ring of formal power series of t with the coefficients belonging to R is denoted by $R[[t]]$.

2. Main Result

Let Ω be a bounded domain in $\boldsymbol{C}^n$ that contains the origin 0. Let T be a positive real number. Consider a linear partial differential operator (1.1). For simplicity, we assume that

$$b_{m-l} \in \mathcal{O}(\Omega), \quad c_{j,\alpha} \in C^\infty([0,T];\mathcal{O}(\Omega)).$$

As for formal solutions when the condition (A) is satisfied, it is easy to prove the following.

Proposition 1. *Assume*

(A) $\mathcal{C}^{(P)}(0;\lambda) \neq 0$ *for* $\lambda \in (m-k) + \boldsymbol{N}$.

Take a subdomain Ω_0 *of* Ω *that includes* 0 *and that satisfies*

$(\mathrm{A})_{\Omega_0}$ $\mathcal{C}^{(P)}(x;\lambda) \neq 0$ *on* Ω_0 *for* $\lambda \in (m-k) + \boldsymbol{N}$.

Then, for every $f(x,t) \in \mathcal{O}(\Omega_0)[[t]]$ *and every* $g_j(x) \in \mathcal{O}(\Omega_0)$ $(0 \leq j \leq m-k-1)$, *there exists a unique formal solution* $u(t,x) \in \mathcal{O}(\Omega_0)[[t]]$ *of the characteristic Cauchy problem* (CP).

Remark 2. The condition $(\mathrm{A})_{\Omega_0}$ may look like a collection of infinite number of conditions. But, since $\mathcal{C}^{(P)}(x;\lambda)$ is a polynomial of λ, this is a collection of a finite number of conditions on x, and hence, if (A) is satisfied then we can always take Ω_0.

In this talk, we assume the following condition on $\mathcal{C}^{(P)}(x;\lambda)$.

(E) $\mathcal{C}^{(P)}(x;\lambda) = \prod_{j=1}^{r}(\lambda - \lambda_j(x)) \cdot \mathcal{D}^{(P)}(x;\lambda)$,
where
(a) $\lambda_j \in \mathcal{O}(\Omega)$, $\lambda_j(0) \in (m-k) + \boldsymbol{N}$ $(1 \leq j \leq r)$,
(b) $\mathcal{D}^{(P)}(0,\lambda) \neq 0$ for every $\lambda \in (m-k) + \boldsymbol{N}$.

In [2], the author considered the restricted case when λ_j are all constant. In this restricted case, we can give a formal solution in the form of

$$u = \sum_{j=0}^{\infty} t^j \sum_{l=0}^{r} u_{j,l}(x)(\log t)^l, \quad u_{j,l} \in \mathcal{O}(\Omega_0)\ (0 \leq j; 0 \leq l \leq r).$$

In general case, we can not expect to have a formal solution of this form. We shall give formal solutions in a more complicated form.

In order to see what kind of functions we need, let us consider the following simple example.

Example 3. $P := t\partial_t - \lambda(x)$, where $\lambda(x) \in \mathcal{O}(\Omega)$ and $\lambda(0) = p \in \boldsymbol{N}$. By freezing x, we can easily solve the equation $Pu = t^p$ in $\mathcal{R}(\boldsymbol{C}^*)$:

$$u = \begin{cases} \dfrac{-1}{\lambda(x) - p} t^p + C_x t^{\lambda(x)} & \text{if } \lambda(x) \neq p, \\ (\log t) t^p + C_x t^{\lambda(x)} & \text{if } \lambda(x) = p, \end{cases}$$

where C_x is an arbitrary constant that may depend on x. This is not a good representation since we want solutions that is holomorphic with respect to x. A good one is

$$u = \begin{cases} \dfrac{t^{\lambda(x)-p} - 1}{\lambda(x) - p} t^p & + \quad C(x) t^{\lambda(x)} \quad \text{if } \lambda(x) \neq p, \\ (\log t) t^p & + \quad C(x) t^{\lambda(x)} \quad \text{if } \lambda(x) = p, \end{cases}$$

where $C(x) \in \mathcal{O}(\Omega)$ is arbitrary.

This example suggests that we need a family of functions such as

$$u(t,x) = \begin{cases} \dfrac{t^{\mu(x)} - 1}{\mu(x)} & \text{if } \mu(x) \neq 0 \\ \log t & \text{if } \mu(x) = 0 \end{cases}, \tag{2.2}$$

where $\mu \in \mathcal{O}(\Omega)$.

Definition 4. (1) Put

$$F^{(1)}(z_1; t) := \left\{ \begin{array}{ll} \dfrac{t^{z_1} - 1}{z_1} & (z_1 \neq 0) \\ \log t & (z_1 = 0) \end{array} \right\} = \sum_{p=1}^{\infty} \frac{(\log t)^p}{p!} z_1^{p-1}, \tag{2.3}$$

$$\begin{aligned} & F^{(j+1)}(z_1, \ldots, z_{j+1}; t) \\ & := \begin{cases} \dfrac{F^{(j)}(z_1, \ldots, z_{j-1}, z_{j+1}; t) - F^{(j)}(z_1, \ldots, z_{j-1}, z_j; t)}{z_{j+1} - z_j} & (z_{j+1} \neq z_j) \\ \partial_{z_j} F^{(j)}(z_1, \ldots, z_j; t) & (z_{j+1} = z_j) \end{cases} \\ & = \sum_{p=j+1}^{\infty} \frac{(\log t)^p}{p!} \sum_{k_1 + \ldots + k_{j+1} = p-j-1} z_1^{k_1} \ldots z_{j+1}^{k_{j+1}}. \end{aligned} \tag{2.4}$$

Also put $F^{(0)}(;t) := 1$.

It is easy to see that $F^{(j)}(z_1, \ldots, z_j; t)$ is holomorphic on $\boldsymbol{C}^j \times \mathcal{R}(\boldsymbol{C}^*)$ and symmetric in $(z_1, \ldots, z_j) \in \boldsymbol{C}^j$. Further, $F^{(j)}(0, \ldots, 0; t) = \dfrac{1}{j!}(\log t)^j$.

(2) The function u given by (2.2) can be written as $F^{(1)}(\mu(x); t)$. Thus, we define the following function spaces.

For $\mu_1(x), \ldots, \mu_r(x) \in \mathcal{O}(\Omega)$ and $p \in \boldsymbol{N}$, put

$$\begin{aligned} \mathcal{F}_\Omega^{(p)}[\mu_1, \ldots, \mu_r] \; := \; & \{ \sum_{l=0}^{p} \sum_{I=(i_1, \ldots, i_l)} v_I(x) F^{(l)}(\mu_{i_1}(x), \ldots, \mu_{i_l}(x); t) \\ & : 1 \leq i_1 \leq \ldots \leq i_l \leq r, \;\; v_I(x) \in \mathcal{O}(\Omega) \, \} \\ \subset \; & \mathcal{O}(\Omega \times \mathcal{R}(\boldsymbol{C}^*)). \end{aligned} \tag{2.5}$$

Note that $\mathcal{F}_{\Omega}^{(0)}[\mu_1, \ldots, \mu_r] = \mathcal{O}(\Omega)$ (constant functions with respect to t). Further, if $p \leq q$, then $\mathcal{F}_{\Omega}^{(p)}[\mu_1, \ldots, \mu_r] \subset \mathcal{F}_{\Omega}^{(q)}[\mu_1, \ldots, \mu_r]$.

Remark 5. If $\mu_j(x) \equiv 0$ $(1 \leq j \leq r)$, then

$$\mathcal{F}_{\Omega}^{(p)}[0] = \oplus_{l=0}^{p} \mathcal{O}(\Omega)(\log t)^l := \{\, v(t,x) = \sum_{l=0}^{p} v_l(x)(\log t)^l : v_l \in \mathcal{O}(\Omega) \,\}.$$

By the following proposition, a formal series of the form

$$\sum_{p=0}^{\infty} t^p v_p(t,x), \quad v_p(t,x) \in \mathcal{F}_{\Omega_0}^{(N_p)}[\mu_1, \ldots, \mu_r] \quad (N_p \in \boldsymbol{N})$$

can be considered as an asymptotic series that is an extension of a formal power series $\sum_{p=0}^{\infty} t^p v_p(x)$, $v_p(x) \in \mathcal{O}(\Omega_0)$.

Proposition 6. *If $s > 0$ and $\operatorname{Re} \mu_l(x) + s > 0$ on Ω $(1 \leq l \leq j)$, then for every $g(t,x) \in \mathcal{F}_{\Omega}^{(p)}[\mu_1, \ldots, \mu_r]$, there holds*

$$t^s g(t,x) \longrightarrow 0 \quad (t \to 0).$$

This convergence is considered in an arbitrary sector of $t \in \mathcal{R}(\boldsymbol{C}^*)$, and is uniform on an arbitrary compact set $K \subset\subset \Omega$.

Now, the following is the main theorem.

Main Theorem. *Assume the condition* (E). *Put $\mu_j(x) := \lambda_j(x) - \lambda_j(0)$ $(j = 1, 2, \ldots, r)$, and put $\mu_0(x) \equiv 0$. Take a subdomain Ω_0 of Ω that includes 0 and that satisfies the following three conditions.*

(a) *$\mathcal{D}^{(P)}(x; \mu_j(x) + q) \neq 0$ on Ω_0 for every $q \in (m-k) + \boldsymbol{N}$ and $j = 0, 1, 2, \ldots, r$.*

(b) *If $j \neq l$, then $\mu_j(x) - \mu_l(x) \notin \boldsymbol{Z} \setminus \{0\}$ on Ω_0.*

(c) *$\operatorname{Re} \mu_l(x) + 1 > 0$ on Ω_0 $(1 \leq l \leq r)$.*

Then, for every $f \in \mathcal{O}(\Omega_0)[[t]]$ and every $g_j \in \mathcal{O}(\Omega_0)$ $(0 \leq j \leq m-k-1)$, there exists an asymptotic solution of the Cauchy problem (CP) *in the form*

$$u(t,x) = \sum_{j=0}^{m-k-1} \frac{g_j(x)}{j!} t^j + \sum_{p=0}^{\infty} t^{m-k+p} v_p(t,x), \tag{2.6}$$

where $v_p \in \mathcal{F}_{\Omega_0}^{(r+mp)}[\mu_1, \ldots, \mu_r]$.

If λ_j are all constants in addition, then we can take $v_p \in \mathcal{F}_{\Omega_0}^{(r)}[0]$, that is, $v_p(t,x) = \sum_{l=0}^{r} v_{p,l}(x)(\log t)^l$, where $v_{p,l} \in \mathcal{O}(\Omega_0)$ $(p \geq 0; 0 \leq l \leq r)$.

Remark 7. (1) We can always take Ω_0 satisfying (a)–(c), since $\mu_j(0) = 0$ $(1 \leq j \leq r)$ and since $\mathcal{D}^{(P)}$ is a polynomial of λ.

(2) This formal solution is not unique in general.

(3) Using already known results, we can easily get an exact solution u of (CP) in $C^{m-k-1}([0,T_0];\mathcal{O}(\Omega_0'))$ of which the asymptotic expansion is the formal solution given in the theorem above. When $m-k=0$, we consider $C^{-1}([0,T_0];\mathcal{O}(\Omega_0'))$ as $\{ f(t,x) \in C^0((0,T_0];\mathcal{O}(\Omega_0')) : tf(t,x) \in C^0([0,T_0];\mathcal{O}(\Omega_0')) \}$.

(4) We can obtain a similar theorem for $C^\infty(U)$ instead of $\mathcal{O}(\Omega)$. We omit the detail since the proof is almost the same. We can obtain exact solutions also for Fuchsian hyperbolic operarors considered by H. Tahara.

3. Sketch of the Proof of Main Theorem

We give a sketch of the proof of Main Theorem after some preliminaries.

3.1. Basic Properties of $\mathcal{F}_\Omega^{(p)}[\mu_1,\ldots,\mu_r]$. The following is the basic properties of $\mathcal{F}_\Omega^{(p)}[\mu_1,\ldots,\mu_r]$ that is used in the proof of Main Theorem.

Proposition 8. *Let $\mu_1,\ldots,\mu_r \in \mathcal{O}(\Omega)$.*

(1) $\vartheta(\mathcal{F}_\Omega^{(p)}[\mu_1,\ldots,\mu_r]) \subset \mathcal{F}_\Omega^{(p)}[\mu_1,\ldots,\mu_r]$.

(2) $\mathcal{O}(\Omega)\cdot\mathcal{F}_\Omega^{(p)}[\mu_1,\ldots,\mu_r] \subset \mathcal{F}_\Omega^{(p)}[\mu_1,\ldots,\mu_r]$.

(3) $\partial_{x_i}(\mathcal{F}_\Omega^{(p)}[\mu_1,\ldots,\mu_r]) \subset \mathcal{F}_\Omega^{(p+1)}[\mu_1,\ldots,\mu_r] \quad (1 \leq i \leq n)$.

If $\mu_1(x),\ldots,\mu_r(x)$ are all constants, in addition, then

$$\partial_{x_i}(\mathcal{F}_\Omega^{(p)}[\mu_1,\ldots,\mu_r]) \subset \mathcal{F}_\Omega^{(p)}[\mu_1,\ldots,\mu_r] \quad (1 \leq i \leq n).$$

3.2. Key Equation. In the proof of Main Theorem, it is the key part to solve the following ordinary differential equation with holomorphic parameter x.

$$\mathcal{C}(x;\vartheta)v = g(t,x), \tag{3.7}$$

where $\mathcal{C}(x;\lambda) \in \mathcal{O}(\Omega)[\lambda]$ and $g \in \mathcal{F}_\Omega^{(p)}[\mu_1,\ldots,\mu_r]$.

Proposition 9. *Let $\mu_1,\ldots,\mu_r \in \mathcal{O}(\Omega)$ and let $p \in \boldsymbol{N}$. Put $\mu_0(x) \equiv 0$. Assume that*

$$\mathcal{C}(x;\lambda) = \prod_{l=1}^{r'}(\lambda - \mu_{j_l}(x)) \cdot \mathcal{D}(x;\lambda),$$

where

(i) $\{j_1, \ldots, j_{r'}\} \subset \{1, \ldots, r\}$,

(ii) $\mathcal{D}(x;\lambda) \in \mathcal{O}(\Omega)[\lambda]$,

(iii) $\mathcal{D}(x;\mu_l(x)) \neq 0$ *on* Ω $(0 \leq l \leq r)$.

Then, for every $g(t,x) \in \mathcal{F}_\Omega^{(p)}[\mu_1, \ldots, \mu_r]$, *there exists a solution* $v(t,x) \in \mathcal{F}_\Omega^{(p+r')}[\mu_1, \ldots, \mu_r]$ *of the equation* (3.7).

3.3. Sketch of the Proof. Put $G(t,x) := \sum_{j=0}^{m-k-1} \frac{g_j(x)}{j!} t^j$, $u =: G(t,x) + t^{m-k}\tilde{u}$, $\tilde{f} := f - P(G)$, and $\tilde{P}(\tilde{u}) := P(t^{m-k}\tilde{u})$. Then, we have

$$Pu = f \quad \Longleftrightarrow \quad \tilde{P}\tilde{u} = \tilde{f}.$$

Since $\tilde{P}$ has the same form as (1.1) with $k = m$, and since $\mathcal{C}^{(\tilde{P})}(x;\lambda) = \mathcal{C}^{(P)}(x;\lambda + m - k)$, we may assume $k = m$ without loss of generality.

If $k = m$, then we can expand P with respect to t formally as

$$P = \mathcal{C}^{(P)}(x;\vartheta) + \sum_{l=1}^{\infty} t^l B_l(x, \partial_x; \vartheta),$$

where

$$B_l(x, \partial_x; \vartheta)\Big(\mathcal{F}_\Omega^{(p)}[\mu_1, \ldots, \mu_r]\Big) \subset \mathcal{F}_\Omega^{(p+m)}[\mu_1, \ldots, \mu_r].$$

By substituting $f = \sum_{p=0}^{\infty} t^p f_p(x)$ and $u = \sum_{p=0}^{\infty} t^p v_p(t,x)$ into $Pu = f$, we get the following recursive equations.

$$(\mathrm{R})_p \qquad \mathcal{C}^{(P)}(x;\vartheta + p)v_p(t,x) = f_p(x) - \sum_{l=1}^{p} B_l(x, \partial_x; \vartheta + p - l)v_{p-l}(t,x)$$

$(p = 0, 1, \ldots)$. (Note that $\vartheta(t^p v) = t^p(\vartheta + p)v$.)

Put $r_p := \#\{\, j \in \{1, \ldots, r\} : \lambda_j(0) = p \,\}$, where $\#A$ denotes the cardinal of a set A, and put $R_p := \sum_{l=0}^{p} r_l \leq r$.

By the repeated use of Proposition 9, we can get solutions $v_p(t,x) \in \mathcal{F}_{\Omega_0}^{(R_p+mp)}[\mu_1, \ldots, \mu_r]$ $(p \in \boldsymbol{N})$.

If $\lambda_j(x)$ are all constants, then we have $\mu_j(x) \equiv 0$, and we can take $v_p \in \mathcal{F}_{\Omega_0}^{(R_p)}[0]$.

References

[1] Baouendi, M. S. and Goulaouic, C., Cauchy problems with characteristic initial hypersurface, *Comm. Pure Appl. Math.* **26** (1973), 455–475.

[2] Mandai, T., On exceptional cases of Cauchy problems for Fuchsian partial differential operators, *Publ. Res. Inst. Math. Sci.* **20** (1984), 1007–1019.

[3] Tahara, H., Fuchsian type equations and Fuchsian hyperbolic equations, *Japan. J. Math. (N.S.)* **5** (1979), 245–347.

[4] ———, Singular hyperbolic systems, III. On the Cauchy problem for Fuchsian hyperbolic partial differential equations, *J. Fac. Sci. Univ. Tokyo Sect. IA Math.* **27** (1980), 465–507.

Department of Mathematics, Faculty of General Education, Gifu University, Yanagido 1-1, Gifu 501-11, JAPAN

e-mail address : g00325@sinet.ad.jp

S PISKAREV AND S-Y SHAW

Perturbation of cosine operator functions by step response and cumulative output[1]

1. Introduction

Recently, in [10] we introduced two operator families, namely cosine step response and cosine cumulative output, and discussed some properties of them. In [9] we have established some multiplicative perturbation theorems for cosine operator functions, which also contain the classical additive perturbation theorems as corollaries. The purpose of this paper is to discuss how some mixed-type perturbations are caused by those cosine step responses and cosine cumulative outputs whose local semivariations vanish at zero. The results obtained in [9], [10], and the present paper constitute a theory of cosine step response and cosine cumulative output, which is in many respects parallel to the corresponding theory of step response and cumulative output for semigroups (cf. [7], [8], [12]).

We first recall some definitions and basic properties. Let X be a Banach space, and $B(X)$ denote the space of all bounded linear operators on X. Throughout this paper, $\{C(t); t \in (-\infty, \infty)\}$ is a strongly continuous cosine operator function on X. By definition, it is a family of operators in $B(X)$ satisfying

(a) $C(0) = I$;

(b) $C(t+s) + C(t-s) = 2C(t)C(s)$ for t, $s \in (-\infty, \infty)$;

(c) the function $C(\cdot)x$ is continuous on $(-\infty, \infty)$ for every $x \in X$.

The associated sine operator function $S(\cdot)$ is defined by $S(t) = \int_0^t C(s)ds$ for all $t \in (-\infty, \infty)$. It is well known that $C(\cdot)$ is exponentially bounded, i.e. $\|C(t)\| \leq Me^{\omega t}$ for some $M \geq 1, \omega \in (-\infty, \infty)$ and all $t \geq 0$. We may assume that such choice of constants M and ω also satisfy $\|S(t)\| \leq Me^{\omega t}$ for $t \in (-\infty, \infty)$. The infinitesimal generator A of $C(\cdot)$ is defined as $Ax = \lim_{t \to 0} 2(C(t) - I)x/t^2$, with the natural domain (see [2], [4], [14]). It is a densely defined closed operator, and if $\lambda > \omega$, then $\lambda^2 \in \rho(A)$ and

$$\lambda(\lambda^2 - A)^{-1}x = \int_0^\infty e^{-\omega t} C(t)x dt, \; x \in X. \tag{1.1}$$

[1]Research supported in part by the National Science Council of Taiwan.

It is known ([11, Theorem 2.3], [6, Proposition 2], or [5]) that an exponentially bounded, strongly continuous operator function $C(\cdot)$ is a cosine operator function with generator A if and only if there is ω such that $\lambda^2 \in \rho(A)$ and (1.1) hold for all $\lambda > \omega$.

A strongly continuous family $\{F(t); -\infty < t < \infty\}$ of operators in $B(X)$ is called a C_0*-cosine step response* for the cosine function $C(\cdot)$ if $F(0) = 0$ and

$$F(t+s) - 2F(t) + F(t-s) = 2C(t)F(s) \quad \text{for} \quad t,\, s \in (-\infty, \infty). \tag{1.2}$$

A strongly continuous family $\{G(t); -\infty < t < \infty\}$ in $B(X)$ is called a C_0*-cosine cumulative output* for $C(\cdot)$ if $G(0) = 0$ and

$$G(t+s) - 2G(t) + G(t-s) = 2G(s)C(t) \quad \text{for} \quad t,\, s \in (-\infty, \infty). \tag{1.3}$$

The *infinitesimal operator* A_s of the pair $(C(\cdot), F(\cdot))$ is defined as

$$A_s x := \lim_{h \to 0+} \frac{2}{h^2}(C(h) + F(h) - I)x,$$

with the natural domain. The infinitesimal operator A_c of the pair $(G(\cdot), C(\cdot))$ is defined as

$$A_c x := \lim_{h \to 0+} \frac{2}{h^2}(C(h) + G(h) - I)x.$$

These are respective analogues of step response and cumulative output for C_0-semigroups, the latter being studied in [1], [7], [8], and [12].

Section 2 consists of some preliminary results. Section 3 is concerned with perturbation which is caused by a C_0-cosine step response $F(\cdot)$, and Section 4 is concerned with perturbation which is caused by a C_0-cosine cumulative output $G(\cdot)$.

The existence of the perturbed cosine functions depends on the behavior of the local semivariation or variation of $F(\cdot)$ or $G(\cdot)$ at 0. An operator-valued function $F(\cdot)$ is said to be *locally of bounded semivariation* if for some $t > 0$

$$SV(F(\cdot), t) := \sup\{\|\sum_{j=1}^{n}[F(t_j) - F(t_{j-1})]x_j\|\,;\, x_j \in X, \|x_j\| \le 1\} < \infty,$$

where the supremum is taken over all subdivisions of $[0, t]$. The function $F(\cdot)$ is said to be *locally of bounded strong variation* if for some $t > 0$ and all $x \in X$

$$\mathrm{Var}(F(\cdot)x, t) := \sup\{\sum_{j=1}^{n} \|(F(t_j) - F(t_{j-1}))x\|\,;\, 0 = t_0 < t_1 < \cdots < t_n = t, n \ge 1\}$$

is finite, and is said to be *locally of bounded uniform variation* if for some $t > 0$

$$\mathrm{Var}(F(\cdot),t) := \sup\{\sum_{j=1}^{n} \|(F(t_j)-F(t_{j-1}))\| \,;\, 0 = t_0 < t_1 < \cdots < t_n = t, n \geq 1\} < \infty.$$

2. Preliminary results

In this section we collect some basic properties of C_0-cosine step response $F(\cdot)$ and C_0-cosine cumulative output $G(\cdot)$ (see [10]), and two multiplicative perturbation theorems (see [9]).

Proposition 2.1 ([10, Proposition 3.1]). *The following properties are satisfied:*

(i) $(C(t)-I)F(s) = (C(s)-I)F(t)$ *and* $G(s)(C(t)-I) = G(t)(C(s)-I)$ *for* $t, s \geq 0$.

(ii) *The functions* $F(\cdot)$ *and* $G(\cdot)$ *are exponentially bounded.*

(iii) $\frac{d^2}{dt^2}[\lambda(\lambda^2 - A)^{-1}F(t)x] = C(t)\lambda^2\hat{F}(\lambda)x$ *and*

$\frac{d^2}{dt^2}[G(t)\lambda(\lambda^2 - A)^{-1}x] = \lambda^2\hat{G}(\lambda)C(t)x$ *for* $x \in X$, $\lambda > \omega$, *and* $t > 0$.

(iv) $F(t)x = (\lambda^2 - A)\int_0^t S(s)\lambda\hat{F}(\lambda)x\,ds$

$= \int_0^t S(s)\lambda^3\hat{F}(\lambda)x\,ds - (C(t) - I)\lambda\hat{F}(\lambda)x$ *for* $x \in X$, $t \geq 0$.

(v) $G(t)x = \lambda\hat{G}(\lambda)(\lambda^2 - A)\int_0^t S(s)x\,ds$

$= \lambda^3\hat{G}(\lambda)\int_0^t S(s)x\,ds - \lambda\hat{G}(\lambda)(C(t) - I)x$ *for* $x \in X$, $t \geq 0$.

Proposition 2.2 (from [10, Propositions 3.3, 3.4]). (i) *The infinitesimal operator* A_s *of the pair* $(C(\cdot), F(\cdot))$ *is closed and* $A_s = A(I - \lambda\hat{F}(\lambda)) + \lambda^3\hat{F}(\lambda)$ *for* $Re\,\lambda > \omega$.

(ii) *The infinitesimal operator* A_c *of the pair* $(G(\cdot), C(\cdot))$ *satisfies* $D(A) \subset D(A_c)$ *and* $A_c x = (I - \lambda\hat{G}(\lambda))Ax + \lambda^3\hat{G}(\lambda)x$ *for* $x \in D(A)$. *Moreover, if* $G(t)$ *is uniformly continuous in* t, *then* A_c *is closed,* $D(A_c) = D(A)$, *and* $A_c = (I - \lambda\hat{G}(\lambda))A + \lambda^3\hat{G}(\lambda)$ *for large* λ.

Theorem 2.3 ([9, Theorem 2.3]). *Let* A *be the infinitesimal generator of a cosine operator function* $C(\cdot)$ *on* X. *Suppose an operator* B *satisfies the condition:*

(*) *For all continuous functions* $f \in C([0,t],X)$, $\int_0^t S(t-s)Bf(s)ds \in D(A)$ *and* $\|A\int_0^t S(t-s)Bf(s)ds\| \leq \gamma_B(t)\|f\|_{[0,t]}$, *where* $\gamma_B(\cdot)$ *is some locally bounded function with* $\overline{\lim_{t\to 0+}}\,\gamma_B(t) < 1$.

Then both $A(I+B)$ and $(I+B)A$ are generators of cosine operator functions. Moreover, the cosine operator function $C_1(\cdot)$ generated by $A(I+B)$ satisfies $\|C_1(t) - C(t)\| = O(\gamma_B(t))\,(t \to 0^+)$.

Theorem 2.4 ([9, Theorem 2.6]). *Let A be the infinitesimal generator of a cosine operator function $C(\cdot)$ on X. Suppose an operator B satisfies the condition:*
(**) $\delta_B(t) := \sup\{\int_0^t \|BS(s)Ax\|ds;\ x \in D(A), \|x\| \le 1\} < 1$ *for some* $t > 0$.
Then both $(I+B)A$ and $A(I+B)$ are generators of cosine operator functions. Moreover, the cosine operator function $C_2(\cdot)$ generated by $(I+B)A$ satisfies $\|C_2(t) - C(t)\| = O(\delta_B(t))\,(t \to 0^+)$.

3. Perturbation caused by a C_0-cosine Step Response

Let $F(\cdot)$ be a C_0-step response for a cosine operator function $C(\cdot)$ with generator A. In this section we study conditions under which the infinitesimal operator A_s of the pair $(C(\cdot), F(\cdot))$ generates a cosine operator function $C_s(\cdot)$, and also the infinitesimal difference between $C_s(\cdot)$ and $C(\cdot)$.

Theorem 3.1. *Let $F(\cdot)$ be a C_0-cosine step response. The following assertions hold:*

(i) *If $SV(F(\cdot),t) = o(1)\,(t \to 0^+)$, then the infinitesimal operator $A_s = A(I - \lambda\hat{F}(\lambda)) + \lambda^3\hat{F}(\lambda)$ of the pair $(C(\cdot), F(\cdot))$ generates a cosine operator function $C_s(\cdot)$. Moreover, if $SV(F(\cdot),t) = O(\gamma(t))\,(t \to 0^+)$, then $\|C_s(t) - C(t)\| = O(\gamma(t))\,(t \to 0^+)$.*

(ii) *The perturbed semigroup $C_s(\cdot)$ satisfies the equation:*

$$C_s(t) = C(t) + (dF * C_s)(t) \equiv C(t) + \int_0^t F(d\tau)C_s(t-\tau),\ t \ge 0. \tag{3.1}$$

Proof. It is proved in [9, Theorem 3.2] that $SV(F(\cdot),t) = o(1)\,(t \to 0^+)$ if and only if $\gamma_{(-\lambda\hat{F}(\lambda))}(t) = o(1)\,(t \to 0^+)$, and they have the same order of convergence. Hence (i) follows from Theorem 2.3. To prove (ii) let $\mathcal{Q}(\cdot)$ denote the function on the right hand side of (3.1). Then, taking Laplace transform, we have $\hat{\mathcal{Q}}(\lambda) = \lambda(\lambda^2 - A)^{-1} + \lambda\hat{F}(\lambda)\lambda(\lambda^2 - A_s)^{-1}$, so that for $x \in D(A_s)$

$$\begin{aligned}\frac{1}{\lambda}\hat{\mathcal{Q}}(\lambda)(\lambda^2 - A_s)x &= (\lambda^2 - A)^{-1}(\lambda^2 - A_s)x + \lambda\hat{F}(\lambda)x \\ &= (\lambda^2 - A)^{-1}[\lambda^2 - A(I - \lambda\hat{F}(\lambda)) - \lambda^3\hat{F}(\lambda)]x + \lambda\hat{F}(\lambda)x \\ &= (\lambda^2 - A)^{-1}(\lambda^2 - A)(I - \lambda\hat{F}(\lambda))x + \lambda\hat{F}(\lambda)x = x.\end{aligned}$$

Hence $\hat{Q}(\lambda) = \lambda(\lambda^2 - A_s)^{-1}$, and it follows that $Q(\cdot)$ is a cosine operator function with generator A_s. Thus $Q(\cdot) = C_s(\cdot)$, that is, (3.1) holds.

In particular, $SV(F(\cdot), t) = O(t^\alpha)\,(t \to 0^+)$ implies $\|C_s(t) - C(t)\| = O(t^\alpha)\,(t \to 0^+)$. Clearly, the converse of this statement is true for $\alpha > 2$ because $\|C_s(t) - C(t)\| = o(t^2)\,(t \to 0^+)$ implies $C_s(\cdot) \equiv C(\cdot)$, which then implies $\lambda^2 - A = \lambda^2 - A_s = (\lambda^2 - A)(I - \lambda\hat{F}(\lambda))$, and consequently $F(\cdot) \equiv I$ and $SV(F(\cdot), t) \equiv 0$. The rest of this section is to show that this is also true for the case $\alpha = 2$.

We need the notion of Favard class

$$Fav_{C(\cdot)} := \{x \in X; \|C(t)x - x\| = O(t^2) \text{ as } t \to 0\},$$

which, equipped with the norm

$$\|x\|_{F_C(\cdot)} := \|x\| + \sup_{0<t\leq 1} \frac{2}{t^2}\|C(t)x - x\|$$

becomes a Banach space.

Theorem 3.2. *Let $F(\cdot)$ be a C_0-cosine step response for a cosine operator function $C(\cdot)$, and let A_s be the infinitesimal operator of the pair $(C(\cdot), F(\cdot))$. The following statements are equivalent.*

(i) *$F(\cdot)$ is locally of bounded uniform variation and* $Var(F(\cdot), t) = O(t^2)$, $(t \to 0^+)$.

(ii) *$F(\cdot)$ is locally of bounded strong variation and* $\sup\{Var(F(\cdot)x, t); x \in X, \|x\| \leq 1\} = O(t^2)\,(t \to 0^+)$.

(iii) *$F(\cdot)$ is locally of bounded semivariation and* $SV(F(\cdot), t) = O(t^2)$ $(t \to 0^+)$.

(iv) $\|F(t)x\| = O(t^2)\,(t \to 0^+)$ *for all* $x \in X$.

(v) $\|F(t)\| = O(t^2)\,(t \to 0^+)$.

(vi) $R(\hat{F}(\lambda)) \subseteq Fav_{C(\cdot)}$.

(vii) *A_s generates a cosine operator function $C_s(\cdot)$ such that* $\|C_s(t) - C(t)\| = O(t^2)$ $(t \to 0^+)$.

Proof. The implications "(i)$\Rightarrow$ (ii)+(iii)," "(ii)$\Rightarrow$ (iv)," and "(iii)$\Rightarrow$ (iv)" are direct consequences of the definitions of $Var(F(\cdot), t)$, $Var(F(\cdot)x, t)$ and $SV(F(\cdot), t)$. "(iv)$\Leftrightarrow$ (v)" holds because of the uniform boundedness principle.

To show "(v) $\Rightarrow$ (i)", suppose $\|F(s)\| \leq Ks^2$ for $0 \leq s \leq \tau$. For any subdivision $\{0 = t_0, t_1, \cdots, t_n = t\}$ of $[0, t] \subset [0, 1]$ with $h_i = t_i - t_{i-1} \leq \tau$, one has

$$F(t_i) - F(t_{i-1}) = F(t_{i-1}) - F(t_{i-1} - (t_i - t_{i-1})) + 2C(t_{i-1})F(t_i - t_{i-1})$$

and so

$$\|F(t_i) - F(t_{i-1})\| \leq 2KMe^{\omega}\ m_i h_i^2,$$

where m_i is such that $(m_i - 2)h_i \leq t_i < (m_i - 1)h_i$. Therefore

$$\sum_{i=1}^{n} \|F(t_i) - F(t_{i-1})\| \leq 2KMe^{\omega} \sum_{i=1}^{n} m_i h_i^2 \leq 2KMe^{\omega} \sum_{i=1}^{n} (t_n - \sum_{j=i+1}^{n} h_j + 2h_i)h_i$$
$$\leq 2KMe^{\omega} \sum_{i=1}^{n} 3t_n h_i = 6KMe^{\omega}t^2.$$

(iv) $\Leftrightarrow$ (vi). Using the first identity in (i) of Proposition 2.1 one has for any $x \in X$

$$\begin{aligned} h^{-2}(C(h) - I)\hat{F}(\lambda)x &= \int_0^{\infty} e^{-\lambda t} h^{-2}(C(h) - I)F(t)x dt \\ &= \int_0^{\infty} e^{-\lambda t}(C(t) - I)h^{-2}F(h)x dt \\ &= \lambda(\lambda^2 - A)^{-1}h^{-2}F(h)x - \lambda^{-1}h^{-2}F(h)x. \end{aligned}$$

As was shown in Proposition 2.1 (iii), $\lambda(\lambda^2 - A)^{-1}h^{-2}F(h)x$ converges to $\frac{1}{2}\lambda^2\hat{F}(\lambda)x$. Hence $\|F(h)x\| = O(h^2)(h \to 0)$ if and only if $\|(C(h) - I)\hat{F}(\lambda)x\| = O(h^2)(h \to 0)$, i.e. $\hat{F}(\lambda)x \in Fav_{C(\cdot)}$.

"(iii) $\Rightarrow$ (vii)" follows from Theorem 3.1(i), and "(vii) $\Rightarrow$ (vi)" is proved in "(i) $\Rightarrow$ (ii)" of Theorem 5.1 in [9].

In particular, Theorem 3.2 shows that if an operator A_1 is the infinitesimal operator A_s of the pair $(C(\cdot), F(\cdot))$ for some locally square Lipschitz continuous step response $F(\cdot)$ for $C(\cdot)$, then A_1 generates a cosine operator function $C_1(\cdot)$ that satisfies $\|C_1(t) - C(t)\| = O(t^2)(t \to 0^+)$. That the converse of the above statement is also true is contained in the following theorem from [9, Theorem 5.1].

Theorem 3.3. *Let $C(\cdot)$ be a cosine operator function with generator A, and let A_1 be a linear operator. The following statements are equivalent.*

(i) A_1 *generates a cosine operator function* $C_1(\cdot)$ *that satisfies* $\|C_1(t) - C(t)\| = O(t^2)\,(t \to 0^+)$.

(ii) *For some (each)* $\lambda > \omega$ *there exists a* $B_\lambda \in B(X, Fav_{C(\cdot)})$ *such that* $A_1 = A(I - B_\lambda) + \lambda^2 B_\lambda$.

(iii) $A_1 = A(I - \lambda\hat{F}(\lambda)) + \lambda^3\hat{F}(\lambda)$ *for some square Lipschitz continuous* C_0*-cosine step response* $F(\cdot)$.

(iv) A_1 *generates a cosine operator function* $C_1(\cdot)$, $D(A_1^*) = D(A^*)$, *and* $A_1^* - A^*$ *is a bounded operator from* $D(A^*)$ *to* X^*.

(v) A_1 *generates a cosine operator function* $C_1(\cdot)$ *and* $\|(\lambda^2 - A_1)^{-1} - (\lambda^2 - A)^{-1}\| = O(\lambda^{-4})\,(\lambda \to \infty)$.

4. Perturbation caused by a C_0-cosine Cumulative Output

In this section we give a condition on the variation of a cumulative output $G(\cdot)$ for $C(\cdot)$ which ensures that the infinitesimal operator A_c of the pair $(G(\cdot), C(\cdot))$ generates some cosine operator function. Let us denote

$$\beta(G(\cdot), t) := \sup\{Var(G(\cdot)x, t);\ x \in X, \|x\| \le 1\}. \tag{4.1}$$

Theorem 4.1. *Let* $G(\cdot)$ *be a* C_0*-cosine cumulative output for a cosine operator function* $C(\cdot)$ *with generator* A*. The following statements hold.*

(i) $\beta(G(\cdot), t) = o(1)\,(t \to 0^+)$ *if and only if* $\delta_{(-\lambda\hat{G}(\lambda))}(t) = o(1)\,(t \to 0^+)$, *and* $\beta(G(\cdot), t) = O(t^\alpha)\,(t \to 0^+)$ *if and only if* $\delta_{(-\lambda\hat{G}(\lambda))}(t) = O(t^\alpha)\,(t \to 0^+)$, *where* $(0 \le \alpha \le 2)$.

(ii) *If* $\beta(G(\cdot), t) = o(1)\,(t \to 0^+)$, *then* $A_c = (I - \lambda\hat{G}(\lambda))A + \lambda^3\hat{G}(\lambda)$ *(for all large* λ*) generates a cosine operator function* $C_c(\cdot)$*. Moreover, we have* $\|C_c(t) - C(t)\| = O(\beta(G(\cdot), t))\,(t \to 0^+)$.

(iii) *The cosine operator function* $C_c(\cdot)$ *satisfies the equation:*

$$\begin{aligned}(4.2)\quad C_c(t)x &= C(t)x + \int_0^t C_c(t-\tau)G'(\tau)x\,d\tau \\ &= C(t)x + \int_0^t C_c(t-\tau)\lambda\hat{G}(\lambda)S(\tau)(\lambda^2 - A)x\,d\tau,\quad x \in D(A),\ t \ge 0.\end{aligned}$$

Proof. (i) With $B_\lambda = -\lambda\hat{G}(\lambda)$, (v) of Proposition 2.1 becomes

$$B_\lambda C(t)x = G(t)x + \lambda^2 B_\lambda \int_0^t S(s)x\,ds + B_\lambda x.$$

Since $\lambda\|B_\lambda\| \int_0^t \|S(s)x\|ds \leq \lambda\|B_\lambda\| \, Me^{\omega t}t^2 \, \|x\|$, which has order $O(t^2)$ as $t \to 0$, and since

$$Var(B_\lambda C(\cdot)x, t) = \int_0^t \|\frac{d}{ds} B_\lambda C(s)x\|ds = \int_0^t \|BS(s)Ax\|ds,$$

one sees that $\beta(G(\cdot), t)$ and $\delta_{B_\lambda}(t)$ have the same order of convergence as $t \to 0$.

(ii) Since $\|G(t)x\| \leq Var(G(\cdot)x, t) \leq \beta(G(\cdot), t)$ for all x with $\|x\| \leq 1$, $\|G(t)\| \leq \beta(G(\cdot), t) \to 0$ as $t \to 0$. It follows from Proposition 2.2(ii) that $A_c = (I + B_\lambda)A - \lambda B_\lambda$. Since, by (i), we have $Var(B_\lambda C(\cdot)x, t) \leq \delta_{B_\lambda}(t)\|x\|$, with $\delta_{B_\lambda}(t) \to 0$, we can apply Theorem 2.4 to conclude that $(I + B_\lambda)A$ is a generator of a cosine operator function $\mathcal{T}(\cdot)$, and so its bounded perturbation A_c generates a cosine operator function $C_c(\cdot)$.

Since $\mathcal{T}(\cdot)$ and $C(\cdot)$ are related by the equation (see [9, (2.3)]):

$$\mathcal{T}(t)x = C(t)x + \int_0^t \mathcal{T}(t-s)B_\lambda S(s)Ax \, ds, \; x \in D(A),$$

it follows that

$$\begin{aligned} \|\mathcal{T}(t)x - C(t)x\| &\leq M' e^{\omega' t} \int_0^t \|B_\lambda S(s)Ax\| \, ds \\ &= M' e^{\omega' t} \; Var(B_\lambda C(\cdot)x, t) \\ &\leq M_1 e^{\omega' t} \; \beta(G(\cdot), t)\|x\| \end{aligned}$$

for all $x \in D(A)$. Thus we have $\|\mathcal{T}(t) - C(t)\| \leq M_1 e^{\omega' t}\beta(G(\cdot), t)$. As a bounded perturbation of $\mathcal{T}(\cdot)$, $C_c(\cdot)$ also satisfies $\|C_c(t) - C(t)\| = O(\beta(G(\cdot), t)) \; (t \to 0^+)$.

To prove (iii) let $\Upsilon(\cdot)$ denote the function on the right hand side of (4.2). Then, taking Laplace transform, we have $\hat{\Upsilon}(\lambda) = \lambda(\lambda^2 - A)^{-1} + \lambda(\lambda^2 - A_c)^{-1}\lambda\hat{G}(\lambda)$, so that

$$\begin{aligned} (\lambda^2 - A_c)\hat{\Upsilon}(\lambda) &= \lambda(\lambda^2 - A_c)(\lambda^2 - A)^{-1} + \lambda^2\hat{G}(\lambda) \\ &= \lambda[\lambda^2 - (I - \lambda\hat{G}(\lambda))A - \lambda^3\hat{G}(\lambda)](\lambda^2 - A)^{-1} + \lambda^2\hat{G}(\lambda) \\ &= \lambda(I - \lambda\hat{G}(\lambda))(\lambda^2 - A)(\lambda^2 - A)^{-1} + \lambda^2\hat{G}(\lambda) = \lambda. \end{aligned}$$

Hence $\hat{\Upsilon}(\lambda) = \lambda(\lambda^2 - A_c)^{-1}$, and it follows that $\Upsilon(\cdot)$ is a cosine operator function with generator A_c. Thus $\Upsilon(\cdot) = C_c(\cdot)$, that is, (4.2) holds.

Remark. The statement (i) for the case $\alpha > 2$ is trivial because then $G(\cdot) \equiv 0$.

The next theorem is the counterpart of Theorem 3.2 for cosine cumulative outputs.

Theorem 4.2. *Let $G(\cdot)$ be a C_0-cosine cumulative output for a cosine operator function $C(\cdot)$, and let A_c be the infinitesimal operator of the pair $(G(\cdot), C(\cdot))$. The following conditions are equivalent.*

(i) *$G(\cdot)$ is locally of bounded uniform variation and*

$Var(G(\cdot), t) = O(t^2)(t \to 0^+)$.

(ii) *$G(\cdot)$ is locally of bounded strong variation and*

$\sup\{Var(G(\cdot)x, t); x \in X, \|x\| \le 1\} = O(t^2)(t \to 0^+)$.

(iii) *$G(\cdot)$ is locally of bounded semivariation and*

$SV(G(\cdot), t) = O(t^2)(t \to 0^+)$.

(iv) $\|G(t)x\| = O(t^2)(t \to 0^+)$ *for all* $x \in X$.

(v) $\|G(t)\| = O(t^2)(t \to 0^+)$.

(vi) $R(\hat{G}^*(\lambda)) \subseteq F_{C^*(\cdot)}$.

(vii) *A_c generates a cosine operator function $C_c(\cdot)$ such that* $\|C_c(t) - C(t)\| = O(t^2)$.

Proof. The proof of the equivalence of conditions (i)-(v) is exactly the same as the proof in Theorem 3.2.

(v) $\Leftrightarrow$ (vi). Using the second identity in (i) of Proposition 2.1 one has for any $x \in X$

$$\begin{aligned} h^{-2}\hat{G}(\lambda)(C(h) - I)x &= \int_0^\infty e^{-\lambda t} h^{-2} G(t)(C(h) - I)x dt \\ &= \int_0^\infty e^{-\lambda t} h^{-2} G(h)(C(t) - I)x dt \\ &= h^{-2}G(h)\lambda(\lambda^2 - A)^{-1}x - \lambda^{-1}h^{-2}G(h)x. \end{aligned}$$

Hence $h^{-2}(C^*(h) - I^*)\hat{G}^*(\lambda)x^* = [h^{-2}G(h)\lambda(\lambda^2 - A)^{-1}]^*x^* - \lambda^{-1}h^{-2}G^*(h)x^*$ for all $x^* \in X^*$. As was shown in Proposition 2.1(iii), $h^{-2}G(h)(\lambda^2 - A)^{-1}$ converges strongly to $\frac{1}{2}\lambda\hat{G}(\lambda)$. Hence it is uniformly bounded for h near 0, so that $\|G^*(h)x^*\| = O(h^2)(h \to 0)$ if and only if $\|(C^*(h) - I^*)\hat{G}^*(\lambda)x^*\| = O(h^2)(h \to 0)$, i.e. $\hat{G}^*(\lambda)x^* \in Fav_{C^*(\cdot)}$. This being true for all $x^* \in X^*$, we obtain the equivalence of (v) and (vi).

The implication "(ii) $\Rightarrow$ (vii)" follows from (ii) of Theorem 4.1. To complete the proof we show "(vii) $\Rightarrow$ (iv)". It follows from Proposition 2.2(ii) that $D(A) \subseteq D(A_c)$

and $A_c x = Ax + \lambda \hat{G}(\lambda)(\lambda^2 - A)x$ for $x \in D(A)$, which together with Proposition 2.1(v) shows that $G(t)x = (A_c - A)\int_0^t S(s)x\,ds$ for all $x \in X$. Now condition (vii) implies that $\|(A_c - A)x\| \le \overline{\lim}_{h\to 0^+} \frac{2}{h^2}\|(C_c(h) - C(h))x\| \le K\|x\|$ for $x \in D(A)$, so that $\|G(t)x\| \le \|(A_c - A)\int_0^t S(s)x\,ds\| \le K\,\|\int_0^t S(s)x\,ds\| = O(t^2)$ for all $x \in X$.

Hence if an operator A_2 is the infinitesimal operator A_c of the pair $(G(\cdot), C(\cdot))$ for some locally square Lipschitz continuous C_0-cosine cumulative output $G(\cdot)$, then $D(A_2)$ contains $D(A)$, and A_2 generates a cosine operator function $C_2(\cdot)$ that satisfies $\|C_2(t) - C(t)\| = O(t^2)\,(t \to 0^+)$. The next theorem tells that these two statements are actually equivalent, and this kind of perturbed cosine functions are just the additive perturbations by bounded operators.

Theorem 4.3 ([9, Theorem 5.2]). *Let $C(\cdot)$ be a cosine operator function with the generator A. For any operator A_2, the following statements are equavalent.*

(i) *$A_1 = A + Q$ for some $Q \in B(X)$.*

(ii) *$D(A) \subseteq D(A_2)$, and A_2 generates a cosine operator function $C_2(\cdot)$ such that $\|C_2(t) - C(t)\| = O(t^2)\,(t \to 0^+)$.*

(iii) *There exists an operator $B \in B(X)$ such that $R(B^*) \subseteq F_{C^*(\cdot)}$ and $A_1 = (I + B)A - \lambda^2 B$ for some $\lambda > \omega$.*

(iv) *There is a locally square Lipschitz continuous C_0-cosine cumulative output $G(\cdot)$ for $C(\cdot)$ such that $A_2 = (I - \lambda\hat{G}(\lambda))A + \lambda^3\hat{G}(\lambda)$.*

References

1. O. Diekmann, M. Gyllenberg, and H. Thieme, *Perturbing semi-groups by solving Stieltjes renewal equations,* Differential and Integral Equations **6** (1993), 155-181.
2. H. O. Fattorini, *Ordinary differential equations in linear topological spaces, I.*, J. Differential Eq. **5** (1968), 72-105.
3. H. O. Fattorini, *Un teorema de perturbacion para generadores de functiones coseno,* Revista de la Unión Msthemática Argentina **25** (1971), 199-211.
4. J. A. Goldstein, "Semigroup of linear operators and applications", Oxford 1985.
5. Y.-C. Li and S.-Y. Shaw, *Integrated C-cosine functions and the abstract Cauchy problem,* 1991, preprint.

6. Y.-C. Li and S.-Y. Shaw, *On generators of integrated C-semigroups and C-cosine functions*, Semigroup Forum **47** (1993), 29-35.
7. S. Piskarev and S.-Y. Shaw, *On some properties of step responses and cumulative outputs,* Chinese J. Math. (1994), in press.
8. S. Piskarev and S.-Y. Shaw, *Multiplicative perturbations of semigroups and some applications to step responses and cumulative outputs*, J. Funct. Anal., to appear.
9. S. Piskarev and S.-Y. Shaw, *Perturbation and comparison of cosine operator functions*, Semigroup Forum, to appear.
10. S. Piskarev and S.-Y. Shaw, *On certain operator families related to cosine operator functions*, Preprint, 1993.
11. S.-Y. Shaw and Y.-C. Li, *On n-times integrated C-cosine functions*, in Evolution Equations, Marcel Dekker, New York, 1994.
12. S.-Y. Shaw and S. Piskarev, *Asymptotic behavior of some semigroup-related functions,* Saitama Math. J. (1994), to appear.
13. M. Shimizu and I. Miyadera, *Perturbation theory for cosine families on Banach spaces,* Tokyo J. Math. **1** (1978), 333-343.
14. M. Sova, *Cosine operator functions*, Rozprawy. Mat. **49** (1966), 1-47.
15. T. Takenaka and N. Okazawa, *A Phillips-Miyadera type perturbation theorem for cosine functions of operators*, Tôhoku Math. J. **30** (1978), 107-115.
16. C. C. Travis and G. F. Webb, *Perturbation of strongly continuous cosine family generators,* Colloq. Math. **45** (1981), 277-285.
17. M. Watanabe, *A Perturbation theory for abstract evolution equations of second order*, Proc. Japan. Acad. **58**, Ser. A, (1982), 143-146.

Department of Mathematics, National Central University, Chung-Li, Taiwan

W M RUESS

Asymptotic behavior of solutions to delay functional differential equations

1 Statement of the problem

The object of this paper is the study of solutions $x_\varphi : \mathbb{R} \to X$ to the following partial functional differential equation with infinite delay:

$$\text{(FDE)} \quad \begin{cases} \dot{x}(t) + (\alpha I + B)x(t) \ni F(x_t)\,, & t \geq 0 \\ x|_{\mathbb{R}^-} = \varphi \in \hat{E}. \end{cases}$$

Here, α is a real constant, X a Banach space, $B \subset X \times X$ a (generally) nonlinear and multivalued accretive operator in X, E a "suitably" chosen Banach space of continuous initial history functions $\varphi : \mathbb{R}^- \to X$, and $F : \hat{E} \subset E \to X$ a Lipschitz continuous mapping from a subset $\hat{E}$ of E into X with Lipschitz constant $M > 0$. As usual, for a function $x : \mathbb{R} \to X$ and $t \geq 0$, the function $x_t : \mathbb{R}^- \to X$ is defined by $x_t(s) := x(t+s)\,,\ s \in \mathbb{R}^-$. The initial history spaces E will be restricted to the following class of weighted sup-norm spaces of continuous functions (spaces of fading memory type [1, 17]):

$E = E_v = \{\varphi \in C(\mathbb{R}^-, X) \mid v\varphi$ is bounded and uniformly continuous$\}$, endowed with the norm $\|\varphi\|_v = \sup_{s \leq 0} v(s)\,\|\varphi(s)\|$. The function $v : \mathbb{R}^- \to (0,1]$ is supposed to satisfy

(v1) v is continuous, nondecreasing, and $v(0) = 1$;

(v2) $\lim_{u \to 0^-} \dfrac{v(s+u)}{v(s)} = 1$ uniformly over $s \in \mathbb{R}^-$.

The following problems will be addressed: Existence of global solutions to (FDE) (section 3), and compactness, almost periodicity properties and asymptotic stability of solutions (section 4).

Part of the results presented here is out of joint work with W.H. Summers [35, 36].

The literature on (FDE) and its nonautonomous counterpart – with B and F depending on "time" t as well, is enormous; compare the survey articles [7] and [18]. There is a variety of papers on direct methods for proving existence of – at least local – solutions to (FDE) for B single-valued but under less restrictive conditions on the history-controlling operator F (continuity as opposed to Lipschitz-continuity, for instance), cf. [19, 20, 21, 23, 24, 26].

In contrast, in this paper we take up the semigroup approach of associating with (FDE) in X a Cauchy problem in the initial history space E as developed for the case of globally Lipschitz-continuous operators $F : E \to X$ (and, mainly, for B single-valued) in [2, 3, 4, 5, 12, 13, 16, 22, 28, 29, 38, 39, 40], and extend it to the more general context as given above [35]. (For a local approach of a somewhat different kind, cf. [14, 15].) This approach can be structured into three steps: Associate with (FDE) an operator A in E by

$$\begin{cases} D(A) = \{\varphi \in E \mid \varphi' \in E, \varphi(0) \in D(B), \varphi'(0) \in F(\varphi) - (\alpha I + B)\varphi(0)\} \\ A\varphi := -\varphi' \, , \varphi \in D(A), \end{cases}$$

and consider the following statements:

(S1) $-A$ generates a strongly continuous semigroup $(S(t))_{t\geq 0}$ on $cl\, D(A) \subset \hat{E}$ of type γ, with $\gamma = max\{0, M - \alpha\}$: $\|S(t)\varphi - S(t)\psi\| \leq e^{\gamma t}\, \|\varphi - \psi\|$, $t \geq 0$, $\varphi, \psi \in cl\, D(A)$.

(S2) If $\varphi \in cl\, D(A)$, and $x_{\varphi} : \mathbb{R} \to X$ is defined by

$$x_{\varphi}(t) = \begin{cases} \varphi(t) & t \leq 0 \\ (S(t)\varphi)(0) & t \geq 0, \end{cases}$$

then $S(t)\varphi = (\, x_{\varphi})_t$ (i.e., $(S(t))_{t\geq 0}$ acts as a translation).

(S3) For $\varphi \in cl\, D(A)$, the function x_{φ} from **(S2)** solves (FDE).

Obviously, once **(S1)** – **(S3)** hold for (possibly, a subset of) $\varphi \in cl\, D(A)$, one not only is assured of global solutions to (FDE) but, at the same time, has the full machinery of semigroup theory at hand for a study of stability and further results on asymptotic behavior of solutions to (FDE).

In this respect, the following result has been known for quite a while (though formulated mainly for the finite delay case, or simply for $E = (BUC(\mathbb{R}^-, X), sup - norm)$ in the infinite-delay case, and for B single-valued).

Theorem 1.1 *Assume that $B \subset X \times X$ is m-accretive and $F : E \to X$ is globally (defined and) Lipschitz continuous, i.e., $\hat{E} = E$. Then we have:*

(a) (S1) and (S2) hold.

(b) If, in addition, $B : D(B) \subset X \to X$ is single-valued, X^* is uniformly convex, and

$\varphi \in \hat{D}(A)$ (the generalized domain of A [8]), then the function x_φ from ***(S2)*** *is the unique (strong) solution to (FDE): x_φ is locally absolutely continuous on $\mathbb{R}^+$, differentiable a.e. and fulfills (FDE) a.e. on $\mathbb{R}^+$.*

Compare [2, 3, 4, 5, 16, 29, 38, 39, 40], and, for the nonautonomous analog, see [12, 13, 22, 28]. Concrete model examples, however, indicate that the assumptions on the operators B (single-valued) and F (globally Lipschitz or Lipschitz on balls [5]) in proposition (b) of Theorem 1.1 are far too restrictive for applications, see section 2. The main point of the results in section 3 will be to show that this proposition actually holds in our original more general context and, most importantly, that, in the sense of integral solutions , the function x_φ from **(S2)** always is the unique solution to (FDE) for any $\varphi \in cl\, D(A)$.

2 Concrete model examples

2.1 Thermostat problem

One of the prototype examples of diffusion-absorption processes is the regulation of temperature through heat injection/extraction controlled by a thermostat. In the setting of a bounded open subset Ω of $\mathbb{R}^n, n \in \{2,3\}$, this process is governed by the operator $B \subset L^2(\Omega) \times L^2(\Omega)$,

$$B = (-\Delta + \tilde{\beta}) \qquad \text{with} \qquad D(B) = W^{2,2}(\Omega) \cap W_0^{1,2}(\Omega) \cap D(\tilde{\beta}),$$

where

$$\begin{cases} D(\tilde{\beta}) &= \{u \in L^2(\Omega) \mid \exists v \in L^2(\Omega) : v(\omega) \in \beta(u(\omega)) \text{ a.e. } \omega \in \Omega\}, \\ \tilde{\beta}(u) &= \{v \in L^2(\Omega) \mid v(\omega) \in \beta(u(\omega)) \text{ a.e. } \omega \in \Omega\}, \qquad u \in D(\tilde{\beta}), \end{cases}$$

with $\beta \subset \mathbb{R} \times \mathbb{R}$ a maximal monotone graph with $0 \in \beta(0)$.

Specifically, the particular choice of

$$\beta(r) = \begin{cases} (-\infty, 0] & , \quad r = h_1 \\ 0 & , \quad h_1 < r < h_2 \\ [0, \infty) & , \quad r = h_2 \end{cases}$$

would correspond to the ideal case of keeping the temperature within the prescribed interval $[h_1, h_2]$ at all times uniformly over Ω. (For details and further examples, cf. [6, 11].) Specialized to this particular m-accretive multivalued operator B, (FDE) would thus describe temperature control for materials with a suitable thermal memory.

2.2 Goodwin-oscillator with delay

This model for biochemical reaction sequences with end product inhibition (cf. [27]) is described by the system

$$\begin{aligned}\dot{x}_1(t) + a_1 x_1(t) &= b_1\left[1 + \left(\int_{-\infty}^0 k(-s)x_n(t+s)\,ds\right)^m\right]^{-1}\\ \dot{x}_i(t) + a_i x_i(t) &= b_i \int_{-\infty}^0 k(-s)x_{i-1}(t+s)\,ds, \quad i \in \{2, ..., n\}\end{aligned}$$

where $a_i, b_i > 0$ for $i \in \{1, ..., n\}$, $k \in L^1(\mathbb{R}^+)$ with $k \geq 0$, and $m \in \mathbf{N}$.

2.3 Spatially diffused population with delay in the birth process

For $x \in [0,1]$ and $t \geq 0$, this process is modeled by the equation

$$\dot{u}(t,x) - c\frac{d^2}{dx^2}u(t,x) = au(t,x)[1 - bu(t,x) - \int_{-1}^0 u(t + r(s), x)d\eta(s)],$$

where $a, b, c \in \mathbb{R}^+$, $\eta \in M^+[-1, 0]$ and $r : [-1, 0] \to \mathbb{R}^-$ is a continuous delay function (cf. [9, 25, 37]).

While example 2.1 indicates the need for allowing the operator B in (FDE) to be multi-valued, restricting the domain of definition and of Lipschitz continuity, respectively, of the history-controlling operator F to proper subsets $\hat{E}$ of E is forced by example 2.2 ($\hat{E}$ = cone of $n-$vectors of nonnegative functions on $\mathbb{R}^-$) and example 2.3 ($\hat{E}$ = truncated cone of nonnegative functions bounded by an arbitrary, but fixed, positive constant), respectively.

3 Existence of global solutions

In the context of (FDE) as in section 1, the following assumptions make the "local approach" to global solutions work [35].

(A1) $\hat{X} \subset X$ and $\hat{E} \subset E$ are closed subsets of X and E, respectively, such that, for $x \in \hat{X}$, $\psi \in \hat{E}$, and $\lambda > 0$ with $\lambda\gamma < 1$, if $\varphi_x \in E$ is the solution to

$$\varphi - \lambda\varphi' = \psi, \qquad \varphi(0) = x,$$

then $\varphi_x \in \hat{E}$.

(**A2**) If $\psi \in \hat{E}$ and $\lambda > 0$ with $\lambda\gamma < 1$, then

$$\frac{1}{1+\lambda\alpha}(\psi(0) + \lambda F(\varphi_x)) \in \Big(I + \frac{\lambda}{1+\lambda\alpha} B\Big)(D(B) \cap \hat{X})$$

for each $x \in \hat{X}$.

The subsequent results on (FDE) are all subject to assumptions (**A1**) and (**A2**) being in effect.

Theorem 3.1 ([31, 32, 35]) *(a) (S1) and (S2) hold.*

(b) If $\hat{E}$ is convex, and (i) if $B \subset X \times X$ is maximal accretive, and X is reflexive with its norm Fréchet-differentiable on $X \setminus \{0\}$ (for instance, if X^* is uniformly convex), or (ii) if $D(B)$ is closed, $B : D(B) \subset X \to X$ is continuous, and $\varphi \in \hat{D}(A)$, then the function x_φ from (S2) is the unique (strong) solution to (FDE).

The analog of Theorem 3.1, (b) (i), for the nonautonomous case of time dependent operators $B(t)$ and $F(t)$, $t \geq 0$, is contained in [31, Thm. 3.1].

The method of proof of Theorem 3.1, (b) (i) ([31, 32, 35]) is built upon approximating the semigroup $(S(t))_{t\geq 0}$ by the semigroups $(S_\lambda(t))_{t\geq 0}$, $\lambda > 0$, generated by the Yosida approximants A_λ of A. (Recall that $A_\lambda = \lambda^{-1}(I - J_\lambda)$, with $J_\lambda = (I + \lambda A)^{-1}$, $\lambda > 0$.) This explains the particular assumptions on X in Theorem 3.1, (b) (i), typical for this approach.

However, the problem of (global) solutions to (FDE) in our context can be resolved in complete generality by using the general Crandall-Liggett formula $S(t)\varphi = \lim_{n\to\infty}(I + \frac{t}{n}A)^{-n}\varphi$, and explicitly computing this limit (evaluated at $0 \in \mathbb{R}^+$) in the present particular context ([33]).

Theorem 3.2 ([33]) *For all $\varphi \in cl\, D(A)$, the function x_φ from (S2) is the unique integral solution to (FDE), i.e.,*

$$\begin{aligned}\| x_\varphi(t) - x\|^2 - \| x_\varphi(r) - x\|^2 \;\leq\; & -2\alpha \int_r^t \| x_\varphi(\tau) - x\|^2\, d\tau \\ & +2\int_r^t \langle F((x_\varphi)_\tau) - y,\, x_\varphi(\tau) - x\rangle_s\, d\tau\end{aligned}$$

for all $[x,y] \in B$ *and all* $0 \le r \le t$.

As a consequence, the following substantial extension of Theorem 3.1, (b) (i), on strong solutions can be achieved.

Theorem 3.3 ([33]) *If* $B \subset X \times X$ *is maximal accretive in cl* $D(B)$, *and* X *has the Radon-Nikodym property, then, for every* $\varphi \in \hat{D}(A)$, *the function* x_φ *from* (**S2**) *is the unique strong solution to (FDE).*

4 Asymptotic behavior of solutions to (FDE)

We consider two questions directed by the following *leitmotif*: If, in (FDE), the damping dominates the influence of the history, i.e., if $\alpha \ge M$ = Lipschitz constant of F on $\hat{E}$, then solutions to (FDE) qualitatively enjoy the same asymptotic properties as the solutions to its undelayed counterpart, namely, the Cauchy problem

$$\text{(CP)} \quad \begin{cases} \dot{x}(t) + Bx(t) \ni 0\,, \quad t \ge 0 \\ x(0) = \varphi(0). \end{cases}$$

(**Q1**) Assume that the resolvents $J_\lambda^B = (I + \lambda B)^{-1}$, $\lambda > 0$, are compact. Then bounded solutions to (CP) have relatively compact range and are asymptotically almost periodic (cf. [10, 34]). Does the same hold true for solutions to (FDE) if $\alpha \ge M$?

(**Q2**) If α is strictly larger than M, are solutions to (FDE) asymptotically stable or even exponentially asymptotically stable ?

Concerning the initial history spaces E_v, in addition to the basic assumptions (v1) and (v2), the following special properties of the weight functions v will play a role in answering these questions.

$$(v3) \quad \lim_{s\to-\infty} v(s) = 0; \qquad\qquad (v3^*) \quad \lim_{t\to\infty} \sup_{s\le -t} \frac{v(s)}{v(s+t)} = 0.$$

Clearly, (v3*) implies (v3). If, for $\mu \ge 0$, $v_1(s) = e^{\mu s}$, and $v_2(s) = (1 + |s|)^{-\mu}$, $s \le 0$, then (a) v_1 and v_2 fulfill (v1) and (v2), and (b) if $\mu > 0$, v_1 and v_2 both fulfill (v3), v_1

even fulfills (v3*), but v_2 fails to satisfy (v3*).

As for (**Q1**), the answer is positive in complete generality.

Theorem 4.1 ([32]) *Assume that the resolvents J_λ^B of B are compact [respectively, weakly compact], and that $\alpha \geq M$. Then we have :*

(a) For $\varphi \in cl\, D(A)$, if $x\varphi|_{\mathbb{R}^+}$ is bounded, it has relatively compact [respectively, weakly relatively compact] range.

(b) If the resolvents J_λ^B of B are compact and if, in addition, v satisfies (v3), then, for $\varphi \in cl\, D(A)$, if either $\lim_{s\to-\infty} v(s)\, \|\varphi(s)\| = 0$, or v satisfies (v3), we have: If $x\varphi|_{\mathbb{R}^+}$ is bounded, then $x\varphi|_{\mathbb{R}^+} : \mathbb{R}^+ \to X$ is asymptotically almost periodic, and there exists a unique element $\psi \in cl\, D(A)$ such that (i) $x_\psi|_{\mathbb{R}^+}$ is almost periodic, and (ii) $\lim_{t\to\infty} \left\| x\varphi(t) - x_\psi(t)\right\| = 0$. Moreover, if $\varphi \in \hat{D}(A)$, then so is ψ.*

As for (**Q2**), one has to find a way in between two extreme cases:
1. For $v(s) \equiv 1$, i.e., $E_v = (BUC(\mathbb{R}^-, X), sup-norm)$, $\alpha > M$ does not imply asymptotic stability for (FDE) [36, Example 4.1.A].
2. For $v(s) = e^{\mu s}, \mu > 0$, and $\alpha > M$, solutions to (FDE) are exponentially asymptotically stable [30, Cor. 4.1].

For weights v "reasonably" vanishing at $-\infty$, problem (**Q2**) can be answered positively in full generality.

Theorem 4.2 ([33]) *Assume that v satisfies (v3), and that $\alpha > M$, and let $\varphi, \psi \in cl\, D(A)$. If (i) $\lim_{s\to-\infty} v(s)\varphi(s) = 0 = \lim_{s\to-\infty} v(s)\psi(s)$, or (ii) v satisfies (v3*), then* $\lim_{t\to\infty} \left\| x\varphi(t) - x_\psi(t)\right\| = 0$.

As for Theorem 3.2 above, the proof of Theorem 4.2 requires explicit formulas for the (difference of the) $n-$th powers $J_{\frac{t}{n}}^n$ (acting on φ, ψ and evaluated at $0 \in \mathbb{R}^-$) of the resolvents $J_\lambda = (I + \lambda A)^{-1}$ of A; for details, see [33].

In closing this paper, we note that (a) conditions (**A1**) and (**A2**) are trivially fulfilled with $\hat{X} = X$ and $\hat{E} = E$ in case $B \subset X \times X$ is m-accretive and $F : E \to X$ is globally (defined and) Lipschitz continuous, but that (b) "local" examples like those of section 2 above can as well be put in the setting of these basic assumptions, so that all results recorded in this paper may be applied accordingly. For a sample of applications, cf. [35, 36].

References

[1] F.V. Atkinson and J. R. Haddock, *On determining phase spaces for functional differential equations*, Funkcial. Ekvac. 31 (1988), 331-347

[2] V. Barbu and S.I. Grossman, *Asymptotic behavior of linear integrodifferential systems*, Trans. Amer. Math. Soc. 173 (1972), 277-288

[3] D.W. Brewer, *A nonlinear semigroup for a functional differential equation*, Trans. Amer. Math. Soc. 236 (1978), 173-191

[4] D.W. Brewer, *A nonlinear contraction semigroup for a functional differential equation.* In: Volterra equations, S.-O. Londen and O.J. Staffans (Eds.), Lecture Notes Math.737, Springer 1979, 35-44

[5] D.W. Brewer, *Locally Lipschitz continuous functional differential equations and nonlinear semigroups*, Illinois J. Math. 26 (1982), 374-381

[6] H. Brézis, *Monotonicity methods in Hilbert spaces and some applications to nonlinear partial differential equations.* In: Contributions to nonlinear functional analysis, E. Zarantonello (Ed.), Acad.Press 1971, 101-156

[7] C. Corduneanu and V. Lakshmikantham, *Equations with unbounded delay: a survey*, Nonlinear Anal. 4 (1979), 831-877

[8] M.G. Crandall, *A generalized domain for semigroup generators*, Proc. Amer. Math. Soc. 37 (1973), 434-440

[9] J.M. Cushing, *Integrodifferential equations and delay models in population dynamics*, Lecture Notes in Biomathematics 20, Springer, Berlin 1977

[10] C.M. Dafermos and M. Slemrod, *Asymptotic behavior of nonlinear contraction semigroups*, J. Funct. Anal. 13 (1973), 97-106

[11] G. Duvaut and J.L. Lions, *Inequalities in Mechanics and Physics*, Grundlehren math. Wiss., vol. 219, Springer, Berlin, Heidelberg, New York, 1976

[12] J. Dyson and R. Villella-Bressan, *Functional differential equations and non-linear evolution operators*, Proc. Royal Soc. Edinburgh 75A (1975/76), 223-234

[13] J. Dyson and R. Villella-Bressan, *Semigroups of translation associated with functional and functional differential equations*, Proc. Royal Soc. Edinburgh 82A (1979), 171-188

[14] J. Dyson and R. Villella-Bressan, *Integral solutions of locally Lipschitz continuous functional differential equations*, preprint

[15] J. Dyson and R. Villella-Bressan, *Nonautonomous locally Lipschitz continuous functional differential equations in spaces of continuous functions*, preprint

[16] H. Flaschka and M.J. Leitman, *On semigroups of nonlinear operators and the solution of the functional differential equation* $\dot{x}(t) = F(x_t)$, J. Math. Anal. Appl. 49 (1975), 649-658

[17] J.K. Hale and J. Kato, *Phase space for retarded equations with infinite delay*, Funkcial. Ekvac. 21 (1978), 11-41

[18] F. Kappel and W. Schappacher, *Some considerations to the fundamental theory of infinite delay equations*, J. Differential Equations 37 (1980), 141-183

[19] F. Kappel and W. Schappacher, *Nonlinear functional differential equations and abstract integral equations*, Proc. R. Soc. Edinb. 84A (1979), 71-91

[20] A.G. Kartsatos, *A direct method for the existence of evolution operators associated with functional evolutions in general Banach spaces*, Funkc. Ekvacioj 31 (1988), 89-102

[21] A.G. Kartsatos and M.E. Parrott, *Global solutions of functional evolution equations involving locally defined Lipschitzian perturbations*, J. London Math. Soc. (2), 27 (1983), 306-316

[22] A.G. Kartsatos and M.E. Parrott, *Convergence of the Kato approximants for evolution equations involving functional perturbations*, J. Differential Equations 47 (1983), 358-377

[23] A.G. Kartsatos and M.E. Parrott, *A method of lines for a nonlinear abstract functional evolution equation*, Trans. Amer. Math. Soc. 286 (1984), 73-89

[24] A.G. Kartsatos and M.E. Parrott, *The weak solution of a functional differential equation in a general Banach space*, J. Differential Equations 75 (1988), 290-302

[25] W. Kerscher and R. Nagel, *Positivity and stability for Cauchy problems with delay.* In: Partial Differential Equations, Lecture Notes in Mathematics 1324, Springer 1987.

[26] K. Kunisch and W. Schappacher, *Variation of constants formulas for partial differential equations with delay*, Nonlinear Analysis 5 (1981), 123-142

[27] N. MacDonald, *Time lag in a model of a biochemical reaction sequence with end product inhibition*, J. Theoret. Biol. 67 (1977), 549-556

[28] M.E. Parrott, *Representation and approximation of generalized solutions of a nonlinear functional differential equation*, Nonlinear Analysis 6 (1982), 307-318

[29] A.T. Plant, *Nonlinear semigroups of translations in Banach space generated by functional differential equations*, J. Math. Anal. Appl. 60 (1977), 67-74

[30] A.T. Plant, *Stability of nonlinear functional differential equations using weighted norms*, Houston J. Math. 3 (1977), 99-108

[31] W.M. Ruess, *The evolution operator approach to functional differential equations with delay*, Proc. Amer. Math. Soc. 119 (1993), 783-791

[32] W.M. Ruess, *Compactness and asymptotic stability for solutions of functional differential equations with infinite delay.* In: Evolution Equations (G. Ferreyra, G. Ruiz Goldstein, F. Neubrander, Eds.), Lecture Notes Pure Appl. Math. 168, Marcel-Dekker 1994

[33] W.M. Ruess, *Existence and asymptotic stability of solutions to partial functional differential equations with infinite delay*, in preparation.

[34] W.M. Ruess and W.H. Summers, *Minimal sets of almost periodic motions*, Math. Ann. 276 (1986), 145-186

[35] W.M. Ruess and W.H. Summers, *Operator semigroups for functional differential equations with delay*, Trans. Amer. Math. Soc. 341 (1994), 695-719

[36] W.M. Ruess and W.H. Summers, *Almost periodicity and stability for solutions to functional differential equations with infinite delay*, to appear

[37] H.W. Steech, *The effect of time lags on the stability of the equilibrium state of population growth equations*, J. Math. Biology 5 (1978), 115-120

[38] G.F. Webb, *Autonomous nonlinear functional differential equations and nonlinear semigroups*, J. Math. Anal. Appl. 46 (1974), 1-12

[39] G.F. Webb, *Functional differential equations and nonlinear semigroups in L^p- spaces*, J. Differential Equations 20 (1976), 71-89

[40] G.F. Webb, *Asymptotic stability for abstract nonlinear functional differential equations*, Proc. Amer. Math. Soc. 54 (1976), 225-230

W.M. RUESS

Fachbereich Mathematik
Universität Essen

D-45117 Essen
Germany

K SAXTON

Global existence for a quasilinear heat equation

The aim of this paper is the analysis of global existence for a second order wave equation with first order dissipation. The equation describes the propagation of finite speed thermal waves.

1. Introduction. The classical model of heat conduction fails to take into account that heat can propagate as a wave called second sound. Second sound was observed for the first time in superfluid helium II for very small temperatures, around one or two degrees Kelvin. Recently the same heat pulse phenomenon has been observed in high purity crystals of NaF and Bi at low temperature. The heat conductivity of these crystals peaks near fifteen degrees Kelvin, at which the second sound is most easily detectable.

There are many theories dealing with finite speed of heat disturbances. In the present paper, the model of heat conduction replaces Fourier's law with a modification introduced by Kosinski [3], and physically motivated by Cimmelli and Kosinski [1]. The difference between this approach and others in the literature is that the heat flux becomes proportional to the gradient of a "new" so-called semi-empirical temperature β. This temperature β is related to the absolute temperature $\vartheta > 0$ by the initial-value problem

$$\begin{aligned} &\beta_t = f(\vartheta, \beta) \\ &\beta(0) = \beta_0 \quad , \quad \beta \in (0, \infty) \, . \end{aligned} \tag{1.1}$$

The heat flux $\mathbf{q}$, is related to $\nabla\beta$ by

$$\mathbf{q} = k_0 \nabla \beta \, , \tag{1.2}$$

where k_0 represents the thermal conductivity at constant temperature.

We employ the following constitutive equations for nonlinear heat conductors,

Partially supported by LEQSF Grant (1991–94) RD–A–22.

$$\psi = \widehat{\psi}(\vartheta, \nabla\beta)$$

and

$$\eta = -\partial_\vartheta \widehat{\psi}(\vartheta, \nabla\beta) \ , \tag{1.3}$$

where ψ is the Helmholtz free energy and η the specific entropy. For an explicit function f in Eq. (1.1), the Maxwell-Cattaneo relation can be obtained: $\tau \mathbf{q}_t + \mathbf{q} = -k\nabla\vartheta$,with τ the relaxation time depending on ϑ and β (cf. [3]). We will assume that the free energy $\widehat{\psi}$ is given by

$$\widehat{\psi}(\vartheta, \nabla\beta) = \widehat{\psi}_1(\vartheta) + \frac{1}{2}\widehat{\varepsilon}_2(\nabla\beta)^2 \tag{1.4}$$

where $\widehat{\varepsilon}_2 = k_0\tau_0/(\rho_0\vartheta^0)$, and τ_0, k_0 are characteristic material constants representing the relaxation time and thermal conductivity at constant reference temperature ϑ^0, β^0. ρ_0 is a mass density.

From the energy balance law, one obtains a second order hyperbolic equation. In the case of a one-dimensional rigid body (cf. Kosinski and Saxton [4]) this becomes

$$-C(\beta, \beta_t)\beta_{tt} + b\beta_x\beta_{xt} + a\beta_{xx} + H(\beta, \beta_t) = 0 \tag{1.5}$$

where

$$\begin{aligned} C(\beta, \beta_t) &= \rho_0\tau_0\vartheta c_V(\vartheta) \\ H(\beta, \beta_t) &= \rho_0\vartheta c_V(\vartheta) f_0'(\beta)\beta_t \\ a &= k_0\vartheta^0 \\ b &= \tau_0 k_0 \ , \end{aligned} \tag{1.6}$$

and

$$\begin{aligned} \tau_0 f(\vartheta, \beta) &= \vartheta^0 \log(\vartheta^0/\vartheta) + f_0(\beta) \\ f_0(\beta) &= \tau_0 f(\vartheta^0, \beta) \ . \end{aligned} \tag{1.7}$$

Here $c_V = \vartheta\partial_\vartheta\eta > 0$ is the specific heat at constant volume, and the relations (1.7) are derived by Kosinski in [3]. Eq. (1.5) can be rewritten as a system of hyperbolic equations by introducing new dependent variables

$$\begin{aligned} \beta_t &= w \\ \beta_x &= p \ , \end{aligned} \tag{1.8}$$

then

$$
\begin{aligned}
-C(\beta,w)w_t + bpw_x + ap_x + H(\beta,w) &= 0 \\
\beta_t - w &= 0 \\
p_t - w_x &= 0\ .
\end{aligned} \tag{1.9}
$$

Kosinski and Saxton [4] showed that the amplitude of acceleration waves satisfied a Bernoulli type equation. As a consequence, a sufficiently large initial jump in the first order derivatives of the data leads to the finite time blow up of the resulting acceleration waves.

In the following section we give an existence result for global in time, small amplitude solutions.

2. Global smooth solution.

We obtain sufficient conditions for global existence of small amplitude solutions of the Cauchy problem for Eq. (1.5). This equation is different from the equation considered by Nishida, [6], and Matsumara, [5], since the role of time and space variables are interchanged.

To overcome this difficulty we assume that the function f_0 in Eqs. (1.7) is linear, that is

$$
f_0(\beta) = -(\beta - \beta^0)\ . \tag{2.1}
$$

A local existence theorem was established by Cimmelli and Kosinski [2]. The initial conditions for the heat propagation (1.5) can be given physically for β and ϑ. Using the evolution equation (1.1), and (1.7) with (2.1), these can be obtained as:

$$
\begin{aligned}
\beta(0,x) &= \beta_0(x) \\
\beta_t(0,x) &= \frac{\vartheta^0}{\tau_0}\log\frac{\vartheta^0}{\vartheta(0,x)} - \frac{1}{\tau_0}(\beta_0(x) - \beta^0) = \beta_1(x)\ .
\end{aligned} \tag{2.2}
$$

We calculate ϑ from Eq. (1.7), using (1.1) and (2.1), then

$$
\vartheta = \vartheta^0 \exp\left\{\frac{\tau_0}{\vartheta^0}\left(-\beta_t - \frac{1}{\tau_0}\beta + \frac{1}{\tau_0}\beta^0\right)\right\} = \vartheta(\xi), \tag{2.3}
$$

where for convenience we introduce the variable ξ defined by

$$
\xi = \beta_t + \frac{1}{\tau_0}\beta\ . \tag{2.4}
$$

We express the coefficients in (1.6) in terms of ξ, so that

$$
\begin{aligned}
C(\beta,\beta_t) &= \tau_0 g(\xi)\ , \\
H(\beta,\beta_t) &= -g(\xi)\beta_t
\end{aligned} \tag{2.5}
$$

and

$$g(\xi) = \rho_0 \vartheta(\xi) c_V(\vartheta(\xi)) > 0 \ .$$

Next we substitute these coefficients into the Eq. (1.5) and multiply by τ_0^{-1}, which gives

$$-g(\xi)\beta_{tt} + k_0 \beta_x \beta_{xt} + \frac{k_0 \vartheta^0}{\tau_0} \beta_{xx} - \frac{1}{\tau_0} g(\xi) \beta_t = 0 \ . \tag{2.6}$$

If we define $P(\xi)$ such that

$$P'(\xi) = g(\xi) > 0 \ , \tag{2.7}$$

then

$$P(\xi)_t = g(\xi)\xi_t \ . \tag{2.8}$$

On using (2.8), Eq. (2.6) can be written in the form

$$\left(\frac{k_0 \vartheta^0}{\tau_0} \beta_x \right)_x = \left(P(\xi) - \frac{1}{2} k_0 (\beta_x)^2 \right)_t \ . \tag{2.9}$$

Let $\phi(t, x)$ be a potential function, such that

$$\begin{aligned} \phi_x &= P(\xi) - \frac{1}{2} k_0 (\beta_x)^2 \\ \phi_t &= \frac{k_0 \vartheta^0}{\tau_0} \beta_x \ . \end{aligned} \tag{2.10}$$

Our next step is to obtain an equation for $\phi(t, x)$. First, on differentiating (2.10), we obtain

$$\phi_{tt} - \frac{k_0 \vartheta^0}{\tau_0 g(\xi)} \phi_{xx} - \frac{\tau_0}{2\vartheta^0 g(\xi)} (\phi_t)_x^2 + \frac{1}{\tau_0} \phi_t = 0 \tag{2.11}$$

The variable ξ can be expressed in terms of ϕ_x, and ϕ_t from Eq. (2.10), namely,

$$\xi = P^{-1}(\nu) \tag{2.12}$$

where

$$\nu = \phi_x + \frac{1}{2} \frac{\tau_0^2}{k_0 (\vartheta^0)^2} (\phi_t)^2 \ , \tag{2.13}$$

since the function P is invertible (cf. (2.7)).

We define a function $\sigma(\nu)$, satisfying the relation

$$\sigma'(\nu) = \frac{k_0 \vartheta^0}{\tau_0 g(P^{-1}(\nu))} > 0 \ . \tag{2.14}$$

As a result, we have the following second order quasi-linear wave equation containing first order dissipation,

$$\phi_{tt} - \sigma(\nu)_x + \frac{1}{\tau_0}\phi_t = 0 \tag{2.15}$$

with the initial conditions

$$\phi(0,x) = \phi_0(x)\ , \quad \phi_t(0,x) = \phi_1(x)\ ,$$

where ν is given by (2.13).

The initial data for $\phi(t,x)$ can be obtained from (2.2) via (2.10).

This equation is now in the form analyzed by Nishida [6]. Eq. (2.15) will possess (cf. Nishida [6], Matsumara [5]), global smooth solutions $\phi(t,x) \in C^2(I\!R^+ \times I\!R)$ for $|\phi_0(\cdot)|_{C^2} + |\phi_1(\cdot)|_{C^1}$ sufficiently small provided the function σ satisfies the conditions (cf. (2.14))

$$\begin{aligned} &\sigma(0) = 0 \\ &\sigma'(\nu) \geq \sigma_0 = \text{const} > 0\ , \quad \text{and} \ \ \sigma(\cdot) \in C^4\ . \end{aligned} \tag{2.16}$$

The solutions $\phi(t,x)$ have the property

$$\|\phi(t)\|_2 \equiv |\phi(t,\cdot)|_{C^2} + |\phi_t(t,\cdot)|_{C^1} + |\phi_{tt}(t,\cdot)|_{C^0} < \infty \ \text{ for all } t > 0\ . \tag{2.17}$$

This result is obtained using energy estimates in spaces of L^2-function together with Sobolev's Lemma in one-space dimension.

We next show that a solution of Eq. (1.5) (equivalently (2.6) with (2.4)) satisfies, similarly,

$$\|\beta(t)\|_2 \equiv |\beta(t,\cdot)|_{C^2} + |\beta_t(t,\cdot)|_{C^1} + |\beta_{tt}(t,\cdot)|_{C^0} < \infty \ \text{ for all } t > 0\ . \tag{2.18}$$

Using (2.17), and Eq. $(2.10)_2$ we have that β_x, β_{xx}, and β_{xt} are bounded in C^0.

Now we will see that β_t, and β_{tt} are also bounded, which will give (2.18), under the additional assumption,

$$P'(\xi) \geq P_0 = \text{const} > 0\ . \tag{2.19}$$

Eq. $(2.10)_1$ implies $P(\xi)$ is bounded, which in turn implies that ξ is bounded, under the assumption (2.19). Thus there is a constant $\mathcal{D}$ for $t \in [0,T]$, $T < \infty$, such that (cf. Definition (2.4))

$$-\mathcal{D} < \beta_t + \frac{1}{\tau_0}\beta < \mathcal{D}\ . \tag{2.20}$$

This implies the following inequality

$$-\mathcal{D}e^{\frac{t}{\tau_0}} < \left(\beta e^{\frac{t}{\tau_0}}\right)_t < \mathcal{D}e^{\frac{t}{\tau_0}} \ . \tag{2.21}$$

By integration of (2.21) we obtain upper and lower bounds for β, and by (2.20), for β_t.

Let us differentiate $(2.10)_1$ with respect to t, to give

$$\phi_{xt} = P'(\xi)\xi_t - k_0\beta_x\beta_{xt} \ ,$$

which shows that in view of (2.19), and the boundedness of ϕ_{xt}, β_x, and β_{xt}, $\xi_t = \beta_{tt} + \tau_0^{-1}\beta_t$ is bounded. This together with our prior result gives the boundedness of β_{tt}.

Similar arguments can be applied to show that from the initial conditions, $|\beta_0(\cdot)|_{C^2} + |\beta_1(\cdot)|_{C^1}$ is small if and only if $|\phi_0(\cdot)|_{C^2} + |\phi_1(\cdot)|_{C^1}$ is small. The above arguments imply the following:

LEMMA: The Cauchy problem for Eq. (1.5) with (1.7) and (2.1) together with data (2.2) will have global smooth solutions $\beta(t,x) \in C^2(\mathbb{R}^+ \times \mathbb{R})$ for $|\beta_0(\cdot)|_{C^2} + |\beta_1(\cdot)|_{C^1}$ sufficiently small. Then

$$\|\beta(t)\|_2 \equiv |\beta(t,\cdot)|_{C^2} + |\beta_t(t,\cdot)|_{C^1} + |\beta_{tt}(t,\cdot)|_{C^0} < \infty \quad \text{for} \quad t > 0 \ ,$$

provided that

$$A_0 \le \vartheta(\xi)c_V(\vartheta(\xi)) \le B_0 \ , \quad \text{and} \quad c_V(\cdot) \in C^3 \ ,$$

where

$$A_0 = \text{const} > 0 \quad , \quad B_0 = \text{const} > 0 \ ,$$

and ϑ is given by (2.3) with (2.4).

The constants A_0 and B_0 are determined such that the assumptions $(2.16)_2$ and (2.19) are satisfied,

$$A_0 = P_0/\rho_0$$
$$B_0 = k_0\vartheta^0/(\tau_0\sigma_0\rho_0) \ .$$

3. Remarks on the breakdown of smooth solutions.

The breakdown of smooth solutions under the assumption that the function $g(\xi) = c_0 = \text{const}$ was obtained by Saxton in [7]. Then the function $\sigma(\nu)$ defined by (2.14) with condition $(2.16)_1$ is linear,

$$\sigma(\nu) = a_0 \nu \tag{3.1}$$

and Eq. (2.15) takes the form

$$\phi_{tt} - \left(a_0 \phi_x + \frac{b_0}{2a_0} \phi_t^2 \right)_x + \frac{1}{\tau_0} \phi_t = 0 \,, \tag{3.2}$$

where the constants a_0, and b_0 are introduced for convenience,

$$\begin{aligned} a_0 &= \frac{k_0 \vartheta^0}{\tau_0 c_0} \\ b_0 &= \frac{k_0}{c_0} \,. \end{aligned} \tag{3.3}$$

The corresponding system of equations for w, and p (cf. (1.9)) can be reduced to 2×2 form

$$\begin{aligned} w_t - b_0 p w_x - a_0 p_x + \frac{1}{\tau_0} w = 0 \\ p_t - w_x = 0 \,, \end{aligned} \tag{3.4}$$

with initial data

$$\begin{aligned} w(x,0) &= w_0(x) \\ p(x,0) &= p_0(x) \,. \end{aligned} \tag{3.5}$$

The above system is of nonconservative form and includes dissipation. By using Riemann invariants along characteristics, breakdown of solutions can be established for large initial data, [7].

REFERENCES

[1] V. A. Cimmelli and W. Kosinski, *Heat waves in continuous media at low temperatures*, Quaderni del Dipartimento di Matematica, n.6/1992, Universaita della Basilicata, Potenza.

[2] V. A. Cimmelli and W. Kosinski, *Well posedness for a nonlinear hyperbolic heat equation*, Ricerche di Matematica **42**, 49–68, (1993).

[3] W. Kosinski, *Elastic waves in the presence of a new temperature scale, Elastic wave propagation*, M. F. McCarthy and M. A. Hayes, eds., Elsevier Science (North–Holland), Amsterdam, 1989, p. 629.

[4] W. Kosinski and K. Saxton, *The effect on finite time breakdown due to modified Fourier laws*, Quart. Appl. Math. **51**, 55–68 (1993).

[5] A. Matsumara, *Global existence and asymptotic solutions of the second order quasi-*
linear hyperbolic equations with first order dissipation, Publ. Res. Inst. Math. Sci. Kyoto Univ. **A 13**, 349–379, (1977).

[6] T. Nishida, *Nonlinear hyperbolic equations and related topics in fluid dynamics*, Publications Mathematiques D'orsay, Universite de Paris-Sud, Departement de Mathematique, **78.02** (1978).

[7] K. Saxton, *Global existence and singularity formation in solutions of a modified Fourier law*, Quart. Appl. Math., to appear.

Katarzyna Saxton

Department of Mathematics and Computer Science

6363 St. Charles Avenue

Loyola University

New Orleans

Louisiana 70118

R SAXTON AND V VINOD

Nonstrictly hyperbolic systems of partial differential equations

We examine the influence of regions of nonstrict hyperbolicity on the Cauchy problem for two $n \times n$ systems of quasilinear, hyperbolic equations. The first of these systems takes the form

$$\mathbf{u}_t + (\chi(\tfrac{1}{2}|\mathbf{u}|^2)\mathbf{u})_x = \mathbf{0},$$

where χ is a scalar-valued function. The second reads

$$\mathbf{u}_t + \Lambda(\mathbf{u})\mathbf{u}_x = \mathbf{0},$$

where Λ is a diagonal matrix, $\Lambda = diag\{\chi^1, \ldots, \chi^n\}$, with ith entry, χ^i, independent of u^i.

1 Introduction

The system of first order partial differential equations

$$\mathbf{u}_t + \mathcal{A}(\mathbf{u})\mathbf{u}_x = \mathbf{0}, \ \mathbf{u} \in \mathbb{R}^n, \ (t, x) \in \mathbb{R}_+ \times \mathbb{R}, \tag{1.1}$$

is hyperbolic if the eigenvalues, λ_i, of $\mathcal{A}$ are real. It is *strictly* hyperbolic at a point $\mathbf{u}$ if all the eigenvalues are distinct there and *nonstrictly* hyperbolic if any pair becomes equal.

We consider two systems which become nonstrictly hyperbolic on some region in phase space and investigate how this can influence the formation of singularities. The first of these is a conservation law

$$\mathbf{u}_t + \mathcal{F}_x(\mathbf{u}) = \mathbf{0} \tag{1.2}$$

which has $\mathcal{F}(\mathbf{u}) = \chi(\frac{1}{2}|\mathbf{u}|^2)\mathbf{u}$, where χ is a scalar-valued function. We consider this in Section **2**.

In Section **3**, we consider the second, diagonal, system

$$\mathbf{u}_t + \Lambda(\mathbf{u})\mathbf{u}_x = \mathbf{0}. \tag{1.3}$$

Here Λ is a matrix-valued function of $\mathbf{u}$,

$$\Lambda = diag\{\chi^1, \ldots, \chi^n\}, \tag{1.4}$$

where the entries, χ^i, of Λ will be assumed to be independent of the corresponding components, u^i, of $\mathbf{u}$.

Each system has n eigenvalues some of which become equal on a set $\overline{\Sigma}$ in phase space. They are therefore nonstrictly hyperbolic there. The principal distinguishing feature of the two systems turns out to be that, while in (1.2), finite time breakdown takes place *off* Σ (an analogue of $\overline{\Sigma}$ which is identical in the 2×2 case), in (1.3) this can only take place *on* $\overline{\Sigma}$. Furthermore, although the first system generally possesses a direction of genuine nonlinearity which dictates when breakdown can take place off Σ, the second system is linearly degenerate throughout phase space.

The 2×2 counterpart of (1.2) has been examined earlier in [1], while that of (1.3) has been analysed from a different perspective in [2].

2 Formation of Singularities, I

Consider the system of equations

$$\mathbf{u}_t + \mathcal{F}_x(\mathbf{u}) = \mathbf{0}, \tag{2.1}$$

where the flux function takes the form $\mathcal{F}(\mathbf{u}) = \chi(\frac{1}{2}|\mathbf{u}|^2)\mathbf{u}$, and assume $\chi(z),\ \chi'(z)z,\ \chi''(z)z^2 \in C(\mathbb{R}_+;\mathbb{R})$. Setting $\mathcal{A}(\mathbf{u}) = \nabla_u(\chi(\frac{1}{2}|\mathbf{u}|^2)\mathbf{u})$, gives

$$\mathcal{A} = \chi'\mathbf{u}\otimes\mathbf{u} + \chi\mathbf{I}, \tag{2.2}$$

which has characteristic polynomial

$$|\lambda\mathbf{I} - \mathcal{A}| \ = \ (\lambda - \chi)^{n-1}(\lambda - (\chi + \chi'|\mathbf{u}|^2)). \tag{2.3}$$

Labelling the characteristic speeds,

$$\lambda_i = \begin{cases} \chi, & 1 \le i \le n-1, \\ \chi + \chi'|\mathbf{u}|^2, & i = n, \end{cases} \tag{2.4}$$

means that the corresponding right eigenvectors, $\mathbf{r}_i$, satisfy $(\chi\mathbf{I} - \mathcal{A})\mathbf{r}_i = -\chi'(\mathbf{u}\otimes\mathbf{u})\mathbf{r}_i = -\chi'\mathbf{u}(\mathbf{u}.\mathbf{r}_i), 1 \le i \le n-1$, and $((\chi + |\mathbf{u}|^2\chi')\mathbf{I} - \mathcal{A})\mathbf{r}_n = \chi'(|\mathbf{u}|^2\mathbf{I} - \mathbf{u}\otimes\mathbf{u})\mathbf{r}_n = \chi'(|\mathbf{u}|^2\mathbf{r}_n - \mathbf{u}(\mathbf{u}.\mathbf{r}_n))$. Assuming $\mathbf{u} \ne \mathbf{0}$, then for $\chi' \ne 0$ and $1 \le i \le n-1$, the $\mathbf{r}_i$'s can therefore be chosen from a set, $\mathbf{u}^\perp$, of mutually orthogonal vectors perpendicular to $\mathbf{u}$. $\mathbf{r}_n$ is then proportional to $\mathbf{u}$. The left eigenvectors, $\mathbf{l}_i$, $1 \le i \le n$, are the same as the right.

By (2.4), the set $\overline{\Sigma}$ where the system is nonstrictly hyperbolic consists of $\mathbb{R}^n$ if $n > 2$, or $\overline{\Sigma} = \{\mathbf{u} \in \mathbb{R}^2,\ \chi'|\mathbf{u}|^2 = 0\}$ if $n = 2$, ([1]).

Let $\Sigma = \{\mathbf{u} \in \mathbb{R}^n,\ \chi'|\mathbf{u}|^2 = 0\},\ n \ge 2$.

Since the first $n-1$ characteristic fields satisfy $\mathbf{r}_i.\nabla_u\lambda_i \propto (\mathbf{u}^\perp)_i.\mathbf{u}\chi' = 0$, they are linearly degenerate, while the nth characteristic field satisfies $\mathbf{r}_n.\nabla_u\lambda_n \propto \mathbf{u}.\nabla_u(\chi + \chi'|\mathbf{u}|^2)$. Setting $\Upsilon = \{\mathbf{u} \in \mathbb{R}^n,\ (3\chi' + \chi''|\mathbf{u}|^2)|\mathbf{u}|^2 = 0\}$, the nth characteristic field becomes genuinely nonlinear only outside of Υ.

Lemma 2.1 *Let* $\mathbf{u} \in C^1([0,T];C^1(\mathbb{R}))$ *be a solution to (2.1). Then given data* $\mathbf{u}_0(x) = \mathbf{u}(0,x)$, $||\mathbf{u}_x||_\infty(t) \to \infty$ *occurs in finite time if* $\chi'|\mathbf{u}_0|^2 \ne 0$ *and* $(3\chi' + \chi''|\mathbf{u}_0|^2)\mathbf{u}_0.\mathbf{u}_{0x} < 0$.

Proof Dotting (2.1) with $\mathbf{u}$ gives

$$(\frac{1}{2}|\mathbf{u}|^2)_t + \mathbf{u}\mathcal{A}\mathbf{u}_x = 0. \tag{2.5}$$

Since $\mathbf{u}$ is proportional to $\mathbf{l}_n$, by (2.4)

$$(\frac{1}{2}|\mathbf{u}|^2)_t + (\chi + \chi'|\mathbf{u}|^2)\,(\frac{1}{2}|\mathbf{u}|^2)_x = 0, \tag{2.6}$$

and since χ has dependence only on $\frac{1}{2}|\mathbf{u}|^2$, (2.6) shows that $|\mathbf{u}|^2$ is constant in the characteristic direction $\frac{dx}{dt} = \lambda_n = (\chi + \chi'|\mathbf{u}|^2)$. Taking the partial derivative of (2.6) with respect to x and simplifying implies that in the same direction

$$(\frac{1}{2}|\mathbf{u}|^2)_x = \frac{\mathbf{u}_0.\mathbf{u}_{0x}}{1 + (3\chi' + \chi''|\mathbf{u}_0|^2)\mathbf{u}_0.\mathbf{u}_{0x}t} \tag{2.7}$$

unless $\chi'|\mathbf{u}|^2 = 0$ ($\mathbf{u} \in \Sigma$), in which case $(\frac{1}{2}|\mathbf{u}|^2)_x = \mathbf{u}_0.\mathbf{u}_{0x}$. For $\mathbf{u} \notin \Upsilon$, $3\chi' + \chi''|\mathbf{u}_0|^2 \neq 0$ and the result clearly follows if $(3\chi' + \chi''|\mathbf{u}_0|^2)\mathbf{u}_0.\mathbf{u}_{0x} < 0$. ∎

Remark. Although the breakdown above takes place for data off Σ, it is possible to see that there exist situations in which no natural counterpart for Σ exists. For instance, the initial value problem for the system $\mathbf{u}_t + \chi(\frac{1}{2}|\mathbf{u}|^2)\mathbf{u}_x = \mathbf{0}$ is everywhere nonstrictly hyperbolic for $n > 1$, $\overline{\Sigma} = I\!R^n$, but since the eigenvalues of $\mathcal{A} = \chi\mathbf{I}$ are all identical, Σ cannot be defined as before. Indeed, singularity formation is controlled in this case simply by the data condition $\chi'\mathbf{u}_0.\mathbf{u}_{0x} < 0$, which requires only genuine nonlinearity, $\mathbf{u} \notin \Upsilon = \{\mathbf{u} \in I\!R^n,\ \chi'\mathbf{u} = \mathbf{0}\}$.

3 Formation of Singularities, II

Next we examine a system for which it is only possible for breakdown to take place *on* $\overline{\Sigma}$. Unlike the previous Section where the term $\frac{1}{2}|\mathbf{u}|^2$ remains constant along characteristics, ensuring that Σ is invariant with respect to the Lagrangian flow, now the set $\overline{\Sigma}$ will be defined through the intersection of distinct characteristic fields. Given data off $\overline{\Sigma}$, a singularity will form providing a time exists at which $\overline{\Sigma}$ becomes nonempty.

Consider the system

$$\mathbf{u}_t + \Lambda(\mathbf{u})\mathbf{u}_x = \mathbf{0}, \tag{3.1}$$

where

$$\Lambda = diag\{\chi^1(P_1\mathbf{u}), \ldots, \chi^n(P_n\mathbf{u})\}, \tag{3.2}$$

and

$$P_i\mathbf{u} = \{u^1, \ldots, u^{i-1}, u^{i+1}, \ldots, u^n\},\ 1 \leq i \leq n. \tag{3.3}$$

We will make the hypothesis that there exist C^1-functions χ_i, $1 \leq i \leq n$, such that

$$\chi^i(P_i\mathbf{u}) = \chi(\mathbf{u}) - \chi_i(u^i) \tag{3.4}$$

where

$$\chi(\mathbf{u}) \equiv \sum_{j=1}^{n} \chi_j(u^j). \tag{3.5}$$

The set $\overline{\Sigma}$ is defined by the equality in phase space of any pair of eigenvalues $\chi^i(P_i\mathbf{u})$, $\chi^j(P_j\mathbf{u})$, hence through the equality of $\chi_i(u^i)$ and $\chi_j(u^j)$. Since the right eigenvectors, $\mathbf{r}_i$ of Λ are the standard basis vectors for $I\!R^n$, it follows immediately that $\mathbf{r}_i.\nabla_u \chi^i(P_i\mathbf{u}) = 0$, $1 \le i \le n$. So the set Υ over which the system is linearly degenerate is $I\!R^n$.

Lemma 3.1 *Let $\Lambda : I\!R^n \to I\!R^{n^2}$ be a C^1-map. Suppose $\mathbf{u}(t,x) \in C^1([0,t^*); C^1(I\!R^n))$ is a solution to (3.1), with $\mathbf{u}(0,x) = \mathbf{u}_0(x) \in I\!R^n \setminus \overline{\Sigma}$, $x \in I\!R$, defined for some maximal interval $[0,t^*)$. Then, under hypothesis (3.4) with (3.5), $t^* < \infty$ if and only if $\mathbf{u} : I\!R^n \setminus \overline{\Sigma} \to \overline{\Sigma}$ as a map from $\mathbf{u}_0 \to \mathbf{u}(t,.)$.*

Proof Define the characteristic Γ_i by $x^i = x^i(t)$, $\frac{dx^i}{dt} = \chi^i(P_i\mathbf{u})$, $x^i(0) = \alpha^i$, $1 \le i \le n$. Differentiation along Γ_i will be written as $D_i \equiv \partial/\partial t + \chi^i \partial/\partial x$, from which it is immediate by (3.1) that $D_i u^i = 0$, $1 \le i \le n$, *i.e.* $u^i(t, x^i(t)) = u_0^i(\alpha^i)$, where $u_0^i(x) \equiv u^i(0,x)$. Differentiating (3.1) with respect to x implies

$$D_i u_x^i + u_x^i \sum_{j\ne i}^{n} \frac{\partial \chi^i}{\partial u^j} u_x^j = 0. \tag{3.6}$$

Also, for $i \ne j$,

$$D_i u^j = D_j u^j + (\chi^i - \chi^j) u_x^j = (\chi^i - \chi^j) u_x^j. \tag{3.7}$$

Consequently, unless $\chi^i = \chi^j$,

$$D_i u_x^i + u_x^i \sum_{j\ne i}^{n} \frac{\partial \chi^i}{\partial u^j} \frac{D_i u^j}{\chi^i - \chi^j} = 0. \tag{3.8}$$

By (3.4), (3.5), equation (3.8) reduces to

$$D_i u_x^i + u_x^i \sum_{j\ne i}^{n} \chi_j(u^j)' \frac{D_i u^j}{\chi_j(u^j) - \chi_i(u^i)} = 0 \tag{3.9}$$

implying

$$D_i u_x^i + u_x^i D_i \sum_{j\ne i}^{n} \ln |\chi_j(u^j) - \chi_i(u^i)| = 0 \tag{3.10}$$

or

$$D_i (u_x^i \prod_{j\ne i}^{n} |\chi_j(u^j) - \chi_i(u^i)|) = 0. \tag{3.11}$$

As a result, the expression $u_x^i \prod_{j\ne i}^{n} |\chi_j(u^j) - \chi_i(u^i)|$ remains constant along the ith characteristic, Γ_i. This constant is nonzero by the assumption $\mathbf{u}_0(x) \in I\!R^n \setminus \overline{\Sigma}$, which leads to the desired result, using (3.4). ∎

Remark. The previous result establishes necessary conditions on solutions for blow-up to take place in finite time. Sufficient conditions on the data can also be obtained in the 2×2 case.

References

[1] Keyfitz, B. L., and Kranzer, H. C., *Non-strictly hyperbolic systems of conservation laws: formation of singularities*, Contemporary Mathematics, 17, 1983, 77-90.

[2] Serre, D., *Large oscillations in hyperbolic systems of conservation laws*, Rend. Sem. Mat. Univ. Pol. Torino, Fascicolo Speciale 1988, Hyperbolic Equations.

Ralph A. Saxton and Vaidyanath Vinod
Department of Mathematics
University of New Orleans
New Orleans
LA 70148, U.S.A.
email: rsaxton@math.uno.edu, vvinod@math.uno.edu

R J WILTSHIRE

Non-linear coupled diffusion and classical Lie symmetry

The importance of the simultaneous flow of moisture and heat in a column of soil has been discussed by many soil scientists (for example, Philip & De Vries,1957; De Vries, 1958; Jackson, 1973) and has particular significance when applied to semi-arid climates where moisture transport occurs essentially in the vapour phase. In such conditions the one dimensional flow when applied to a homogeneous soil may be represented by a pure system of coupled diffusion equations (Jury *et al* , 1981) given as follows:

$$\begin{aligned} \frac{\partial T}{\partial t} &= \frac{\partial}{\partial z}\{\kappa_T(T,\theta)\frac{\partial T}{\partial z} + \kappa_\theta(T,\theta)\frac{\partial \theta}{\partial z}\}, \\ \frac{\partial \theta}{\partial t} &= \frac{\partial}{\partial z}\{D_T(T,\theta)\frac{\partial T}{\partial z} + D_\theta(T,\theta)\frac{\partial \theta}{\partial z}\}, \end{aligned} \qquad (1)$$

where $T(z,t)$ and $\theta(z,t)$ are the respective soil temperature and moisture content at depth, z and time, t. It should also be noted that a straightforward extension would allow for the inclusion of solute flow.

From a mathematical point of view it is the highly non-linear form of the diffusion coefficients, dependent on both T and θ, which is of particular interest, together with the requirement which many soil scientists have for analytic/semi-analytic and qualitative information concerning the transport properties of soil (Philip,1988).

Indeed it is the quest for analytic solutions which is of particular interest here. Such solutions will be found by means of the method of one parameter Lie point symmetries (Chester, 1977; Olver, 1986; Stephani, 1989; Bluman and Kumei, 1989). The approach, highly appropriate for this problem, is further motivated by Hill (1992) who discusses the need to study the symmetries of similar systems involving chemical reaction equations, even when in linear form. It is clear, that in such cases both the classical and the non-classical, sometimes called hidden symmetries, Guo and Shrauner (1993), are not well understood.

In the following, the Lie approach will be applied to the system of equations written in the form:

$$Z(x,t,Y,\dot{Y},Y',Y'') = \frac{\partial Y}{\partial t} - \frac{\partial}{\partial x}\{\Lambda(Y)\frac{\partial Y}{\partial x}\} = 0 \qquad (2)$$

where it will be supposed that $t \in \mathbf{R}^+$, $x \in \Omega \subset \mathbf{R}$ and also that $Y(x,t) : \Omega \times \mathbf{R}^+ \to \mathbf{R}^n$, with $\boldsymbol{Y} = (y_1, y_2, \ldots, y_n)$. In addition, Λ: $\mathbf{R}^n \to \mathbf{R}^{n\times n}$, is diagonally dominant and it is further assumed that Neumann, Dirichlet or Robin boundary conditions hold on the boundary $\partial\Omega$ of Ω. The particular aim will be to determine the constraints which need to be placed upon $\Lambda(\boldsymbol{Y})$ to enable Classical Lie point symmetries to be used in the determination of analytic solutions of (2).

1. One Parameter Lie Groups

In the following, it will be assumed that if $\boldsymbol{H}(x,t,\boldsymbol{Y},\epsilon) \in \mathbf{R}^n$, is infinitely differentiable in $\mathbf{R}^n$, with $\epsilon \in S \subset \mathbf{R}$, then the particular finite point transformation:

$$x_1 = f(x,t,\boldsymbol{Y},\epsilon); \quad t_1 = g(x,t,\boldsymbol{Y},\epsilon); \quad \boldsymbol{Y}_1 = \boldsymbol{H}(x,t,\boldsymbol{Y},\epsilon) \tag{3}$$

is a one-parameter Lie group of transformations in which the infinitesimal form is:

$$\begin{aligned} x_1 &= x + \epsilon\xi(x,t,\boldsymbol{Y}) + O(\epsilon^2) \equiv x + \epsilon\mathcal{L}x + O(\epsilon^2) \\ t_1 &= t + \epsilon\eta(x,t,\boldsymbol{Y}) + O(\epsilon^2) \equiv t + \epsilon\mathcal{L}t + O(\epsilon^2) \\ \boldsymbol{Y}_1 &= \boldsymbol{Y} + \epsilon\pi(x,t,\boldsymbol{Y}) + O(\epsilon^2) \equiv \boldsymbol{Y} + \epsilon\mathcal{L}\boldsymbol{Y} + O(\epsilon^2), \end{aligned} \tag{4}$$

where $\mathcal{L}$ is the infinitesimal generator defined by:

$$\begin{aligned} \mathcal{L} &= \xi(x,t,\boldsymbol{Y})\tfrac{\partial}{\partial x} + \eta(x,t,\boldsymbol{Y})\tfrac{\partial}{\partial t} + \pi(x,t,\boldsymbol{Y}).\nabla \;; \\ \nabla &= \left(\frac{\partial}{\partial y_1}, \ldots, \frac{\partial}{\partial y_n}\right). \end{aligned} \tag{5}$$

The link between the global and infinitesimal forms is defined by:

$$\tfrac{dx_1}{d\epsilon} = \xi(x_1,t_1,\boldsymbol{Y}_1) \;; \; \tfrac{dt_1}{d\epsilon} = \eta(x_1,t_1,\boldsymbol{Y}_1) \;; \; \tfrac{d\boldsymbol{Y}_1}{d\epsilon} = \pi(x_1,t_1,\boldsymbol{Y}_1) \tag{6}$$

where, $x_1 = x$; $t_1 = t$ and $\boldsymbol{Y}_1 = \boldsymbol{Y}$ when $\epsilon = 0$.

If the solution to the differential equation is given by $\boldsymbol{F}(\boldsymbol{Y},x,t) = \mathbf{0}$ then this must remain invariant under the transformations (4) and so applying the infinitesimal generator, this means that $\mathcal{L}\boldsymbol{F} = 0$. In the particular case when $\boldsymbol{F}(\boldsymbol{Y}, x, t) = \boldsymbol{Y} - \boldsymbol{\phi}(x, t) = \boldsymbol{0}$ it is found that:

$$\mathcal{L}\boldsymbol{F} = 0 \quad \Rightarrow \quad \pi(x, t, \boldsymbol{Y}) = \xi(x, t, \boldsymbol{Y})\boldsymbol{Y}' + \eta(x, t, \boldsymbol{Y})\dot{\boldsymbol{Y}} \ . \tag{7}$$

2. Transformation Formulae

Such expressions for first and second partial derivatives of $\boldsymbol{Y}$ are easily computed using the inverse Lie transformation:

$$\begin{aligned} x &= x_1 - \epsilon\xi(x_1, t_1, \boldsymbol{Y}_1) + O(\epsilon^2) \\ t &= t_1 - \epsilon\eta(x_1, t_1, \boldsymbol{Y}_1) + O(\epsilon^2) \\ \boldsymbol{Y} &= \boldsymbol{Y}_1 - \epsilon\pi(x_1, t_1, \boldsymbol{Y}_1) + O(\epsilon^2) \end{aligned} \tag{8}$$

Thus by defining the following operators:

$$\begin{aligned} \mathcal{G}' &= \mathbf{Y}'.\nabla \ ; \qquad \dot{\mathcal{G}} = \dot{\mathbf{Y}}.\nabla; \quad \mathcal{G}'' = \mathbf{Y}''.\nabla \ ; \\ \mathcal{D}' &= \tfrac{\partial}{\partial x} + \mathcal{G}' \ ; \qquad \dot{\mathcal{D}} = \tfrac{\partial}{\partial t} + \dot{\mathcal{G}} \end{aligned} \tag{9}$$

and using

$$\begin{aligned} \tfrac{\partial x}{\partial x_1} &= 1 - \epsilon\mathcal{D}'\xi + O(\epsilon^2) \ ; \qquad \tfrac{\partial x}{\partial t_1} = -\epsilon\dot{\mathcal{D}}\xi + O(\epsilon^2) \\ \tfrac{\partial t}{\partial x_1} &= -\epsilon\mathcal{D}'\eta + O(\epsilon^2) \ ; \qquad \tfrac{\partial t}{\partial t_1} = 1 - \epsilon\dot{\mathcal{D}}\eta + O(\epsilon^2) \end{aligned} \tag{10}$$

it is easy to see that:

$$\tfrac{\partial \boldsymbol{Y}_1}{\partial x_1} = \boldsymbol{Y}' + \epsilon\pi_x + O(\epsilon^2) \ ; \qquad \tfrac{\partial \boldsymbol{Y}_1}{\partial t_1} = \dot{\boldsymbol{Y}} + \epsilon\pi_t + O(\epsilon^2) \tag{11}$$

where:

$$\pi_x = \mathcal{D}'\pi - \boldsymbol{Y}'\mathcal{D}'\xi - \dot{\boldsymbol{Y}}\mathcal{D}'\eta$$

$$\Rightarrow \qquad \pi_x = \pi' - \dot{\boldsymbol{Y}}\eta' + \boldsymbol{Y}'.(\nabla\pi - \xi') - \boldsymbol{Y}'(\boldsymbol{Y}'.\nabla)\xi - \dot{\boldsymbol{Y}}(\boldsymbol{Y}'.\nabla)\eta \ ; \tag{12}$$

$$\pi_t = \dot{\mathcal{D}}\pi - \boldsymbol{Y}'\dot{\mathcal{D}}\xi - \dot{\boldsymbol{Y}}\dot{\mathcal{D}}\eta$$

$$\Rightarrow \qquad \pi_t = \dot{\pi} - \boldsymbol{Y}'\dot{\xi} + \dot{\boldsymbol{Y}}.(\nabla\pi - \dot{\eta}) - \boldsymbol{Y}'\left(\dot{\boldsymbol{Y}}.\nabla\right)\xi - \dot{\boldsymbol{Y}}\left(\dot{\boldsymbol{Y}}.\nabla\right)\eta \tag{13}$$

Similarly for the second derivatives:

$$\tfrac{\partial^2 \boldsymbol{Y}_1}{\partial x_1^2} = \boldsymbol{Y}'' + \epsilon\pi_{xx} + O(\epsilon^2) \ ; \tag{14}$$

with

$$\pi_{xx} = \mathcal{D}'\pi_x - \boldsymbol{Y}''\mathcal{D}'\xi - \dot{\boldsymbol{Y}}'\mathcal{D}'\eta$$
$$\Rightarrow \pi_{xx} = \mathcal{D}'\left\{\mathcal{D}'\pi - \boldsymbol{Y}'\mathcal{D}'\xi - \dot{\boldsymbol{Y}}\mathcal{D}'\eta\right\} - \boldsymbol{Y}''\mathcal{D}'\xi - \dot{\boldsymbol{Y}}'\mathcal{D}'\eta$$
$$\Rightarrow \pi_{xx} = (\mathcal{D}')^2\pi - \boldsymbol{Y}'(\mathcal{D}')^2\xi - \dot{\boldsymbol{Y}}(\mathcal{D}')^2\eta - 2\boldsymbol{Y}''\mathcal{D}'\xi - 2\dot{\boldsymbol{Y}}'\mathcal{D}'\eta \qquad (15)$$

and

$$(\mathcal{D}')^2 q = q'' + \mathcal{G}''q + 2\mathcal{G}'q' + (\mathcal{G}')^2 q \qquad (16)$$

for any $q = q(x, t, \boldsymbol{Y})$. It therefore, follows that:

$$\pi_{xx} = \pi'' + \boldsymbol{Y}'.(2\nabla\pi' - \xi'') - \dot{\boldsymbol{Y}}\eta'' + (\boldsymbol{Y}'.\nabla)^2\pi$$
$$- 2\boldsymbol{Y}'\,(\boldsymbol{Y}'.\nabla\xi') - 2\dot{\boldsymbol{Y}}(\boldsymbol{Y}'.\nabla\eta') - \boldsymbol{Y}'\,(\boldsymbol{Y}'.\nabla)^2\xi - \dot{\boldsymbol{Y}}(\boldsymbol{Y}'.\nabla)^2\eta$$
$$+ \boldsymbol{Y}''.\left[\nabla\pi - \nabla\xi\boldsymbol{Y}' - \nabla\eta\dot{\boldsymbol{Y}} - 2\xi' - 2\boldsymbol{Y}'\nabla\xi\right] - 2\dot{\boldsymbol{Y}}'(\eta' + \boldsymbol{Y}'.\nabla\eta) \qquad (17)$$

Finally, the transformation rules for $\Lambda\,(\boldsymbol{Y})$ and $\nabla\Lambda\,(\boldsymbol{Y})$ are:

$$\Lambda\,(\boldsymbol{Y}_1) = \Lambda\,(\boldsymbol{Y}) + \epsilon\,(\pi.\nabla)\Lambda\,(\boldsymbol{Y}) = (1 + \pi.\nabla)\,\Lambda\,(\boldsymbol{Y}) + O(\epsilon^2)$$
$$\nabla\Lambda\,(\boldsymbol{Y}_1) = \nabla\Lambda\,(\boldsymbol{Y}) + \epsilon\,(\pi.\nabla)\nabla\Lambda\,(\boldsymbol{Y}) = (1 + \pi.\nabla)\,\nabla\Lambda\,(\boldsymbol{Y}) + O(\epsilon^2) \qquad (18)$$

3. Equations Governing Invariance

The Prolongation Operator and the Condition for Invariance: The prolongation operator may be written in the form:

$$\mathcal{L}_P = \xi\frac{\partial}{\partial x} + \eta\frac{\partial}{\partial t} + \pi.\nabla_Y + \pi_x.\nabla_{Y'} + \pi_t.\nabla_{\dot{Y}} + \pi_{xx}.\nabla_{Y''} \qquad (19)$$

and hence the condition for invariance is:

$$\mathcal{L}_P\mathbf{Z} = \mathbf{0}\ . \qquad (20)$$

Thus applying (20) to (2) it may be shown that:

$$\pi_t = (\pi.\nabla).\Lambda\boldsymbol{Y}'' + \boldsymbol{Y}'\,(\pi.\nabla)\nabla\Lambda\boldsymbol{Y}' + \Lambda\pi_{xx} + (\pi_x.\nabla)\Lambda\boldsymbol{Y}' + (\boldsymbol{Y}'.\nabla)\Lambda\pi_x \qquad (23)$$

and so:

$$\dot{\pi} - \boldsymbol{Y}'\dot{\xi} + \dot{\boldsymbol{Y}}.(\nabla\pi - \dot{\eta}) - \boldsymbol{Y}'\,(\dot{\boldsymbol{Y}}.\nabla)\xi\ - \dot{\boldsymbol{Y}}(\dot{\boldsymbol{Y}}.\nabla)\eta =$$
$$\{(\pi.\nabla)\Lambda\}\Lambda^{-1}\left[\dot{\boldsymbol{Y}} - (\boldsymbol{Y}'.\nabla\Lambda)\boldsymbol{Y}'\right] + \boldsymbol{Y}'\,(\pi.\nabla)\nabla\Lambda\boldsymbol{Y}'$$
$$+ \Lambda\left\{\pi'' + \boldsymbol{Y}'.(2\nabla\pi' - \xi'') - \dot{\boldsymbol{Y}}\eta'' + (\boldsymbol{Y}'.\nabla)^2\pi\right.$$

$$
\begin{aligned}
&- 2\boldsymbol{Y}' (\boldsymbol{Y}'.\nabla\xi') - 2\dot{\boldsymbol{Y}}(\boldsymbol{Y}'.\nabla\eta') - \boldsymbol{Y}' (\boldsymbol{Y}'.\nabla)^2\xi - \dot{\boldsymbol{Y}}(\boldsymbol{Y}'.\nabla)^2\eta - 2\dot{\boldsymbol{Y}}'(\eta' + \boldsymbol{Y}'.\nabla\eta)\big\} \\
&+ \big[\dot{\boldsymbol{Y}} - (\boldsymbol{Y}'.\nabla\Lambda)\boldsymbol{Y}'\big].\big[\nabla\pi - \nabla\xi\boldsymbol{Y}' - \nabla\eta\dot{\boldsymbol{Y}} - 2\xi' - 2\boldsymbol{Y}'\nabla\xi\big] \\
&+ \big\{\big(\pi' - \dot{\boldsymbol{Y}}\eta' + \boldsymbol{Y}'.(\nabla\pi - \xi') - \boldsymbol{Y}'(\boldsymbol{Y}'.\nabla)\xi - \dot{\boldsymbol{Y}}(\boldsymbol{Y}'.\nabla)\eta\big).\nabla\big\}\Lambda\boldsymbol{Y}' \\
&+ (\boldsymbol{Y}'.\nabla)\Lambda\,\{\pi' - \dot{\boldsymbol{Y}}\eta' + \boldsymbol{Y}'.(\nabla\pi - \xi') - \boldsymbol{Y}'(\boldsymbol{Y}'.\nabla)\xi - \dot{\boldsymbol{Y}}(\boldsymbol{Y}'.\nabla)\eta\}
\end{aligned} \tag{22}
$$

Hence the Classical Lie point transformation symmetries may be obtained by equating coefficients in (22) as follows:

1) **Constants:** $\dot{\pi} = \Lambda\pi''$

2) $\boldsymbol{Y}'$: $-\boldsymbol{Y}'\dot{\xi} = \Lambda\boldsymbol{Y}'.[2\nabla\pi' - \xi''] + (\pi'.\nabla)\Lambda\boldsymbol{Y}' + (\boldsymbol{Y}'.\nabla)\Lambda\pi'$

3) $\dot{\boldsymbol{Y}}$: $-\dot{\eta}\dot{\boldsymbol{Y}} = (\pi.\nabla\Lambda)\Lambda^{-1}\dot{\boldsymbol{Y}} - \Lambda\dot{\boldsymbol{Y}}\eta'' - 2\xi'\dot{\boldsymbol{Y}}$

4) $\dot{\boldsymbol{Y}}, \boldsymbol{Y}'$: $-\boldsymbol{Y}'(\dot{\boldsymbol{Y}}.\nabla)\xi = -2\Lambda\dot{\boldsymbol{Y}}(\boldsymbol{Y}'.\nabla\eta') - (\boldsymbol{Y}.\nabla)\xi\boldsymbol{Y}' - 2\dot{\boldsymbol{Y}}(\boldsymbol{Y}'.\nabla)\xi$
$\quad - \dot{\boldsymbol{Y}}\eta'.\nabla\Lambda\boldsymbol{Y}' - \boldsymbol{Y}'\nabla\Lambda\dot{\boldsymbol{Y}}\eta'$

5) $\dot{\boldsymbol{Y}}, \dot{\boldsymbol{Y}}$: $-\dot{\boldsymbol{Y}}(\dot{\boldsymbol{Y}}.\nabla)\eta = -\dot{\boldsymbol{Y}}.\nabla\eta\dot{\boldsymbol{Y}}$

6) $\boldsymbol{Y}', \boldsymbol{Y}'$: $0 = -(\pi.\nabla\Lambda)\Lambda^{-1}(\boldsymbol{Y}'.\nabla)\Lambda\boldsymbol{Y}' + \boldsymbol{Y}'(\pi.\nabla)\nabla\Lambda\boldsymbol{Y}'$
$+\Lambda(\boldsymbol{Y}'.\nabla)^2\pi - 2\Lambda\boldsymbol{Y}'(\boldsymbol{Y}'.\nabla)\xi' + (((\boldsymbol{Y}'.\nabla)\pi).\nabla)\Lambda\boldsymbol{Y}'$
Alternatively:
$0 = \big((\boldsymbol{Y}'.\nabla)\big[(\pi.\nabla\Lambda)\Lambda^{-1}\big]\big)\Lambda\boldsymbol{Y}' + \Lambda(\boldsymbol{Y}'.\nabla)^2\pi$
$\quad - 2\Lambda\boldsymbol{Y}'(\boldsymbol{Y}'.\nabla)\xi'$

7) $\boldsymbol{Y}', \dot{\boldsymbol{Y}}, \dot{\boldsymbol{Y}}$: $0 = -\Lambda\dot{\boldsymbol{Y}}(\boldsymbol{Y}'.\nabla)^2\eta + (\boldsymbol{Y}'.\nabla)\Lambda\boldsymbol{Y}'\nabla\eta\dot{\boldsymbol{Y}}$
$\quad - \dot{\boldsymbol{Y}}(\boldsymbol{Y}'.\nabla)\eta\nabla\Lambda\boldsymbol{Y}' - \boldsymbol{Y}'\nabla\Lambda\dot{\boldsymbol{Y}}(\boldsymbol{Y}'.\nabla)\eta$

8) $\boldsymbol{Y}', \boldsymbol{Y}', \boldsymbol{Y}'$: $0 = -\Lambda\boldsymbol{Y}'(\boldsymbol{Y}'.\nabla)^2\xi + [(\boldsymbol{Y}'.\nabla)\Lambda\boldsymbol{Y}'].\{\nabla\xi\boldsymbol{Y}' + 2(\boldsymbol{Y}'.\nabla)\xi\}$
$- \boldsymbol{Y}'(\boldsymbol{Y}'.\nabla)\xi\nabla\Lambda\boldsymbol{Y}' - (\boldsymbol{Y}'.\nabla)\Lambda\boldsymbol{Y}'(\boldsymbol{Y}'.\nabla)\xi$

9) $\dot{\boldsymbol{Y}}'$: $0 = -2\Lambda\dot{\boldsymbol{Y}}'\eta'$

10) $\dot{\boldsymbol{Y}}', \boldsymbol{Y}'$: $0 = -2\Lambda\dot{\boldsymbol{Y}}'\boldsymbol{Y}'\nabla\eta$

Evaluation of Group Invariants: It is easy to show that conditions 4,5,7,8,9, and 10 are automatically satisfied when $\eta = \eta(t)$ and $\xi = \xi(x,t)$. In addition, condition 3 may be substituted into 6 to give:

$$
\begin{aligned}
\Lambda(\boldsymbol{Y}'.\nabla)^2\pi &= 0 \\
\pi.\nabla\Lambda &= \Lambda(2\xi' - \dot{\eta}).
\end{aligned} \tag{23}
$$

In cases when $(2\xi' - \dot{\eta}) = 0$, and for all $\Lambda\ (\boldsymbol{Y})$, condition 3 gives:

$$
\begin{aligned}
\pi(x,t,\boldsymbol{Y}) &= \mathbf{0} \\
\xi = \xi(x,t,\boldsymbol{Y}) &= \tfrac{\gamma}{2}x + \kappa \\
\eta = \eta(x,t,\boldsymbol{Y}) &= \gamma t + \delta
\end{aligned} \tag{24}
$$

Alternatively when $(2\xi' - \dot{\eta}) \neq 0$ we find:

$$
\pi = \tfrac{1}{m}\ (\boldsymbol{Y} + \boldsymbol{\beta}) \tag{25}
$$

where m and $\boldsymbol{\beta}$ are independent of y_i. Hence using condition 3 it is found that Λ has the particular form:

$$
\Lambda = A\prod_i (y_i + \beta_i)^{R_i} \tag{26}
$$

where A is a constant matrix and

$$
\sum_i R_i = m(2\xi' - \dot{\eta})\,. \tag{27}
$$

In this case it may be shown conditions 1 and 2 will also be satisfied when:

$$
\begin{aligned}
\xi = \xi(x,t,\boldsymbol{Y}) &= \lambda x + \kappa \\
\eta = \eta(x,t,\boldsymbol{Y}) &= \gamma t + \delta \\
\pi &= (\boldsymbol{Y} + \boldsymbol{\beta})\frac{(2\lambda - \gamma)}{\sum_i R_i}
\end{aligned} \tag{28}
$$

4. Examples of Analytic Solutions of (2):

(a) With $\gamma = 0$ equation (24) gives $\pi = 0$, $\xi = \kappa$ and $\eta = \delta$. Hence by (7) a similarity variable can be defined through $\omega = \delta x - \kappa t$, with the result that equation (2) may be integrated immediately with the result that:

$$
\Lambda \boldsymbol{Y}_\omega + \frac{\kappa}{\delta^2}\boldsymbol{Y} = \boldsymbol{a} \tag{29}
$$

where **a** is an arbitrary constant. Although this result holds for any Λ, it is interesting to consider the particular case when Λ is defined by (26) with $\beta_i = 0$, $\forall i$ and $\boldsymbol{a} = \mathbf{0}$. It is straigtforward to show that (29) has the solution:

$$
\boldsymbol{Y} = (r + s\omega)^q \boldsymbol{b} \tag{30}
$$

where r and s are constant, provided that:

$$q\sum_i R_i = 1\,; \qquad \left(A\prod_i b_i^{R_i} sq + \left(\frac{\kappa}{\delta^2}\right)\right)\boldsymbol{b} = 0 \tag{31}$$

(b) In a similar way consider the source solution of:

$$\dot{\boldsymbol{Y}} = \frac{\partial}{\partial x}\left\{A\prod_i (y_i)^{R_i}\frac{\partial Y}{\partial x}\right\} \tag{32}$$

so that

$$\boldsymbol{Y}(x,0) = \boldsymbol{Y}_0\delta(x) \tag{33}$$

where the Dirac delta function satisfies:

$$\delta(\lambda x) = \lambda^{-1}\delta(x) \tag{34}$$

Following Hill (1992), who considers non-linear scalar diffusion, it follows that these equations are invariant whenever:

$$x_1 = e^{\epsilon}x\,; \qquad t_1 = e^{\left(\sum_i R_i + 2\right)\epsilon}t\,; \qquad Y_1 = e^{-\epsilon}\boldsymbol{Y} \tag{35}$$

Thus:

$$\xi(x,t,\boldsymbol{Y}) = x\,; \qquad \eta(x,t,\boldsymbol{Y}) = \left(\sum_i R_i + 2\right)t\,; \qquad \pi = -\boldsymbol{Y} \tag{36}$$

Hence from (7), the functional form of $\boldsymbol{Y}(x,t)$ and associated similarity variable ω is:

$$\boldsymbol{Y}(x,t) = \tfrac{\phi(\omega)}{t^n}\,; \qquad \omega = \tfrac{x}{t^n}\,; \qquad n = \frac{1}{\left(2+\sum_i R_i\right)} \tag{37}$$

Equation (2) therefore becomes :

$$A\prod_i \phi_i^{R_i}\,\tfrac{d\phi}{d\omega} + n\omega\boldsymbol{\phi} = \boldsymbol{a} \tag{38}$$

On setting $\boldsymbol{a} = 0$ it may be shown that:

$$\boldsymbol{\phi}(\omega) = \left[r + \frac{s\omega^2}{2}\right]^p \boldsymbol{b} \tag{39}$$

is a solution of (37) provided that :

$$p\sum_i R_i = 1; \qquad A\prod_i b_i\; s\,\boldsymbol{b} = -\left[\frac{\sum_i R_i}{2+\sum_i R_i}\right]\boldsymbol{b}\,. \tag{40}$$

5.References

Bluman, G.W. & Kumei, S., 1989, *Symmetries and Differential Equations*, Appl. Math. Sci., No. 81, Springer.

Chester, W., 1977, *Continuous Transformations and Differential Equations*, J. Inst. Maths. Applics., **19** , 343-376.

De Vries, D. A., 1958, *Simultaneous Transfer of Heat and Moisture in Porous Media*, Trans Am. Geophys. Un., **39**, 909-16.

Guo, A. & Abraham-Shrauner, B., 1993, *Hidden Symmetries of Energy-Conserving Differential Equations*, IMA Journal of App. Maths., **51** , 147-153.

Hill, J.M., 1992, *Differential Equations and Group Methods*, Ed C.R.C. Press.

Jackson, R.D.,1973, *Diurnal Soil Water Time-Depth Patterns*, Soil Sci. Soc. Am. Proc., **37**, 505-509.

Jury, W. A., Letey, J. and Stolzy, L. H., 1981, *Flow of Water and Energy under Desert Conditions*, Water in Desert Ecosystems, Ed. Evans & Thames, Dowden, Hutchinson and Ross Inc., 92-113.

Olver, P. J., 1986, *Application of Lie Groups to Differential Equations*, Graduate Texts in Mathematics 107, Springer, New York.

Philip J. R., 1988, *Quasianalytic and Analytic Approaches to Unsaturated Flow*
Flow and Transport in the Natural Environment; Advances and Applications. Ed. W.L. Steffen and O.T. Denmead, Springer-Verlag. 30-48.

Philip, J. R. and De Vries, D. A., 1957, *Moisture Movement in Porous Under Temp. Gradients*, Trans. Am. Geophys. Un., **38**, 222-232.

Stephani, H., 1989, *Differential Equations - Their Solution Using Symmetries*, Cambridge University Press.